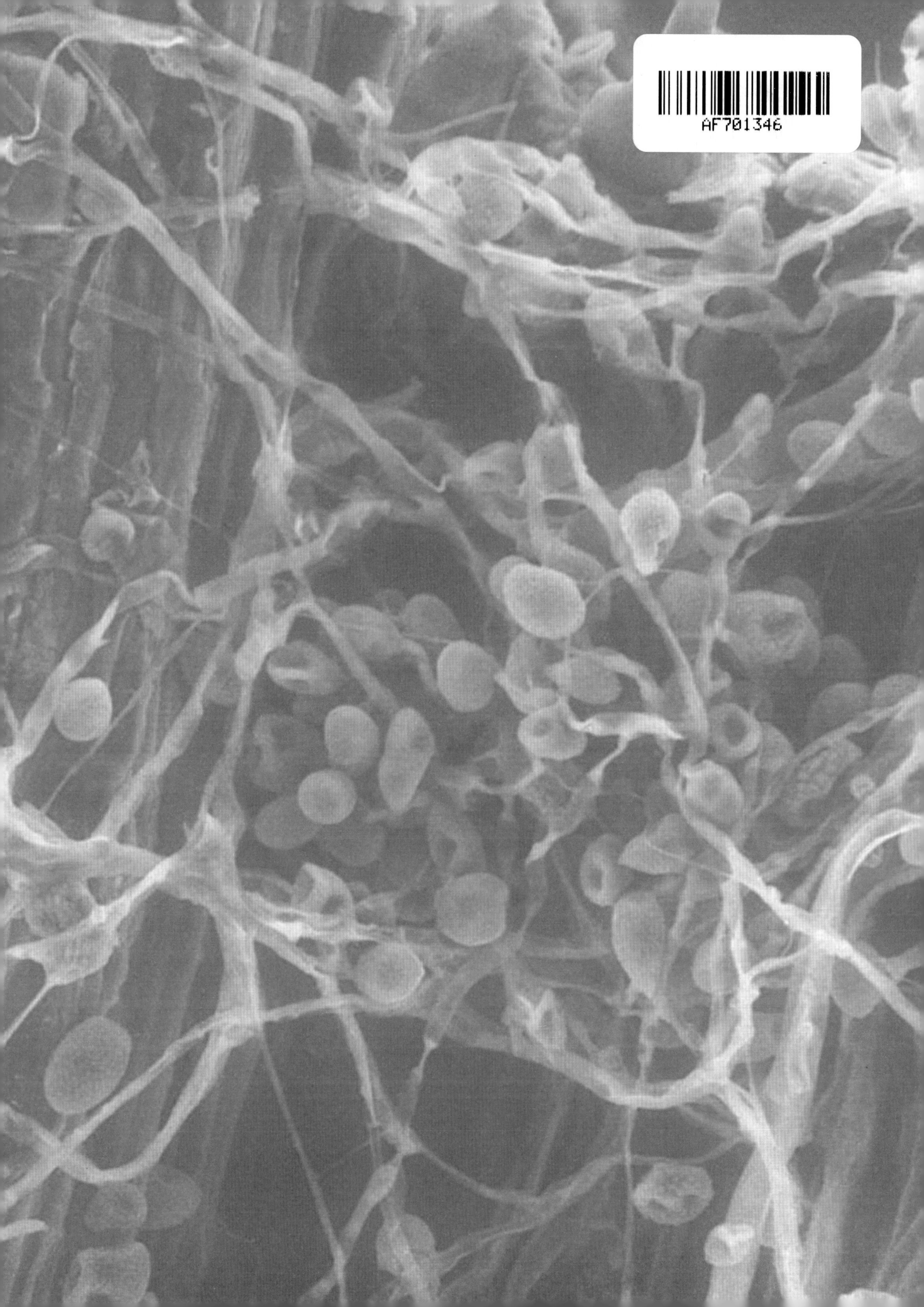

Earth Alive!

FROM MICROBES TO A LIVING PLANET

Earth Alive!

FROM MICROBES TO A LIVING PLANET

Mary E. White

ROSENBERG

First published in Australia in 2003 by Rosenberg Publishing Pty Ltd
PO Box 6125, Dural Delivery Centre NSW 2158, Australia,
Phone 02 9654 1502, Fax 02 9654 1338,
Email rosenbergpub@smartchat.net.au, Web www.rosenbergpub.com.au

National Library of Australia
Cataloguing-in-Publication data

White, M.E. (Mary E.).
Earth alive!: from microbes to a living planet.

Bibliography.
Includes index.
ISBN 1 877058 05 X.

1. Biosphere. 2. Evolution (Biology). 3. Natural history.
4. Paleoecology. I. Title.

577

Cover photos: *Main picture: Like a starry sky, bacteria are the larger illuminated dots; viruses the more numerous, smaller, not so bright dots. The organisms are made visible by microscopy with a dye that stains their DNA. Modern technology has enabled this microcosm in water from the Southern Ocean to be seen and recorded only recently. Image Andrew Davidson, AAD.*
Satellite view of Earth: A greenish tint has been added to the image of our 'blue planet' because it is photosynthesis, performed by green organisms, that provides the nutrients for the webs-of-life that make it a living world. Dean Graetz © CSIRO.
Rubisco molecule: Cornelia Buchen-Osmond, Biosphere 2 Center, Oracle, Arizona

Endpapers: *The surface of crop residues colonised by fungal and actinomycete hyphae and testate amoebae. VVSR Gupta et al. © CSIRO Land & Water, Adelaide*

Cover design: Mary White
Maps and diagrams: Barbara Eckersley
Internal design: Linda Robertshaw

Set in Bembo 10/13pt
Printed in Singapore by Kyodo Printing Co (Singapore) Pte Ltd

10 9 8 7 6 5 4 3 2 1

Sponsorship Acknowledgment

The Publisher and the author gratefully acknowledge the generous sponsorship of two of Australia's leading scientific institutions — CSIRO Land & Water and the Australian National University and its Research School of Biological Sciences. Their financial contributions have enabled the recommended retail price of this book to be not more than $50.

CSIRO
CSIRO LAND and WATER

THE AUSTRALIAN NATIONAL UNIVERSITY

Mary White wishes to express her deep personal gratitude to Dr John Williams, Chief, CSIRO Land & Water, and Professors John Hearn and John Gibson, the Research School of Biological Sciences and the Australian National University, for their interest and support and for facilitating the sponsorship. The writing of a Foreword by John Williams and an account of research programs relevant to this book by John Hearn (page11) have been an extra, much appreciated, bonus. Sincere thanks to you both.

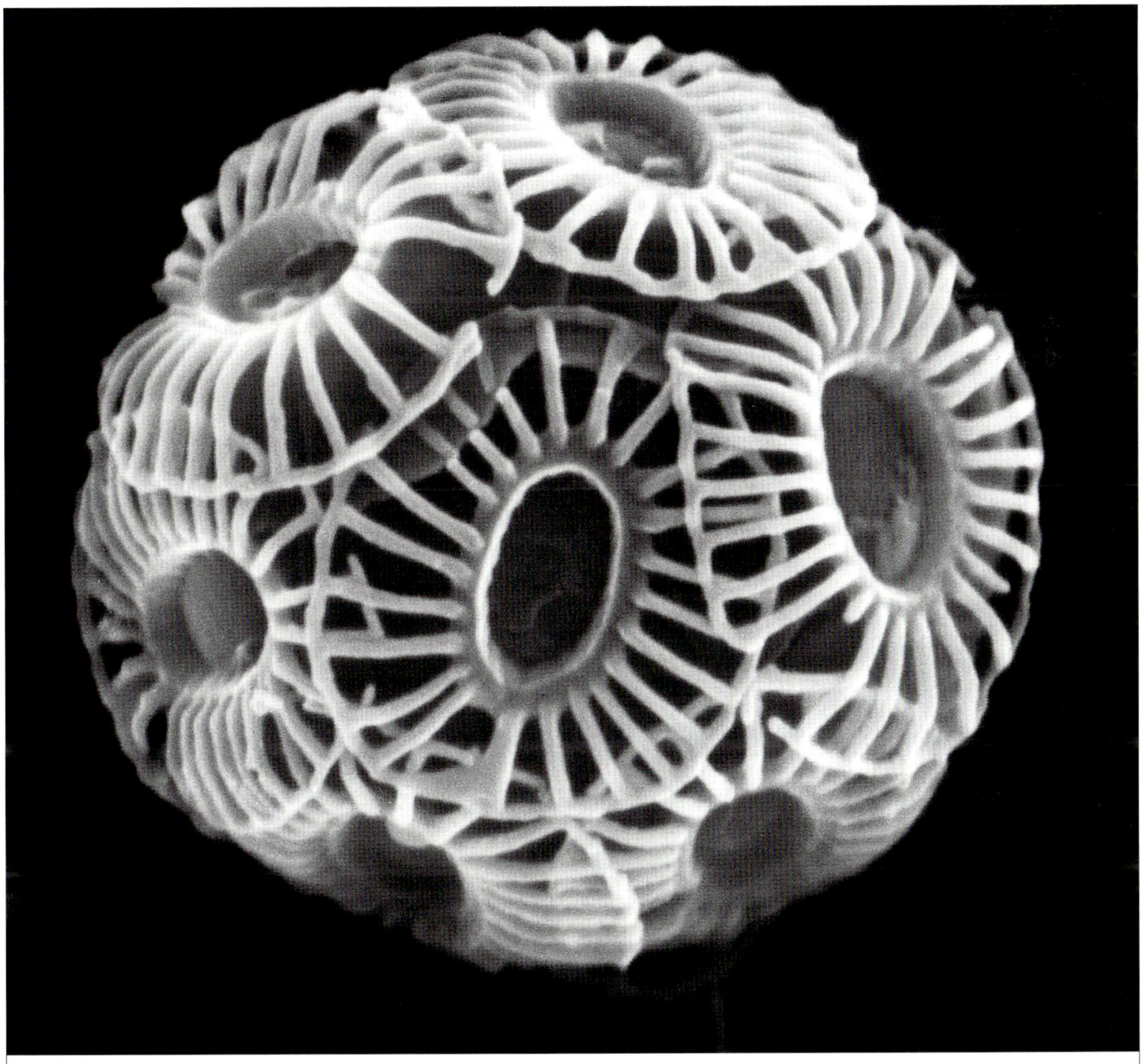

Emiliana, the microscopic organism whose 'blooms' in the oceans influence global weather patterns. *Andrew McEldowney, AAD*

Acknowledgments

The author acknowledges with deep gratitude the help and advice of the many people who generously gave of their time, and supplied scientific information, papers and illustrative material. Where their research became the basis for parts of the text, their review and correction of manuscript was a service of inestimable value and has contributed towards the making of this book. Science communicators would have nothing to communicate if it were not for the availability of information from those who have carried out the original research!

The generosity of CSIRO Land & Water, facilitated by Director Dr John Williams and External Affairs Director Margaret Bryant; and of the Australian Antarctic Division, facilitated by Director Dr Tony Press and Professor Michael Stoddart, require special mention. And it has been a special pleasure to have my daughter Barbara Eckersley again making my maps and diagrams. The encouragement and interest of many friends, particularly Van McCune and Bill Templeman, has helped me at times when the task of assembling this bigger-than-big picture of our living Earth was overwhelming.

Lists of those who have helped in various ways from the various organisations, and others, follow. (Forgive me if I have omitted your name—my records are not always complete—all help was much appreciated.)

AAD: Programme Leader, Biology, Harvey Marchant; Simon Wright; Fiona Scott; Geraldine Nash
Adelaide University: Professor Sally Smith
AGSO: Brian Logan
AIMS: Madeline van Oppen; Charlie Veron
ANU: Cornelia Buchen-Osmond; Penelope Greenslade; Geoffrey Smith
CSIRO: B.M. Doube; Rob Fitzpatrick; Peter Franzmann; Dean Graetz; V.V.S.R. Gupta; D. Jacquier; Margaret McCully; Neil McKenzie; S.M. Neate; Jim Noble; Clive Pankhurst; Joanna Parr; P. Reddell; D.K. Roget; Margaret Roper; A.D. Rovira; M.H. Ryder; Kate Smith; J. Short; Alister Spain; John Walker; Michelle Watt
Earth Sanctuaries Ltd: John Gitsham; Greg Martin; Cheryl McEgan
JCU: Gerry Dickens
Macquarie University: Brian Atwell; Peter Mitchell; Noel Tait; Malcolm Walter
NSW Geological Survey: John Watkins
NT NPW: John Woinarski
SCU Lismore: Peter Harrison
UNE: Andrew Boulton
University of Sydney: Ivan Kennedy; Armstrong Osborne
WA Museum: Bill Humphreys; Ken McNamara
In NSW: Rob Coenraads; Jim Gasteen; David Marsh
In WA: John Beard; Linda Moore; Cliff Ollier

USE OF NAMES AND TERMS

Because of the complications that exist as a result of the many different classifications and the changing concepts, a standard and precise usage throughout this book is as follows:

Three **Super-Kingdoms** are recognised: ARCHAEA, BACTERIA and EUKARYA.

The **five Kingdoms** of living organisms, and the features that qualify their members, are:

1: Kingdom Monera (or Bacteria) *Prokaryotic cells*, without nuclei or organelles, replicating by simply dividing. Some are autotrophic, some are heterotrophic. The subdivisions in this category are: Archaea—methane producers, anaerobic, sulphur bacteria; Bacteria, unspecified, the Eubacteria; and the Cyanobacteria, that have chlorophyll and photosynthesise.

2: Kingdom Protista (or Protists) *Eukaryotic* organisms. Among these are some with lineages ancestral to the Fungal, Plant and Animal Kingdoms. A diverse group of living organisms including the Algae (microscopic single-celled to massive seaweeds) originally classified in the Plant Kingdom; and Protozoa that were classified in the Animal Kingdom.

3: Kingdom Mycota (or Fungi) *Eukaryotic*, saprophytic or parasitic organisms; microscopic individual cells or filamentous hyphae; and macroscopic, as in fruiting bodies like mushrooms and the thallus of lichens.

4: Kingdom Plantae (or Plants) *Eukaryotic* organisms. Starts with land plants, evolved from green algal (Protist) ancestors.

5: Kingdom Animalia (or Animals) *Eukaryotic* organisms. Starts with the big-bang explosion at the start of the Cambrian Period. Evolved from Protists. Invertebrates (including micro-invertebrates) and vertebrates.

Where small organisms of two or more kingdoms are being referred to without separating them according to types, the general term micro-organisms (or microbes) is used.

Contents

Mangroves on the sand flats at Pajinka, Cape York. Tidal ecosystems comprise astronomical numbers of microscopic organisms. *M.E.W.*

Foreword

In her numerous books on the evolution of Australia and its changing landscape, Mary White has applied her extraordinary wealth of knowledge to take on big ideas and even bigger time scales to shed light on why Australia is so different from any other land.

By looking back at the ancient history and evolution of our landscape, she has provided the deep time perspective we need to live sustainably in this ancient and fragile place.

In this latest book, she sets out the life support principles of this planet that underpin our survival.

The living Earth is built on microbes that have transformed the rocks of the earth's crust—the lithosphere—into soil, the foundation for the terrestrial web of life, yet one of the most vulnerable of all natural resources.

Microbial diversity has driven the evolution of all life on Earth as well as the nutrient cycles that are vital to the operation of the biosphere. Microbes play a greater role in carbon cycling and any associated global warming than all the rainforests. Their genetic diversity overshadows the diversity of plants, animals and 'true' fungi. Micro-organisms live in all ecosystems, even those extreme environments that are not habitable to other organisms. In some cases, the same species have been found in seawater, freshwater and soil and recorded in polar regions as well as temperate and tropical areas.

The many functions performed by soil microbes are central to ecological sustainability. Soil is a seething foundry in which matter and energy are in constant flux as it provides the support services for ecosystem primary production. A rich mix of mineral particles, biota, organic matter, gases, water and nutrients, soil constitutes a self-regulating biological factory essential for initiating and maintaining life. Soil determines the partitioning of rainfall, snowmelt or irrigation into overland flow, infiltration, storage, deep drainage and, in turn, groundwater recharge. The way soil accepts, stores and transmits water and associated solute strongly influences the nature of rivers, springs, lakes and wetlands. Organisms in soil recycle residues, converting them to nutrients and other compounds, thereby providing the primary cleaning and recycling function for ecosystems.

The widespread decline in ecosystem processes—the processes that sustain life on earth—reflects our failure to value them. For too long, we have focused on how much we can take from our ecosystems while paying little attention to the essential goods and services they provide—such as climate control and nutrient recycling, food production, provision of pure and sufficient water, storage of atmospheric carbon and maintenance of biodiversity.

The health of ecosystems depends on how we take care of them. If we are to manage them effectively, we need detailed understanding of their condition and how they function.

In this new work, Mary White's science-based, multi-disciplinary approach will play an important role in helping us gain that understanding by enhancing our appreciation of the universe of microbes.

Through numerous physical and chemical processes, energy and matter cycles and flows through the four major subsystems that compose the Earth—the lithosphere (land), biosphere (living things), atmosphere (air) and hydrosphere (water)—and the myriad ecosystems within. The systems themselves, and the complex interactions within and between them, are built on the activity of microbes. Mary White brings them to life in the following pages.

If *Earth Alive!* is able to excite and inform us all about the importance of microbial biodiversity it will be an important step in inspiring the community to value and appreciate the vital ecosystems services supplied by microbes. In turn, this will encourage us to value ecosystem services and functions and to pay for their maintenance. At present we do not value them and we do not pay for them, yet they are the very services on which life depends.

For all these reasons I am delighted to assist in the publication of this book.

John Williams
Chief, CSIRO Land and Water

EARTH TIME SCALE AND THE GEOLOGICAL PERIODS

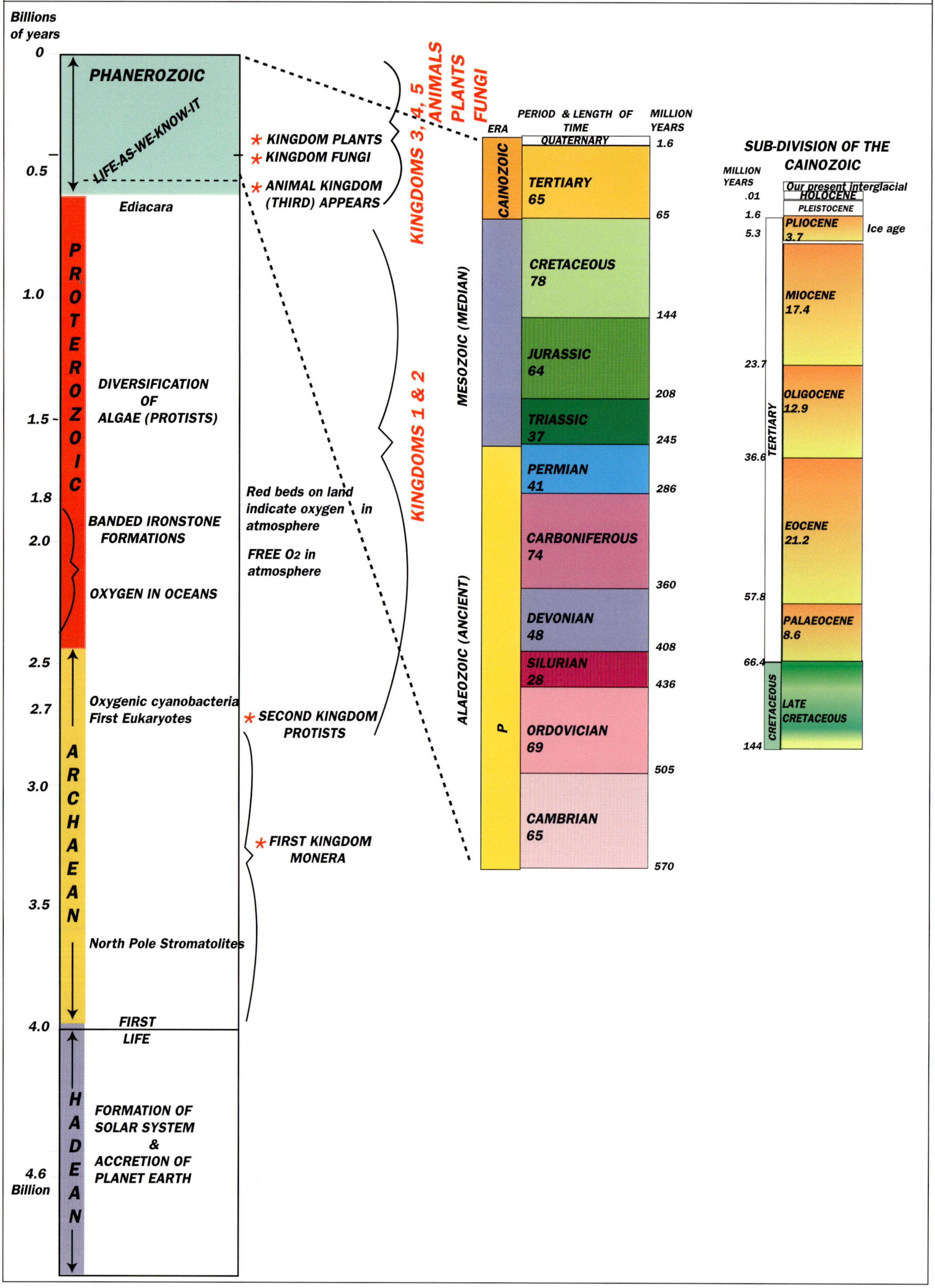

Introduction

For many years now, in all my previous books, I have been presenting the big picture of the co-evolution of the Australian landmass and its biota through great lengths of geological time. The focus in the three most recent has been on how an understanding of that ancient history explains many of the environmental problems afflicting Australia today, and how the parameters for sustainable land and water use have thus been set by the nature of the continent itself.

Now I am venturing into the realms of a bigger picture still—the ultimate 'big picture' that encompasses the Living Planet, the origins of Life, and how the Biosphere, in fact, *is* Life. The research for me has been a revelation of the interconnectedness of everything in the Biosphere; the marvels of bacterial beginnings and the amazing symbioses that produced the *five* Kingdoms of living organisms; the Life-driven control of the atmosphere through time and Life's modification of the inanimate crust of the Earth; and the realisation that this bacterial Earth has marvels and mysteries as complex and as infinite as those of the Universe in which it is spinning ...

I was a Gaian when I started, now I am a proudly Bacterial Gaian living in a Symbiotic Planet!

Because of the 'how do I know what I think till I see what I say' syndrome, that brings some sort of revelation to all writers and synthesisers of interdisciplinary material, the research that has gone into this book has convinced me to adopt a number of innovations. These are: the Five Kingdoms classification of living organisms; bacterial origins and symbiosis as the makers of more complex life forms and different species; and the model for Life's evolution as modified by Dr Lynn Margulis, one of the most inspiring and controversial thinkers of our times. Hers is the Serial Endosymbiosis Theory of Evolution (SET). My background interests in palaeo-environments and prehistoric plants and animals convince me of the logic and wisdom of these advances in evolutionary science.

THE MOST NORTHERLY TERMITE MOUND ON THE TIP OF CAPE YORK PENINSULA

Termites process between a third and a fifth of all phytomass in tropical and subtropical habitats—which amounts to *a quarter of all terrestrial primary production.* Globally they are estimated to add 12 gigatonnes of carbon to the atmosphere annually, and quantities of methane that equal amounts produced by all the world's sheep and cattle—courtesy of the endosymbionts in their guts! They create soil as a by-product of their activities ... invertebrates involved in the running of the world ... *M.E.W.*

There is a practical reason for trying to understand something of this ultimate big picture. The healthy and balanced functioning of the Biosphere that keeps it Life-friendly for our sort of air-breathing creatures requires the maintenance of a web of metabolising, living organisms, from the microbial and invisible to the patently obvious. In particular, I hope to emphasise that it is the 'little things that run the world', as E.O. Wilson so aptly put it[1], because it is the contribution of micro-organisms to the Biosphere that has hardly been recognised by scientists until recently, and remains virtually unknown to the non-scientist. The sheer size of this invisible Kingdom, the biodiversity that it encompasses, and the proportion of living matter on Earth that it comprises, are almost unbelievable.

It is now generally accepted that for the first 3 billion years at least, life on Earth was virtually all microscopic—few organisms achieved a size greater than a millimetre, or a complexity beyond two or three cell types.[1b] It is only in the last 700 million years, or less, that the evolution of much larger and more complex organisms has transformed the Biosphere.

An altered perspective on the modern world emerges when we realise the continuing dominance of microscopic Life. Micro-organisms today constitute about half of the total biomass of the Earth.[220] Ours is, indeed, a living world in which fundamental processes essential for its function are performed by the invisible. Microscopic organisms in the oceans, together in their infinite multitudes, help to control world weather and the composition of the atmosphere. (It has been calculated that 6 billion tonnes of phytoplankton are involved in algal blooms in the cool-temperate seasonal seas annually.) Microscopic soil fungi in symbiotic mycorrhizal relationships with plants enable them to live and thrive in terrestrial environments. It is micro-organisms that recycle Earth's limited and finite budget of the substances essential for Life. And it is not only microscopic living things that maintain the Biosphere—again quoting E.O. Wilson, 'Invertebrates are more important in the maintenance of ecosystems than are vertebrates.'[1]

Against the background of the fascinating glimpses of the shaping of the Biosphere through time, the contents of this book literally 'come down to earth'—to the microbiology of the living soil and how it maintains the green plants that are the basis of all terrestrial food chains; to the symbiotic relationships that, in particular, enable plants to live on the poor soils of our old continent; to the dangers of removing vital links from ecosystem webs, epitomised in the degradation that follows unsustainable land-use practices; and even to Greenhouse warming of our planet.

The Biosphere works on cause and effect, action and reaction, because it is an intricately connected system like a huge machine. But it is a machine whose complexity is almost beyond human comprehension. It is the Biosphere that ultimately sets the rules for the survival of individual organisms of all sorts, including us, here and now. It has been the interconnectedness of everything within this system that has resulted in the evolution of species, and thus of the Biosphere itself, through billions of years. The prevailing conditions within it determined the attributes that were selected for by natural selection. The concept of survival of the fittest has operated since the beginning, and still operates, for all organisms, though chance has a hand in things as well. The natural laws that govern and control population numbers and maintain a dynamic balance in ecosystems are set by the Biosphere and should be called the 'Laws of the Biosphere'.

Through technological and cultural evolution, facilitated by the big brain with the pronounced right hemisphere, the Human species has been able to function outside those constraints. We have enjoyed a honeymoon period during which (because of that highly developed right hemisphere) we have been able to *believe* that we, not the Biosphere, rule the world. Now the honeymoon is over and we are being shown, by the breakdown of the natural life-support systems that are organs of the living Biosphere, that we must understand and heed the Laws or pay for our arrogance.

Between sea and sky—footprints in the sand at Pajinka, on the tip of Cape York. *M.E.W.*

INVITATION to PART ONE

EARTH, AIR, LIFE AND WATER

The Elemental Biosphere

Come on a Voyage of Discovery
to find

Worlds within Worlds ... Microcosms and Macrocosms

Cells within Cells ... Endosymbiosis and Evolution

The Continuum of Life

(each and every individual organism a bacterial masterpiece)

and discover

The Symbiotic Planet ... Earth

Only now, in the twenty-first century, have technology and scientific knowledge reached a stage where many of the mysteries surrounding the creation of our Planet and the coming of Life to make it the Living Earth can be partially unravelled. In all the scientific disciplines an information revolution is taking place, resulting from technological advances. Space-traversing telescopes are demystifying the Universe on one hand. At the other end of the scale, ultra-microscopes and new analytic procedures are enabling the genomes of viruses, bacteria and humans to be studied, bringing evidence for Life's continuum and concepts of evolution that could not have been dreamed of when Charles Darwin was propounding his 'heresies' only a few generations ago.

The mental realms that open up to us as we begin to synthesise these fragments of information are boundless. We can see worlds within worlds—microcosms within macrocosms; we can go back to the Beginning when Life started on a hostile, lifeless young Earth and started to transform it into a Life-friendly planet; trace Life's evolution from bacterial cells that acquired as endosymbionts other bacteria that they had intended to digest; and envisage the development of multicellularity and the arrival of macroscopic life forms. Ultimately we can understand the modern living Biosphere against this background and recognise the interconnectedness of all things, living and inanimate, that comprise the Symbiotic Planet.

SATELLITE VIEW OF EARTH

A greenish tint has been added to the image of our 'blue planet' because it is photosynthesis, performed by green organisms, that provides the nutrients for the webs-of-life that make Earth a living world. *Dean Graetz © CSIRO*

Part One

EARTH, AIR, LIFE and WATER:

The Elemental Biosphere

The word **'Biosphere'** was invented by Austrian geologist Edward Seuss in 1875 in writing his physical descriptions of the Earth's surface, to delineate that portion between atmosphere and lithosphere (hard rock) affected by living things.[3] He described it as 'a specific, life-saturated envelope of the Earth's crust'. The Biosphere has become an infinitely greater concept than its originator could ever have imagined.

Seuss's definition was elaborated by Vladimir Vernadsky (1863–1945), a Russian scientist who saw living matter as a geological force, transforming the Planet's surface. His 1926 book *The Biosphere*[1a] showed how Earth's surface was an ordered transformation of the energies of the sun—that photosynthesis was fuelling the activities of the 'living matter' that effected the transformation.

The Gaia concept of James Lovelock[4,5] in 1979 introduced a theosophical–philosophical element suggesting an almost conscious regulation of all the factors for the benefit of Life. The evolution of Gaian thinking since has found the analogy of the Earth acting like a huge living organism a useful concept—for there can be no doubt that it is Life itself that maintains the Planet as a Life-friendly entity. (James Hutton[6], one of the 'fathers' of modern geology had, in 1788, already suggested that the Earth, 'this planetary body... may be termed a living world'—so the concept of a super-organism has been around for a long time.)

Life keeps global temperature within a suitable range. It has balanced the oxygen levels in the atmosphere over the last 700 million years, and has kept the saltiness of the oceans within satisfactory limits. According to Lovelock, the whole planetary system is 'metastable... stable in its reactive instability'. The persistence of chemical reactivity arises from the combined activities of living organisms. The entire planetary surface, not just the living bodies on it and in it but also the atmosphere that we think of as an inert background, is so far from chemical equilibrium that the Earth's surface is best regarded as alive. (In contrast, the dead planets have stable atmospheres and surfaces where all the chemical reactions between different elements have been completed.)

The 'Life' that maintains the dynamic Life-friendly situation is an infinitely complex and interconnected 'web' of living things that had its origins in bacteria about 4 billion years ago, and is still, believe it or not, bacterially based in the modern world. There is far more diversity and complexity of interactions in the invisible world than in the visible realm. The bacterial world could function quite satisfactorily without visible living organisms, just as it did in the Beginning, but the reverse is not the case.

The Biosphere is about 20 kilometres thick. Its top is the atmosphere and its bottom is continental rock and ocean depths. The whole lower layer of the atmosphere, the troposphere, is involved. It is up to 15 kilometres thick in the tropics, thinning towards the Poles to about 10 kilometres. It is the domain of the world's weather, and its highest reaches have been found to contain living matter—microbes, spores and small invertebrates.

The Biosphere is solar powered. It is an **open energy system** that reflects about a third of incoming radiation back into Space, allowing the rest to be absorbed by the atmosphere, land, waters and living organisms, and to be returned to Space as heat.

The Biosphere is a closed ecological system whose finite matter is, and has been since the beginning of Earth time, recycled again and again. (The cycling of elements and nutrients is described in Part Two.) Elements that are required to build the structures of Life and to maintain their complex metabolic processes are derived from the Earth's crust, from its atmosphere and from its waters.

Living organisms are composed of carbon-based macro-molecules organised in water-based tissues. Their basic composition is thus dominated by hydrogen, oxygen and carbon. The Biosphere's life results from the activities of the living organisms whose activities make its organs—soil, upper lithosphere, hydrosphere and atmosphere—alive.

The Biosphere as a living entity only came into being when the first life forms had reached a stage where their activities were influencing the whole realm—when they had formed a Web of Life and their numbers, wide distribution and activities were altering their environments, and the atmosphere, globally. In the beginning, colonies of microbes were surviving in a hostile world, evolving and changing their immediate locations. A recent review of the habitat and nature of early Life[223] concludes that by 3.5 billion years ago, 'most of the principal biochemical pathways that sustain the modern biosphere had evolved, and were global in scope'.

THE LIVING PLANET—GAIA

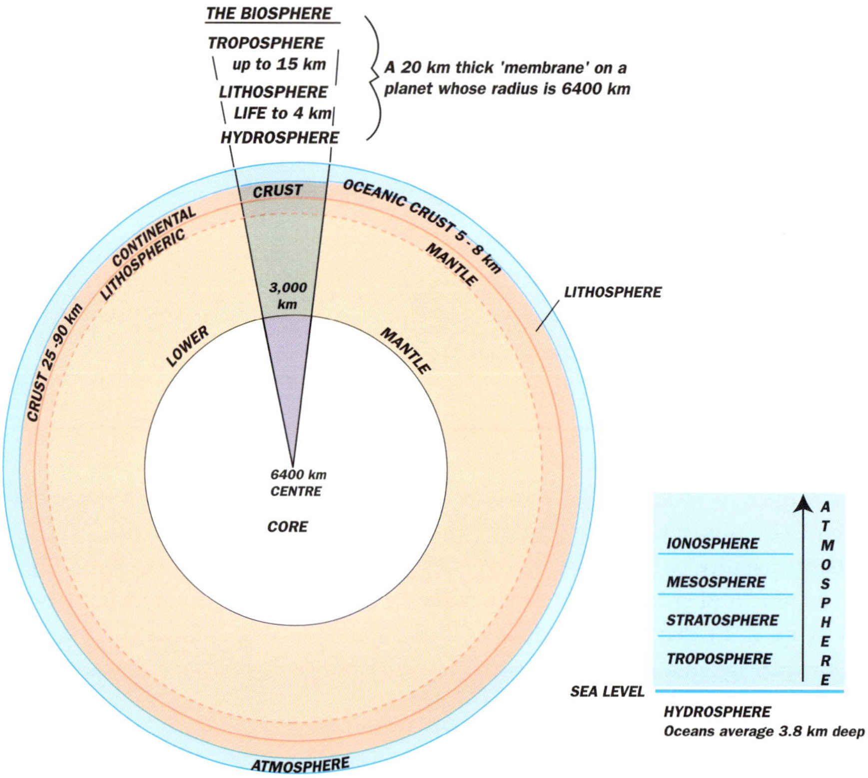

The living skin that keeps the planet Life-friendly is about as invisible in the Universal scheme of things as the microscopic Life that fundamentally runs the Biosphere is to Earth-bound humans. This 20 kilometre thick, Life-impregnated 'membrane' surrounds a heavenly body with a radius of 6400 kilometres—yet, surprisingly, it does the job.

So—is **Gaia the living Planet** perhaps a billion years younger than the Earth?

And if Life and the Biosphere are inseparable, what is Life?

LIFE IS THE BIOSPHERE?

According to Lynn Margulis, when this thinking about Life altering the Earth is taken to its fullest physiological extent, life is the planetary surface.[7] She says that Earth is no more a planet-sized chunk of rock inhabited by Life than your body is merely a skeleton infested with cells.

At every level, from microbe to Biosphere, organic beings use air and water or other organic beings to build their reproducing selves. As Margulis puts it, 'Local ecology becomes global ecology … life does not exist *on* Earth's surface so much as it is Earth's surface.' Life is thus the Biosphere—the zone of growing and self-regulating matter encasing our planet. Life's ***body*** is that whole enveloping skin, analogous to the living outer tissues enclosing dead heartwood in the trunk of a tree. Its ***vital organs*** are the atmosphere, the soil and upper lithosphere, and the hydrosphere. As we shall see, what makes matter alive is its power of self-maintenance (metabolism) and it is in this respect that the Biosphere can legitimately be judged to be alive. Metabolism involves a continual chemical and energy flow—Life produces, repairs and perpetuates itself.

Self-maintenance is known as **autopoiesis**, and in terms of the Planet is, according to Margulis:

> …the aggregate, emergent property of the many gas-trading, gene-exchanging, growing and evolving organisms in it. As human body regulation of temperature and blood chemistry emerges from relations among the body's component cells, so planetary regulation evolved from eons of interactions among Earth's living inhabitants.[7]

An autopoietic entity metabolises continuously; it perpetuates itself through chemical activity, the movement of molecules. Only cells, organisms made of cells, and biospheres made of organisms, are autopoietic and can metabolise. (All processes of metabolism are essentially electric when you break them down to the molecules and atoms that are being exchanged. The negative or positive charges carried by the units that are being metabolised determine whether they enter or exit a cell under whatever conditions prevail at that time.)

Oxygenic photosynthesis forms the nutritional basis for all Life, except for those bacteria (and the food chains that they originate) that still function on chemosynthesis as the earliest bacterial life did in the Beginning. Development of the light-harvesting green pigment chlorophyll by early organisms was the essential breakthrough. Biospheric Life depends on Plants in its terrestrial ecosystems; they depend on the soil to grow and flourish. In marine systems, algae and other protists, and cyanobacteria, are the photosynthesisers. It is no exaggeration to state that Life's most important invention in the history of the Biosphere was the evolution of photosynthesis.

Were the Biosphere to die by breakdown of its systems and a resultant inability to be self-maintaining, the Planet would become just another lifeless globe orbiting the Sun—and how close it came to that situation between about 900 and 600 million years ago in the 'Snowball Earth' episode we shall see as the story of the evolution of the Biosphere unfolds.

WHAT CONSTITUTES A LIVING ORGANISM ?

The **cell**, consisting of a membrane enclosing the 'stuff of Life', DNA and RNA, is the smallest unit of Life. The bacterial cell is this smallest unit. What makes it 'alive' is the metabolic processes by which it uses chemical energy and materials (food) for self-maintenance. The DNA and RNA molecules themselves are not alive. A cell or organism is dead when it ceases to metabolise.

Viruses, according to these criteria, are not strictly alive. They are too small to self-maintain, they do not metabolise, and they must enter a living and metabolising entity, be it a free bacterial cell or a cell of an organism, that will supply them with the products of metabolism, to be able to function. Some come in amazing shapes, now revealed by ultra-microscopes, and may even resemble ultra-microscopic hypodermic needles—why, no one can guess! (Biological viruses reproduce within their hosts in much the same way as digital viruses reproduce within computers.) Margulis[8] thinks that viruses originated as irradiated portions of bacterial cells living in sunlight, and that they spread genes among bacteria in the beginning and have continued to do so to all other organisms, including us, ever since. Populations of virus-infected organisms are then refined by natural selection.

(It is currently being suggested that Viruses should be classified in a separate Kingdom, and regarded as living organisms.)

In 1996, NASA scientists discovered 'nanobes', ultra-small organelle-like particles, in a Martian meteorite.[9] Controversy now rages because their size is but a fraction of the smallest recognised living and metabolising cell, and present classification would exclude them from Bacteria. A similar dilemma exists with filamentous and yeast-like 'cells' of very small size that have been found in sandstones from deep in the lithosphere, brought to the surface by petroleum exploration bores offshore from Western Australia. These show budding and are therefore replicating, but are they (or were they) able to metabolise?

By tracing Life's continuum from the Beginning some of these questions may be answered.

The evolution of Life

A logical sequence of changes to organisms and the environment is revealed when one goes back to the young Earth and first Life emerging, and follows the evolutionary sequence through 4 billion years, right up to the Web of Life as it functions now. The pattern that emerges is amazing and intellectually satisfying. At the outset one might be surprised by Margulis's answer to the question, 'What is Life?' She says 'Bacteria!' is a legitimate answer.[7] But when it can be seen that Bacteria invented multicellularity; that the complex cells that comprise our bodies and those of all other sorts of living things are a symbiosis of bacteria; that bacteria have mastered nanotechnology, controlling specific molecules; along with all the other wonders that will be revealed as you read on, surprise turns to acceptance and new mental horizons open up. The realisation dawns that the bacteria are incredibly highly evolved, not just primitive organisms from which all 'more highly evolved' organisms, including us at the 'top of the evolutionary tree', have evolved.

The bacterial cell does not have a membrane-bound nucleus segregating its genetic material. Cells without nuclei are described as 'prokaryotic' and Bacteria (and Archaea) are known as *prokaryotes*. Their cells also lack other membrane-bound inclusions known as organelles. All other living organisms—which have all evolved from bacteria—have 'eukaryotic' cells with membrane-bound nuclei and organelles, and are classified as *eukaryotes*. The eukaryotic cell is the result of the merger of different bacterial cells that have formed endosymbiotic relationships and operate as a unit, with the different endosymbionts having special functions and cooperating for the benefit of the more complex cell that results.

Fortunately for those who have sought to unravel the mysteries of Life's bacterial origins, modern technology has enabled understanding of the DNA of the different sorts of microorganisms that have persisted, so little changed, through billions of years. This has enabled recognition of the nucleus and other organelles and structures that occur in more highly evolved cells and tissues as derived from early symbiotic events which took place when bacteria ingested others which then became part of a functional unit. Modern research involving genes and proteins has developed to a stage where 'molecular phylogeny' can show relationships and create family trees of groups of organisms. This is revealed by **small sub-unit ribosomal RNA** (SSUrRNA), a genetically specified molecule which is a key constituent of ribosomes, the 'factories' that construct protein in cells. The genes encoding it indicate the relatedness between any life-forms, from the most minute bacterium to the most complex animal. SSUrRNA acts as a 'universal molecular chronometer'.

THE SERIAL ENDOSYMBIOSIS THEORY OF EVOLUTION

The Serial Endosymbiosis Theory of Evolution (SET), as formulated by Lynn Margulis, encapsulates this knowledge, showing that **symbiogenesis**—the formation of new organs and organisms through symbiotic bacterial mergers—has seen the transformation of bacterial Life, through increasingly complex organisms, to all the manifestations of living things on Earth today. The demonstration of this symbiogenetic sequence has enabled a new classification of living organisms involving the erection of a new, fifth, Kingdom for organisms that are not Bacteria (as they have cells with nuclei) and do not fit comfortably into the other Kingdoms.

For these organisms that are generally referred to these days as protists, Margulis has used the Kingdom name Protoctista (the Protoctists) but a more conservative approach, adopted by most scientists, uses the Kingdom name **Protista**. This fifth Kingdom includes the Algae, which were previously accommodated in the Plant Kingdom. The Plant Kingdom now has its genesis in the first land plants that evolved from a green algal ancestor about 450 million years ago.

The conventional classification of organisms into four Kingdoms—Bacteria, Fungi, Plants and Animals—has always presented a few problems, with a large number of 'protozoa', 'ciliates', etc. having characteristics of both Plant and Animal Kingdoms: green cells that photosynthesised but engulfed and digested food; 'animalculi' that were plant-like in some respects; and many organisms that had indeterminate characteristics of several Kingdoms.

ARE THESE STRANGE CREATURES ANCESTORS OF THE ANIMAL KINGDOM?[10]

Scientists at the Howard Hughes Medical Institute at the University of Wisconsin-Madison have found key genetic links between single-celled flagellated organisms classed as choanoflagellates (Protists) and the Animal Kingdom. They believe that this group of protists is the 'Animal Ancestor' that biologists have been searching for since it was realised that Life has microbial beginnings that can be traced through 4 billion years. The choanoflagellates have genetic traits that, until now, had only been found in animals, including a key gene that is involved in communication mechanisms between those organisms and in cellular communication in animals. It has been realised for some time that conservatism is basic to evolutionary processes and it is now apparent that Nature, instead of inventing new genes to accomplish new tasks, has allowed animals to use existing genes in different ways. Evolutionary relationships can thus be traced by looking for common genetic themes.

Professor Sean Carroll, leader of the research group, has said that the choano-flagellates that are a thriving group today:

> are the closest non-animal organisms to animals [and that] they are to animals what chimps are to humans and, by studying some of their genetic characteristics, we can begin to make some strong inferences about how animals evolved.[10]

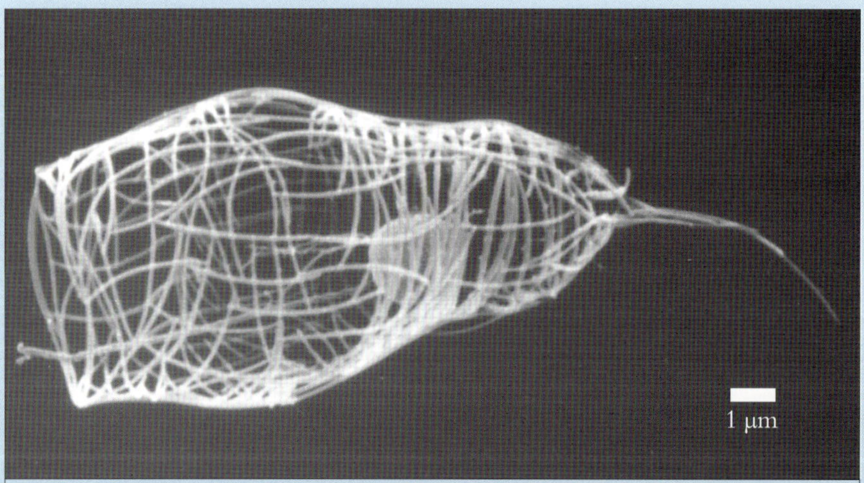

Stephanoeca complexa, a choanoflagellate. Choanoflagellate classification is based on the structure of their lorica—the stick-like house in which the protozoan lives. *Harvey Marchant, AAD*

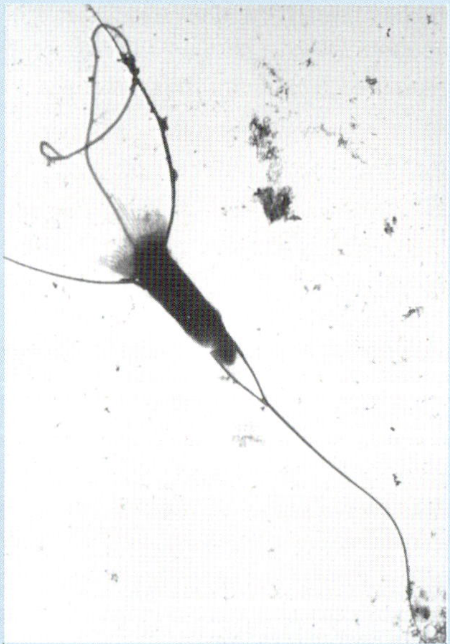

Bicosta spinifera, a choanoflagellate from Antarctic waters. This 5μm long, single-celled organism generates a water current by beating its flagellum amd traps food particles on the tentacles surrounding the base of the flagellum. **Choanoflagellates can filter up to 25 per cent of surface oceanic water per day.** *Harvey Marchant, AAD*

The Protists are a highly diverse group of organisms whose inclusion in a separate Kingdom is based only on their incompatibility with the other well-delineated Kingdoms. As DNA testing proceeds and improves and relationships become clear, the different ancestral lines of evolution from members of this group will be established, clarifying origins of the Fungi, Plant and Animal Kingdoms, and perhaps also showing different ancestry for the brown, red and green algae, for instance. Protists survive today as important organisms in the symbiotic Biosphere.

In addition to the role of endosymbiosis in creating more complex life forms from bacterial beginnings, the Serial Endosymbiosis Theory view of evolution sees endosymbiosis as being the mechanism involved in **the creation of new species.** Acquisition of bacterial or viral genes by organisms has created individuals with different attributes, advantages and disadvantages, and natural selection has worked on these variants over time until eventually their differences became accentuated to a point where they could no longer interbreed—thus becoming new species.

It has been stated that no appearance of a new species has been documented by science, but one recorded example may be pertinent.[7] Theodosius Dobzhansky's experiments with *Drosophila* populations bred at increasingly hotter temperatures resulted in genetic separation of different strains. After two years, hot-bred individuals could not breed with those bred under cold conditions. The difference found in the hot-bred individuals was the absence of a 'parasite', a cellular endosymbiont now found only in those that were cold-bred. (Margulis believes that the 'punctuated equilibrium' documented in evolution, where long static periods are interrupted by bursts of new species, is related to the process seen here—where times of environmental change see selection of individuals in whom new symbiotic relationships better suit them to the changes.)

HOW LIFE BEGAN: THE ACT OF CREATION—A BACTERIUM, THE 'GERM OF LIFE'

It has been shown that simple sugars like fermentable glucose, the ribose of RNA, and even DNA, can be synthesised in a laboratory by passing a lightning-like electrical discharge through a soup of chemicals. This simulates the energised atmosphere of early Earth, when basic chemical requirements were present in abundance. Many rich organic compounds have formed spontaneously in the laboratory experiments. No one has yet been able to create Life, however—that is, imparting the power to metabolise. Some experiments have come close, and perhaps in time they will succeed.

The first minute bacterial cell with the ability to metabolise can be visualised as a bubble-like, oily nano-droplet containing the carbon compounds that were being spontaneously generated in the hot, poisonous and highly volcanic Earth. The oily droplets eventually became membrane-bounded cells, and the power to metabolise that they had acquired has been handed on, so that today all living things can trace a single lineage back to that 'Germ of Life' ancestor—yours, and mine, and that of every bacterium and other organism that has lived since, through 4 billion years, and is living now. So in a very real sense we are part of the Biosphere and **Life is one continuum.**

THE THREE SUPER-KINGDOMS

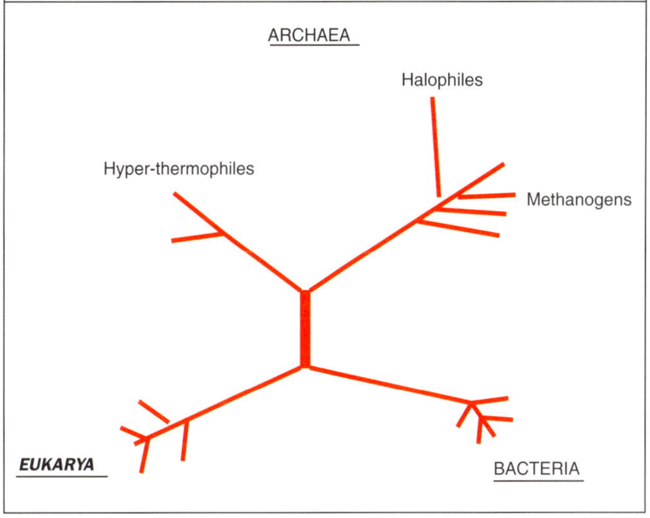

It is suggested[223] that the earliest replicating system, ancestral to the ribosome, was an RNA molecule capable of acting as a ribozyme and in some way self-assembling. Weight is given to this theory by the fact that the ribosomal unit that is basic to cell structure is, in fact, a giant ribozyme supported by a framework of caring proteins. RNA-life may have lived in pores of rock in hot spring environments, most likely those associated with alkaline volcanics, where gases and dissolved nutrients, including phosphorus, were readily available. Such an embryonic environment would account for the unusual basic components that are obligatory to cellular 'housekeeping chemistry'. These include many of the metal proteins, particularly those involving iron, sulphur, nickel, molybdenum, copper, cobalt and zinc.

The closest living organisms to the *first* bacterial cells (Archaea) are today inhabitants of extra-hot and poisonous places like volcanic hot springs and sulphurous vents on mid-ocean ridges. Here, **extremophiles** with bizarre life styles, chemistry and physiology are providing us with a window that allows us to see into the very beginnings of Earth time. They show us the ancestral line from which the **Super-Kingdoms** of Bacteria (Monera: prokaryotes) and the Eukaryotes (Protists, Fungi, Plants and Animals) originated. A widely accepted modern classification elevates these first organisms and their thermophilic, methanogenic, halophytic descendants (heat-loving, methane-producing, salt-tolerant extremophiles) to a third **Super-Kingdom, the Archaea.**[11]

A recent assessment of what the early Earth was like, before there was Life, emphasises just how extreme conditions were in the Beginning.[223] During that 500 to 700 million years, the planet was subjected to bombardment by the numerous

The Burketown bore in Gulf country of northern Australia, where water from the Great Artesian Basin comes to the surface from a depth of more than 600 metres. The water emerges at 68°C and has created a pyramid of sulphur- and mineral-rich sediment on the crusted surface that surrounds the bore pipe. The structure is rich in extremophiles comprising algae and micro-organisms of many sorts, and they abound also in the heated water in the adjacent dam under less extreme conditions. *M.E.W.*

planetesimals that occupied the region now occupied by the inner planets. The largest impacts are believed to have been capable of re-melting the Earth; according to one scenario, at about 4.5 billion years ago the impact of a body as large as Mars spun and tilted Earth and ejected enormous amounts of molten mantle into orbit. This 'big splat' resulted in the formation of the Moon. The spin and tilt of the Earth as it is today evolved from this event, giving day and night, the seasons, and the tides that are governed by the Moon and assist in dissipating heat around the planet and making it habitable. Following that cataclysmic impact, and until about 3.8 billion years ago, other lesser impacts probably heated the planet episodically beyond levels suitable even for hyper-thermophilic organisms.

The early organisms that may have evolved from around 3.8 billion years ago, existed in a hazardous world, probably only in hydrothermal environments. These earliest life forms may have been non-photosynthetic. It was the development of anoxygenic and then oxygenic photosynthesis that would eventually enable Life to move out from those hydrothermal environments. According to some experts who have tried to visualise the primaeval Earth, by about 3.5 billion years ago most of the principal biochemical pathways that sustain the modern Biosphere had evolved and were global in scope.[223]

PROKARYOTES: ABUNDANCE AND BIODIVERSITY[220]

When figures for global biodiversity are presented they seldom include the prokaryotes, yet this essential component of the Earth's biota represents a large proportion of Life's genetic diversity. Estimates of their astronomical numbers are beyond comprehension—it is easier to grasp their abundance in terms of their contribution to global pools of organic carbon, nitrogen and phosphorus.

The total amount of prokaryotic carbon is calculated to be 60–100 per cent of the estimated total carbon in plants. This means that estimates of the amount of stored carbon in living organisms, that up to now have omitted prokaryotes, must be practically doubled. Prokaryotes also represent the largest pools of nitrogen and phosphorus—each about tenfold greater than the amounts present in plants.

Recent estimates for the numbers of prokaryotic species, taking into account their enormous potential to acquire genetic diversity by mutation, suggest that a billion might be a possible number. However, the very definition of a species, which is based on eukaryotes, does not accommodate the situation that is found in prokaryotes. Bacteria can simply swap genetic material; they are haploid and do not have to pair with a compatible partner to reproduce—they have only to divide to produce a new organism.

Algae add colour to the water in the bore dam at Burketown. *M.E.W.*

The Archaea

Accepting that Archaea, extremophiles from extreme habitats in the vicinity of hydrothermal systems in the Beginning, were the common ancestors of all other life forms, may help account for the 'unusual' trace elements that are obligatory in DNA. If Life's housekeeping chemistry evolved in environments where metal sulphides were present and abundant, their continuing requirement by all living things represents an inheritance from the most ancient part of an extraordinarily ancient lineage. Nickel may reflect origins near high-temperature lavas, while the copper, zinc and molybdenum found in proteins are metals typical of hydrothermal deposits. Even the role of heat-shock chaperonins, which play a crucial role in the assembly of complex proteins, may date from the hazardous environment near hydrothermal systems where Life began; the role of manganese ions in the splitting of the water molecule in photosynthesis might be another example. Why otherwise would comparatively rare elements in normal environments be obligatory for basic life processes?

Specialisation for high-temperature environments

Hyper-thermophilic environments in which temperatures are between 40° and 70°C (and much higher in hydrothermal vent environments) require biochemical specialisation in their inhabitants. Their living proteins must remain stable under the conditions of high temperature (and high pressure in deep-sea organisms). Such hot environments are today more widespread than one might at first assume, occurring not just in natural environments such as desert soils, but also in compost, industrial waste dumps and even domestic hot-water supplies, to mention only a familiar few. Not only heat tolerance, but tolerance of strong chemical solutions and toxins may be required. There is a growing belief that an important, possibly vast, thermophilic biosphere exists below the Earth's surface. Hyper-thermophilic Archaea have been found in Alaskan oil reservoirs; microbial biomass is increasingly being detected in deep drilling samples worldwide. Bacteria-like cells have been found in hard sedimentary rock about 15 million years old brought to the surface in bore cores from 850 metres below the sea floor off New Guinea.[12]

The 'Underworld'[13] in which life exists deep in the lithosphere, possibly to 4 kilometres below the surface, at rock temperatures of 130°C or more, is an oxygen-free place where living organisms are believed to be a million times less abundant than in the top layers of soil. They may divide to replicate themselves about once a century, but they still represent one part in a thousand of the entire mass of Life on Earth.

Many Archaea that exist at temperatures at or above 100°C contain novel macro-molecules and metabolic systems which represent a resource for fundamental molecular and physiological studies, and may be a source for potential exploitation in biotechnology.[14] Enzymes isolated from extremely thermophilic micro-organisms show not only resistance to denaturation by heat but resistance to a number of other influences as well—detergents and organic solvents, for instance. The use of hyper-thermophilic DNA polymerases and other DNA-modifying enzymes has already contributed significantly to development of recombinant DNA technology and gene-based diagnostics. (Going back to our own beginnings and the beginnings of all Life for solutions to modern-day health problems!)

When the Archaea first arose they occupied anaerobic aquatic environments rich in sulphur, carbon dioxide and hydrogen. As free oxygen, which came later, is insoluble in boiling water, there was no strong evolutionary pressure for the acquisition of aerobic metabolic systems in high-temperature aquatic environments. Most Archaea are still anaerobic, synthesising their own nutrients in unique metabolic processes, using novel enzymes, co-factors and pathways, and producing methane as a by-product. The Sulfolobales, a group with aerobic metabolism, are a comparatively recent evolutionary divergence.

(A report in *Nature*, referred to in a item in the weekend edition of the *Sydney Morning Herald* of 4–5 May, 2002, announced the discovery of the first member of a new family of microbes, smaller than any known bacteria.[233] It has just been captured in the scalding hot water that emerges under pressure from the seafloor near Iceland and is unlike any other known living organism. Larger than viruses (which are not considered alive), it lives on larger microbes, though whether the relationship is parasitic or symbiotic is not known. It is 'close to the theoretical genome size for a living being' and is believed to comprise a new category within the Archaea; it may represent the first member of a new class of organism. It is an extremophile, living in a high-temperature anoxic environment, like the ancestral Archaea did in the Beginning. Studies of the genome of this ultra-microscopic organism should prove to be most interesting and informative. New frontiers are being opened up at an exciting rate in this origins-of-Life field of science.)

Bacterial evolution from Archaea: the birth of the solar-powered Biosphere

According to current thinking the very earliest living cells did not have to synthesise nutrients.[7] Pre-Life processes such as solar radiation, the restless volcanism of the cooling Earth, and lightning, that had in fact created the first organisms, had also produced organic compounds that they could break down and use as 'food'. 'Primary producers' were only required when the primaeval food source was depleted. With environmental change the driving force in evolution, as it still is today, Life had to develop new strategies. Fermenter bacteria then began to use solar energy to synthesise their own food, and the great lineage of green and purple sulphur bacteria was initiated.

It is believed that photosynthetic life forms were in existence well before 3 billion years ago, possibly as much as 3.5 or 3.7 billion years ago.[15] Purple bacteria and cyanobacteria did then what mitochondria and chloroplasts do today—so Archaean ecology was using the same basic biochemical cycles as are seen on modern Earth.[223] Methane-producing Archaea operated in the same way then as they do in the rumen of cattle today.

Some of the first photosynthesisers were probably green-coloured sulphide scavengers. Such organisms were **autotrophs**, organisms that can synthesise their own food. (**Heterotrophs**, like us, cannot synthesise the food they need—it has to be supplied at the base of a food chain by photosynthetic autotrophs.) The modern anaerobic bacterium *Chlorobium vinosum*, a green sulphur bacterium that today is confined to anoxic places, is similar to its ancestors. Its relatives in the early world, however, were not restricted by the presence of oxygen in the atmosphere. Free hydrogen and CO_2 were abundant, as was hydrogen sulphide, produced by volcanoes. The bacteria used the H_2S, and their waste products were elemental sulphur and methane. (Unlike later photosynthesisers, they did not produce oxygen as a waste product.)

Other early prokaryotes were red. They contained **rhodopsin**, a light-sensitive pigment that could capture and store solar energy, using the same chemical mechanisms as are active in normal photosynthesis, but involving light from a different part of the spectrum. Today, salt-tolerant anaerobic sulphur bacteria still use rhodopsin as their photo-energy receiver, and examples of ancient ancestral inheritance are seen in marine fish that have rhodopsin in the retinal rods of their eyes. Moreover, human vision is aided by its presence in *our* eyes, especially useful when light is dim—a legacy of our fishy ancestry. (All vertebrates evolved from fish.) A further photosynthetic advance came in the form of purple sulphur bacteria that were able to tolerate oxygen, even if only in limited doses. Purple sulphur bacteria today can tolerate oxygen only at night or under cover.

Purple sulphur bacteria that had evolved a unique green chlorophyll system as their photo-energy receptor mutated at some stage to become the **cyanobacteria**. Earliest evidence for cyanobacteria-like organisms is found in 3500 million-year-old stromatolites in Western Australia (see page 22). There is still no way of knowing how they functioned, but at that early stage it would be logical to assume chemosynthesis or photosynthesis of a

BLACK MOUNTAIN

Black Mountain, 26 kilometres south of Cooktown in north Queensland, is an amazing pile of black rocks. It is a granite outcrop whose outer crust has exfoliated into boulders, rocks and stones of all sizes that remain stacked on the largely underground solid core. Almost without vegetation, riddled with caves and crevices, it has an unusual aspect for a granite body—no pile of well-rounded boulders or monolithic plug in a sandplain, just a jumble of black rocks that, on splitting open, reveal pale-coloured granodiorite containing mica and black specks of hornblende. The blackness of the surface is due to three species of cyanobacteria—extremophiles that survive the high temperatures reached by the rock surfaces in summer, and desiccation, presumably having active times when rain reactivates the organisms. *M.E.W.*

THE HISTORY OF CYANOBACTERIA AND OXYGENIC PHOTOSYNTHESIS[16,17,18]

Advances in molecular chemistry have been instrumental in the determination of dates by which different types of organisms had evolved in the distant geological past. The three domains of extant life, the Bacteria, Archaea and Eukarya, have diagnostic lipids in their cell membranes which, in the course of geological time, turn into hydrocarbons. The presence of different molecules in hydrocarbons, derived from the lipids specific to members of the different Super-Kingdoms, becomes a **biomarker**. If the hydrocarbon in a sediment is the product of organisms active at the time of deposition of the strata, and not derived, then the age of the organisms is at least as great as that of the sediments.

In Bacteria, substances known as bacterio-hopanepolyols (BHP) serve a regulating and rigidifying function in the cell membrane. The hydrocarbon skeletons of BHP are extremely resistant and become incorporated into kerogen or bound in sulphur-linked macro-molecules. Both extant cyanobacteria, which perform oxygenic photosynthesis, and their ancestors, have a distinctive 2-methyl-BHP molecule in their cell membranes. Hydrocarbons derived from ancient oxygenic photosynthetic cyanobacteria thus contain diagnostic 2-methyl-BHP biomarker molecules.

In eukaryotes, the lipids that give rigidity to cell membranes are sterols, and methylated steranes found in hydrocarbons are biomarkers for eukaryotic cells.

The first biomarker-determined dates for the presence of both oxygenic cyanobacteria and eukaryote cells come from 2.7 billion-year-old shales from the Pilbara Craton in Western Australia. The kerogen-containing shales of the Roy Hill and Marra Mamba Formations were deposited in a marine continental-slope environment. This is the earliest record of eukaryotes. Much older records of the occurrence of cyanobacterial-type organisms exist—the oldest in stromatolites from 3.5 billion years ago—but up to the present it has not been possible to determine how these organisms, which look like filamentous cyanobacteria, functioned. (See Stromatolites, page 22.) True cyanobacteria presumably arose even before the 2.7 billion year date established by the biomarker.

Cyanobacteria had a pivotal position in the evolution of Life and the Biosphere. The oxygen they produced first oxidised matter in the oceans (the banded ironstone deposits laid down between 2.7 and 2 billion years ago owe their existence to cyanobacteria); then when oxygen started to leak into the atmosphere the world was being prepared for evolution of Life beyond the microbial stage.

Today, cyanobacteria are widely distributed, occurring in all sorts of habitats globally, on land and in the oceans. Some occur as single cells; some form colonies inside gelatinous sheaths; some are filamentous. They are unique in that they are capable of nitrogen fixing as well as oxygenic photosynthesis. (Plants can't fix nitrogen.) Like all prokaryotes, the cells of cyanobacteria are without nuclei or other organelles. The photosynthetic pigments are located within or on membranes (thylakoids) within the cytoplasm, not confined in chloroplasts, and the genetic material is not confined by a membrane.

Cyanobacteria are enormously important in the ecology of the warmer oceans, fixing nitrogen, supplying carbon for the organisms that eat them, and thereby having a major impact on nutrient availability. *Synechococcus*, a species detected only in the last decade, is now known to be responsible for as much as 25 per cent of the oceans' productivity. Other species bloom in the seas as well, *Trichodesmium* being a well-studied, nitrogen-fixing example.

On land, the cyanobacteria may be free living, or form symbiotic relationships, or form communities in soils and in bacterial mats; their activities greatly influence nitrogen and carbon cycling. They are responsible for black stains on paths and in other damp places; and they may form a felt-like surface layer in bare sandy soil or friable loamy soil. They occur as symbionts within the fungal body in some lichens. Some genera, like *Scytonema* and *Calothrix*, are able to be either free living or symbionts in lichens.

Cyanobacteria are important constituents of the biotic crusts that colonise bare soil in arid regions. They are widespread even in the gibber plains of semi-arid and arid Australia, living under pebbles, especially of translucent quartz and chalcedony. In a paddock near Broken Hill, 8 per cent of the surface was found to be covered by cyanobacteria living on the underside of pebbles.[19a]

Microcoleus vaginatus is a common free-living cyanobacterium in crusty soils. It consists of a bunch of finger-like filaments enclosed in a large sticky sheath, rather like a tube worm. When the soil is wet the sheath swells, forcing the filaments out into the soil. As the soil dries out, the filaments manufacture their own sheath material and eventually the soil is crowded with these sheaths, like the fibres in fibreglass. Old sheath material can survive for many years, increasing the organic content of soil and acting as a nutrient source for mites and termites.

In the arid soils of Central Australia, nitrogen-fixing cyanobacteria, free-living, plant-associated and symbiotic, have been found to be contributors to significant nitrate levels in groundwater.[19] Soils near Yulara and in the Ti-Tree Basin, in spinifex plains interspersed with mulga groves, were sampled. Free-living nitrogen-fixing cyanobacteria were found to be widely distributed and thus a potential source of fixed nitrogen in the soil. *Scytonema*

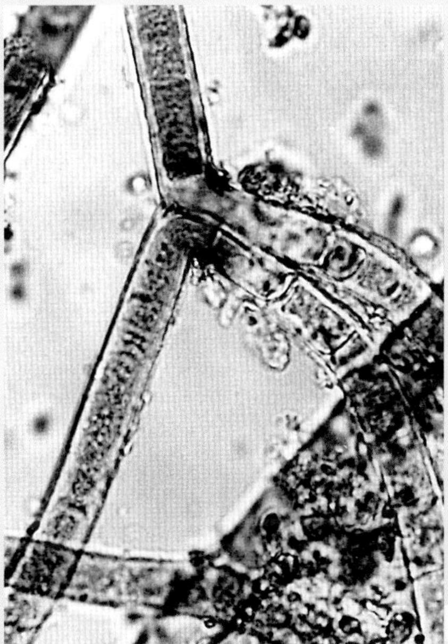

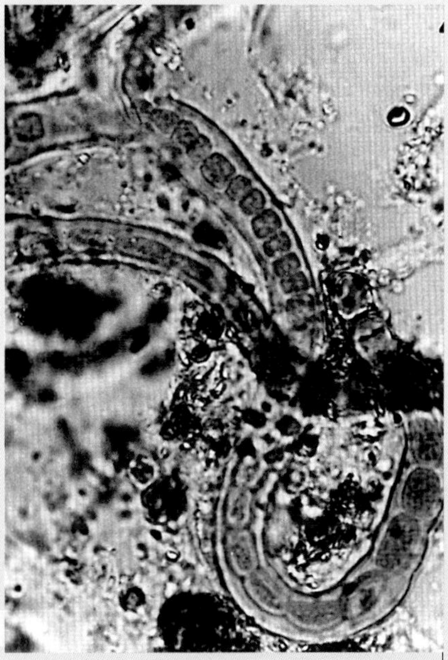

CYANOBACTERIA FROM CENTRAL AUSTRALIA

Chains of cells of cyanobacteria, showing false branching, enclosed in gelatinous sheaths.
Dr Geoffrey Smith, ANU

and *Oscillatoria* were abundant forms. The very low levels of available organic carbon in the sandy desert soils of the region presumably results in diminished rates of nitrogen cycling via denitrification, leading to an accumulation of nitrate in the soil—and ultimately in groundwater through the process of leaching.

The remarkable ability of cyanobacteria to survive the heat and dryness that prevails most of the time in Central Australia, and to come alive and function in brief, infrequent, rainfall events was revealed in these studies. These organisms have metabolic strategies that enable them to be the first colonisers of barren soils when optimal growth conditions return. *Nostoc commune*, also found in these soils, is known to possess a novel water-stress protein that makes it drought resistant.

The presence of cyanobacteria in soils on Antarctica shows adaptation to conditions at the other end of the temperature scale, although they are virtually absent in Antarctic marine phytoplankton.

Cyanobacteria have been found to be a rich source of chemical substances that are by-products of their metabolic processes. Some are highly toxic to animals, including humans, others may prove useful in agriculture and veterinary science, and many are being investigated as possibly useful in the manufacture of pharmaceuticals. (See below for further information on cyanobacterial metabolites, their function and possible medical use.)

ANTI-PHOTOSYNTHETIC CHEMICAL SUBSTANCES PRODUCED BY CYANOBACTERIA[48,49]

It has been discovered that some of the many metabolites that cyanobacteria produce are directed against the oxygenic photosynthetic processes, which in the microbial world are unique to algae and cyanobacteria. Presumably the chemicals are involved in the regulation of natural populations, by either encouraging or discouraging growth and function in neighbouring micro-organisms. Some are algicides, others are poisonous to other species of cyanobacteria. In natural habitats where cyanobacteria are found in communities and in close proximity to other bacteria and micro-organisms, there is competition for limited nutrients, light sources and surface[9] to which they can attach, and having a chemical weapon is an advantage.

The freshwater cyanobacterium *Scytonema hofmanni* is slow-growing. Laboratory experiments to find the effect of its metabolites on other organisms have shown their effectiveness and indicate that without chemicals to restrict other more prolific organisms, *S. hofmanni* could not compete successfully for space or nutrients.

Most of the cyanobacterial chemicals that have been characterised as algicidal are directed against photosynthesis and are therefore natural herbicides. Those whose function is known inhibit electron transport in the cell that they attack, making photosynthesis inoperative. (A flow of electrons is essential for the splitting of the water molecule in oxygenic photosynthesis.) Other toxins inhibit enzymes that facilitate uptake of phosphorus, which is essential for cell function; others directly affect the cell membrane and all the other biochemical processes, including photosynthetic ones, that occur in it.

Cyanobacteria in medicine

Over recent years, research into other metabolites produced by cyanobacteria has been increasing, and cyanobacterial toxicology has become a subject of worldwide significance. Two important and exciting advances have been made by Australian scientists[49] in identifying toxins that show potential for the control of malaria parasites and some cancer cells in humans. This research is ongoing and not yet conclusive.

Cell extracts from cultures of two *Calothrix* species were found to inhibit the growth of a chloroquine-resistant strain of the malaria parasite *Plasmodium falciparum*, and of human HeLa cancer cells. With malaria causing 3 million deaths each year and a cancer cure still elusive, there is naturally much interest in this medical context. In the case of the malarial parasite, it is interesting to note that the *Calothrix* toxin targets a plastid-like organelle within the cells that is a relic of endosymbiosis with a green algal cell or cyanobacterium (like the endosymbiosis that created a chloroplast in plant cells). In other words, the cyanobacterial poison still recognises the chemical signal of the rival cyanobacterial 'ghost' that is now part of an animal cell, and homes in on it.

Many cyanobacteria produce substances poisonous to humans and other animals that are among the most lethal substances known, including nerve, liver, cell and general toxins. The use of some as biocontrol agents (fungicides, herbicides, pesticides), and in many other applications, is the subject of ongoing research by scientists in the Research School of Chemistry and the School of Biochemistry and Molecular Biology at the Australian National University in Canberra.

kind that did not involve splitting the water molecule and releasing oxygen. The big breakthrough in photosynthesis was to occur when ancestral cyanobacteria began splitting the water molecule, H_2O, using its hydrogen and releasing the oxygen as a by-product.

In the early Earth, bacterial mats forming in shallow-water habitats would have had increasingly aerobic cyanobacteria in their surface layers, organisms tolerant of some level of oxygen and low light levels or darkness in the middle layers, and anaerobic, non-photosynthetic organisms in the bottom layer. The anaerobic organisms produced methane, but the aerobic ones liberated oxygen, gradually building up its concentrations in the water, and later in the atmosphere, and preparing the world for Life-as-we-know-it.

It is no exaggeration to say that cyanobacteria, with their oxygen-releasing photosynthetic activities, changed the world and enabled evolution to proceed beyond its anaerobic archaean and bacterial beginnings.

(Ten thousand extant forms of cyanobacteria have so far been identified, but their enormous importance in the functioning of the Biosphere is, like the continuing importance of other micro-organisms, scarcely acknowledged.)

Development of mobility in Bacteria: Spirochaetes and their legacy as endosymbionts

Among the early Bacteria were some that developed **flagellae** that beat rapidly, enabling the organisms (spirochaetes) to swim to new food sources. Flagellae are wonderful examples of nanotechnology—corkscrew-shaped, spinning protein filaments attached to living motors in the membranes of their cells. Complete with rings, tiny bearings and rotors, they are **'proton motors'**, spinning at 150 revolutions per second.[43] (Like an electric fan or an outboard motor!) When such bacteria became endosymbionts, their flagellae became proton motors in the membranes of the mitochondria and chloroplasts which they lost their independence to become, regulating the flow of chemical substances in appropriate directions through the membranes. In this permanent fusion the ingested bacterium lost its independence, and was represented in the composite cell by a small dot or spot, the **centriole**, at the base of the motor.

Another legacy of flagellated organisms that became endosymbionts is seen in the type of cell division that characterises eukaryotic cells. Mitosis involves the 'dance of the chromosomes', an ordered sequence of movements in which chromosomes divide and move apart, following threads of a 'spindle' that guides them to opposite ends of the dividing cell. This ensures that the nuclei in the two new cells that are made from the original one contain the same number of chromosomes with identical gene sequences. This ordered movement originated from the incessant movement of the endosymbiotic motile bacteria that had been taken in by non-nucleated bacterial cells, the process that converted the host cells into eukaryotes. The microtubules of the spindle in mitosis are remnants of the endosymbiotic bacterium.

(Microtubules also make up cilia, and sperm tails. The axons and dendrites in our brains—extensions of nerve cells by which we process information—are underlain by microtubules. Symbiont bacteria with flagellae also became specialised sensory cells in animal tissues, their flagellae becoming sensory hairs, and their inbuilt ability to detect light, specific chemicals, electrical discharges and even magnetism becoming of inestimable value to their recipients.)

Flagellates still swim in oxygen-rich and oxygen-poor environments, in soil pores and in all other watery places, including the guts of wood-eating termites and humans (and in human testicles). They can attach themselves so skilfully to other micro-organisms that their points of attachment are often initially thought by scientists examining them to be centrioles, and their flagellae to be cilia of their host. Such attachments confirm both how endosymbiosis operated and how evolution from a bacterial ancestor proceeded.

CASE STUDY: THE AUSTRALIAN NATIONAL UNIVERSITY

THE AUSTRALIAN NATIONAL UNIVERSITY

The ANU was established by the Federal Parliament in 1946 specifically to lead the development of the intellectual capacity of the nation through research and training in line with the best international standards. As the specially chartered national university, the ANU continues to conduct research at the highest international rankings in all of its Research Schools, Faculties and Centres, offering a unique research-led education to undergraduate and postgraduate students.

ANU has a world reputation for cutting edge research in the humanities, the sciences and the social sciences. The university is broadly organised in twelve National Institutes. Of these, the National Institutes of Bioscience, Earth Sciences, Environment and Health all carry out research and teaching that are relevant to this book—*Earth Alive!*

The Research School of Biological Sciences, part of the National Institute of Biosciences, carries out basic and strategic research in bioscience and biotechnology across plant and animal systems. Programs include Functional Genomics, Global Change Biology, Environmental Interactions and the Visual Sciences. There is special emphasis on the Australian Biota. The research of the National Institute of Bioscience underpins Australia's natural advantages in agriculture and food, biomedicine and health, bioengineering, environment and conservation.

The sea was to become the burgeoning realm of bacterial and viral life as the Earth cooled, with unicellular life evolving there for the first 2 billion years of Earth history. Our planet today is still as much a microbial world as it was in the Beginning, as technology now enables us to begin to understand. The sea is still a bacterial and viral realm, though most of us could never have imagined that this would be found to be the case.

By way of a slight diversion, before we continue this sequential journey through time, tracing the evolution and functioning of the Biosphere, let us enter the Southern Ocean to glimpse an almost unimaginable realm in which microscopic organisms reign supreme. Here we will see webs of Life; the cycling of materials needed for Life; the influence of microscopic Life on the global environment; and the merging of the animate and inanimate in the Hydrosphere where all Life began.

BACTERIA AND VIRUSES CONTROL THE LIVING OCEANS

THE MICROBIAL LOOP

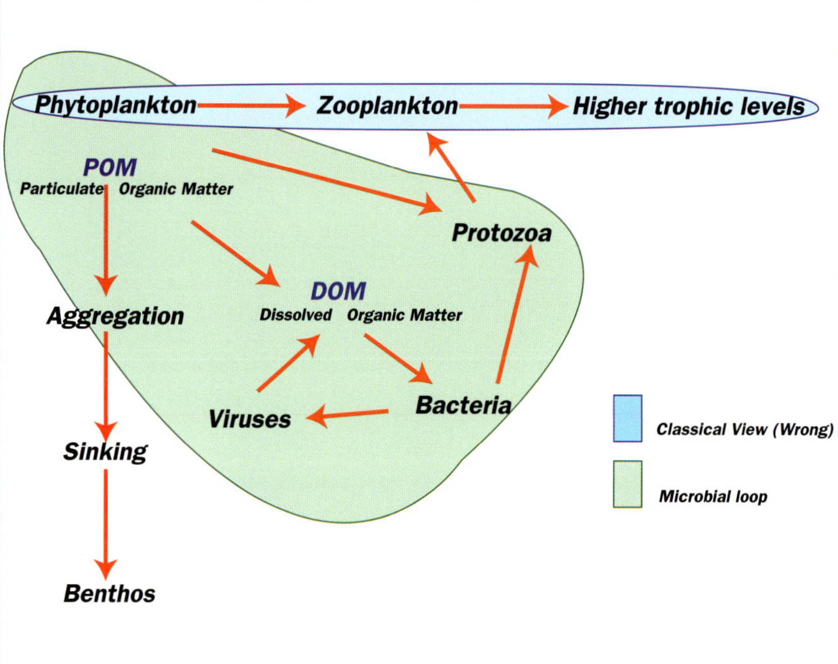

After Marchant[22]

The Microbial Loop represents the modern view of the complex of interactions between phytoplankton, bacteria, protozoa, viruses, DOM and POM, in contrast to the old view that ignores this complexity.

Life in the primaeval oceans when the Earth was young was exclusively bacterial. Microbes were the only life form in the Biosphere for 2 billion years until more complex organisms evolved from them. They modified the evolving world, making it suitable for more highly developed life forms, and they continue to exert control, mainly invisibly and little acknowledged, in the complex living Earth that we know today. The continuum of microbial life from the earliest times to the present day, and its persistence as the foundation for the living Biosphere, is nowhere more clearly revealed than in the modern seas. In recent years scientists of the Australian Antarctic Division have been revealing the secrets of the microbial control of the Southern Ocean.

The Southern Ocean, between the Polar Front and Antarctica, an ocean full of viruses and bacteria? Ultra-microscopic, 10 billion viruses and 1 billion bacteria per litre?[20] And astronomical numbers of other micro-organisms comprising the single-celled plankton in the size range from 0.2–200μm? (μm, a micrometre, is a thousanth of a millimeter.)

In temperate seas, viruses have been found to be consistently the most abundant biological agents; this dominance has now been shown to apply also in the high-latitude Southern Ocean around Antarctica. Viruses consist of genetic material (DNA or RNA) in a protein cover. They are not alive, in that they cannot metabolise or reproduce themselves. They have to invade a host cell and hijack its reproductive system, causing it to produce masses of new virus. When the invaded cell bursts, the newly formed viruses infect other cells and repeat the process. In the ocean's microbial communities they mediate gene transfers, nutrient cycling, the distribution and concentration of particle size, productivity, biodiversity, respiration and sulphur gas emissions.[21] It is believed that they can cause up to 50 per cent mortality in the bacteria that share their realm, and inhibit primary production by up to 78 per cent. The rupture of bacteria and other micro-organisms results in the release of dissolved and colloidal carbon into the sea, where it may be consumed by organisms, thus enhancing planktonic respiration. It may also result in the food available to higher parts of the food chain being reduced. The viruses themselves are grazed by microplankton.

The carbon-cycling food chains involving the viruses, bacteria and other micro-organisms form the 'microbial loop' shown in the diagram. This loop underpins the photosynthesis-based food chains in which phytoplankton provide the materials needed for life by all other creatures that live in the sea or on its produce. (In temperate seas very small cyanobacteria are an abundant element of the microplankton, but they are not known to be as abundant in the Southern Ocean.[22])

Only within the last decade or so has it been realised just how significant are the activities of viruses in the oceans. **Viruses in fact can almost be said to control life in the oceans—more than half of the total oceanic flux of matter and energy passes through the microbial loop in which their activities provide much of the free organic carbon.** The contemporary view of this organic matter is that it comprises a continuum from molecules of amino acids (the building blocks of proteins) and sugars, through colloids, to transparent polymer particles.[20] This so-called 'oceanic dark matter' spans a size range from a few tens of nanometres (millionths of a millimetre) to about a millimetre, and it has been proposed that these organic constituents of seawater interact to form a very dilute gel with a structure that has microscale 'hot spots' where heightened microbial activity takes place. Some bacterial cells from Antarctic Sea ice can be over a quarter of a millimetre long—giants of the microscopic world!

(Feel differently about seawater? Less inviting to dive into?)

Bacteria and viruses not only play a part at the base of the oceanic food web on which all other marine life depends—they also have a role in influencing global climate. The Southern Ocean acts as a major site for the absorption of atmospheric carbon dioxide; it is the activities of micro-organisms that determine whether CO_2 is rapidly released to the atmosphere, or locked up for hundreds or even thousands of years. In addition, the production of sulphur-containing compounds by marine micro-organisms affects global climate. Aerosol particles of sulphides released into the atmosphere form the nuclei for cloud formation; clouds reflect solar energy back into Space, moderating temperature, and they bring rain to the continents; and 17 per cent of the total global production of these sulphur compounds occurs in the Southern Ocean.

(This brief glimpse of Gaia the Microbial Planet is just a scene-setter for what follows in the rest of Part One. A fuller account of ecosystems in the Southern Ocean and how they function is given in Part Four.)

PROTISTS, BACTERIA AND VIRUSES IN THE SOUTHERN OCEAN

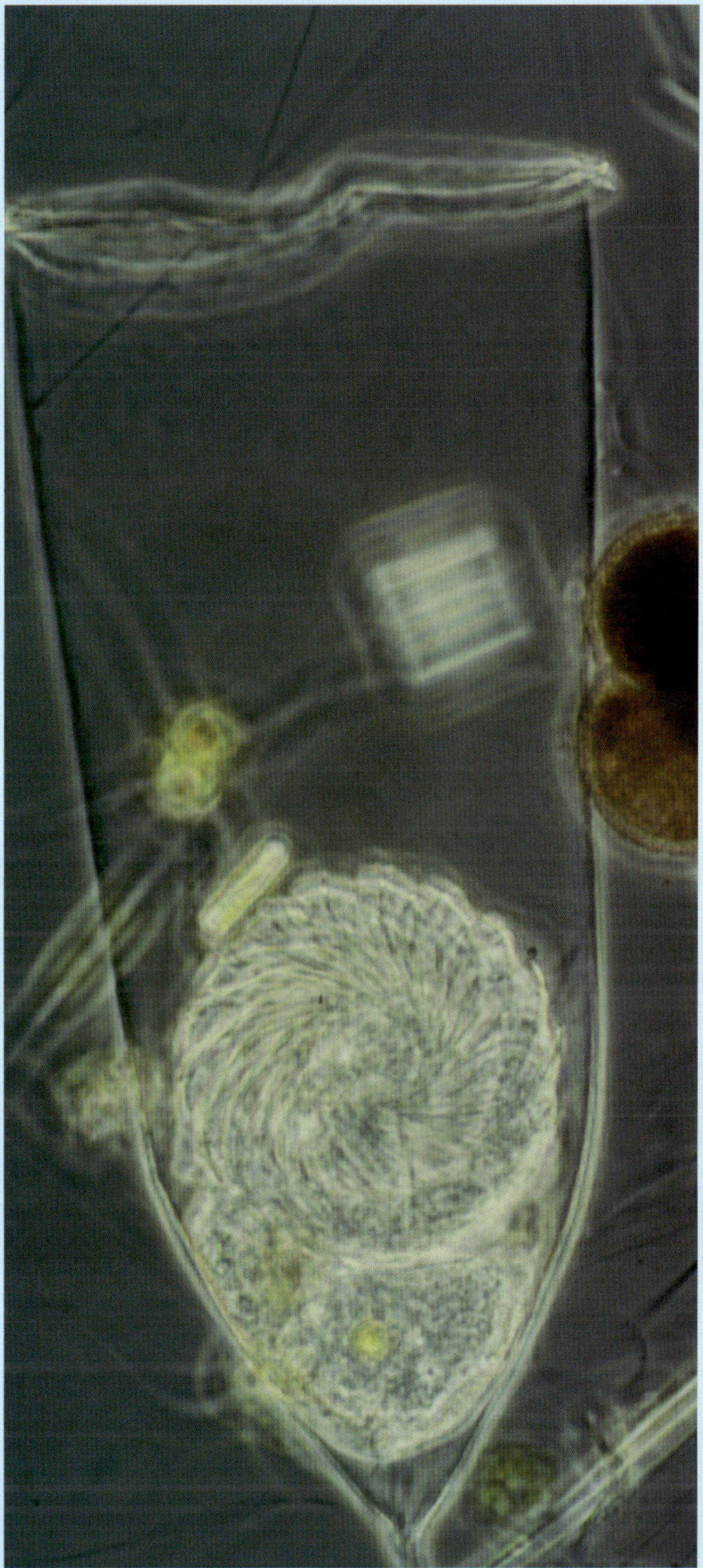

Two marine protozoan grazers. *Left:* A tintinid, contracted into the base of its lorica (flask-shaped gelatinous sheath). *Right:* Brown-pigmented, multi-segmented foraminiferan resting alongside the lorica. The other cells are algae. *Harvey Marchant, AAD*

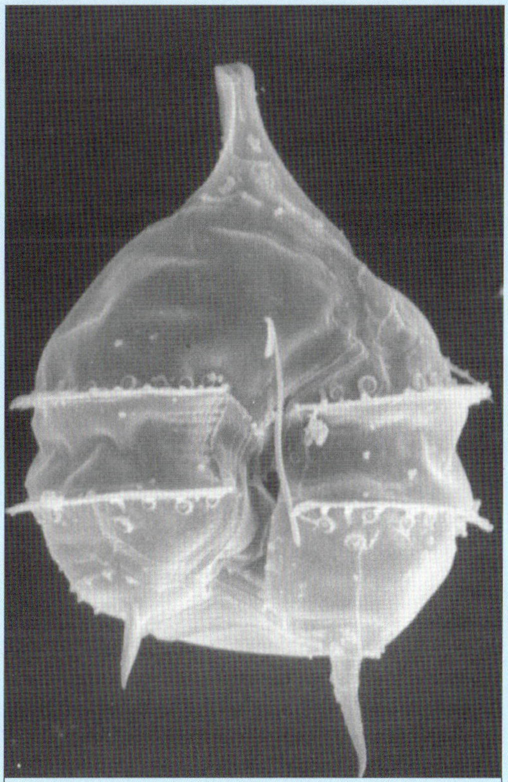

Protopteridinium defectum, a dinoflagellate. Some dinoflagellates are autotrophic, others, like this one, are predators. *Fiona Scott, AAD*

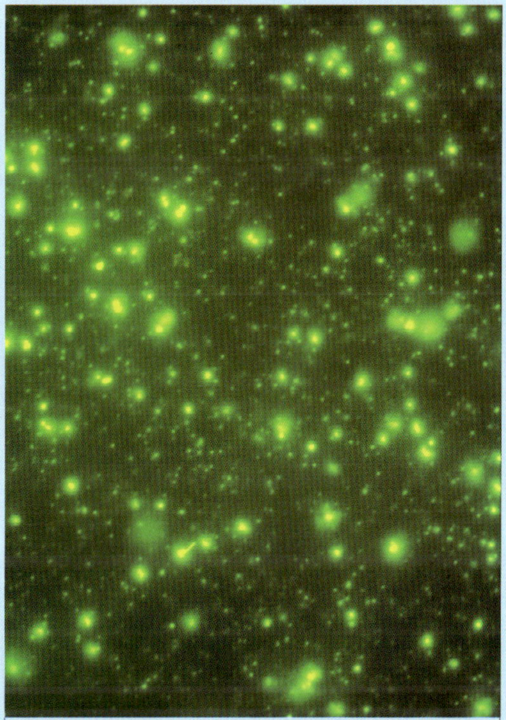

Like a starry sky, bacteria are the larger illuminated dots; viruses the more numerous, smaller, not so bright dots. The organisms are made visible by microscopy with a dye that stains their DNA. Modern technology has enabled this microcosm to be seen and recorded only recently. *Andrew Davidson, AAD*

Cellular evolution according to the SET model

FIRST IT WAS BACTERIA, THEN CAME THE PROTISTS, ANCESTRAL TO THE HIGHER KINGDOMS

The Kingdoms of Plants, Animals and Fungi are all descended from the Protists, which had been around for perhaps 2 billion years before the three other Kingdoms were created. (The earlier assumption that complex cells containing nuclei appeared only about 1 to 1.5 billion years ago has recently been revised on Australian evidence.[16] Oil shales from Western Australia dated at 2.7 billion years show evidence of nucleated cells in the occurrence of 'biomarkers' characteristic of eukaryotes. See page 9 for information on biomarkers for cyanobacteria and eukaryotes.)

The history of evolutionary change within protists, which is the history of the eukaryote cell, has been recorded in the living descendents of this ancient ancestral group. So many weird and wonderful organisms, showing so many variations from the norm, indicate an enormous number of different combinations and permutations in the course of evolution. The simplest organisms, amoebae, and the slime moulds that are just ill-defined blobs of living matter, have proved to be a rich source of information about bacterial origins of Life and protist origins of the higher Kingdoms. Research on cellular slime moulds in particular has led to important findings in animal cell and developmental biology, helping us to understand how multicellular organisms were formed and how they function.

Living cells of single-celled and multicellular organisms are of two types. The simplest are those of Bacteria, of the **Kingdom Monera**. All bacteria are either single **prokaryotic cells** or aggregations of such cells. The prokaryote cell consists of a wall or cell envelope overlying a cell membrane that encloses the cell interior. The genetic information is encoded in DNA dispersed throughout the cytoplasm, not compartmentalised into a nucleus. Respiratory activity is associated with the plasma membrane and the cell envelope. Photosynthetic pigment, when present, is located within or on membranes in the cytoplasm, not confined within organelles. Bacterial cells simply divide to multiply, sharing their genes and cell contents. It can be argued that Bacteria essentially live on for ever, making copies of themselves by cell division; it was only when meiosis (sexual reproduction) evolved in more advanced life forms that death became an inbuilt 'fact of life'.

Bacteria often have spare strands of DNA—extra sets of genes—that can be traded with other bacteria as naked pieces called plasmids, or as protein-coated pieces called viruses, either simply by touching or, in some more advanced forms, by development of a cell bridge between the donor and the recipient.

All other living things (of the Kingdoms of Protists, Fungi, Plants and Animals) have cells that have nuclei and other organelles. The basic **eukaryote cell** evolved when prokaryote ancestors acquired permanent endosymbionts, other prokaryotes that performed their different and useful functions within their hosts' cells and lost their independence. Eukaryote cells divide by **mitosis**—the dance of the chromosomes—choreographed by a symbiont; respiration is carried out by mitochondria, organelles whose DNA is detectably different from the DNA in other parts of the cell; cyanobacteria and their chlorophyll have become the chloroplasts of algae and plant cells. Eukaryotic cells have another prominent feature—a **cytoskeleton**, an intricate system of internal membranes.[29] The processes involved in the different functions within the eukaryotic cell are the same as in prokaryotes; the compartmentalisation has apparently allowed for the evolution of complex multicellular organisms, with extensive division of labour among cells arranged in tissues for optimal efficiency of the whole organism.

Prokaryotic microbes still have an extensive influence on the distribution of matter at or near the Earth's surface.[28] Because their

MITOSIS—THE DANCE OF THE CHROMOSOMES

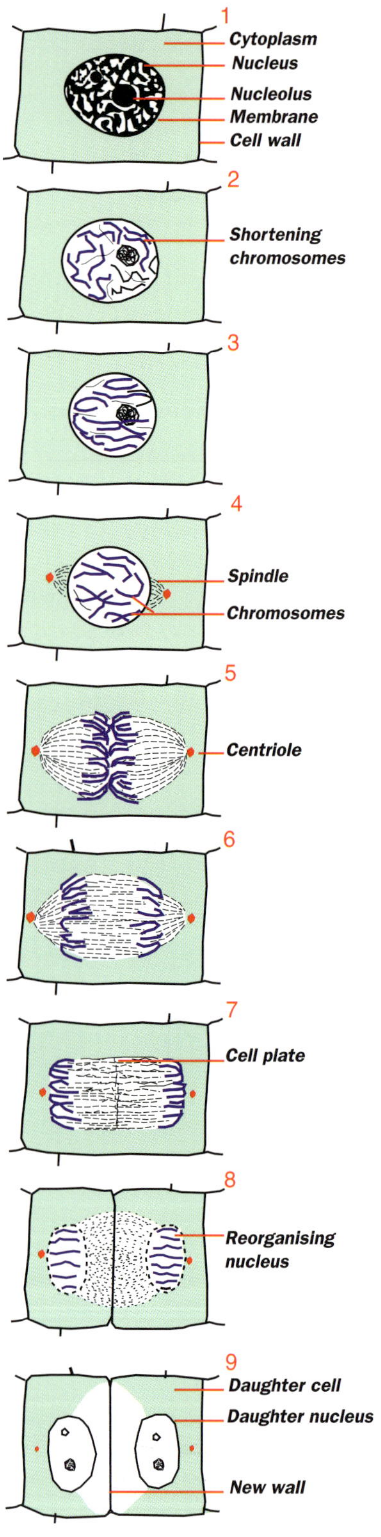

FROM BACTERIAL CELLS TO AUTOTROPHIC AND HETEROTROPHIC PROTISTS

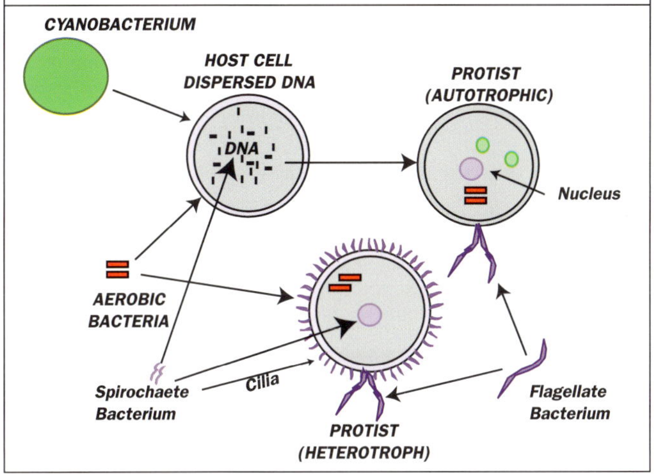

respiratory activity is associated with the plasma membrane and the cell envelope, they can oxidise or reduce relatively large quantities of metals or metalloids with the appropriate enzymes without these substances having to enter the cell interior where they could poison the organism. Very large numbers of different elements are known to be reduced or oxidised by enzymes produced by bacteria; some can even rid their environment of toxic mercury. In eukaryotes, by contrast, because their respiratory activity is located in the mitochondria, the metals and metalloids would have to enter the cell interior for large-scale oxidation or reduction.

The sequence involved in changing a prokaryote bacterial cell to a eukaryote protist is envisaged in this way. At some stage, more than 2 billion years ago, a sulphur- and heat-loving bacterium merged with a swimming bacterium, a spirochaete. This was the first, basic, endosymbiotic merger—the beginning of the series that would lead to the nucleated cell type that characterises organisms of the other four Kingdoms.[8] The two components became the **nucleo-cytoplasm**, the basic substance of the ancestors of animal, plant and fungal cells. This sort of intermediate organism later lost its cell wall and the more flexible membrane underneath began to grow and fold in on itself. This led to formation of a membrane-bounded nucleus, within which the organism isolated its essential DNA, and to other internal membranes, and enabled it to ingest other prokaryotes, folding them inside, originally to digest them.[7] At some stage it took up bacterial cells known as alpha-proteobacteria, which are proficient at respiration. Instead of this 'food' being digested, a symbiotic relationship resulted. The eukaryote sheltered the internalised cells, the endosymbionts provided energy from respiration. Finally the endosymbionts lost the genes they had formerly used for independent growth, transferred others to the host's nucleus, and became mitochondria. (Mitochondria in humans power our metabolism, our muscles, our digestion and even our brains.)

Likewise, chloroplasts derive from cyanobacteria that an early mitochondria-bearing eukaryote took up and kept. Both mitochondria and chloroplasts retain a small number of genes, including those that encode SSUrRNA (small subunit ribosomal

ENDOSYMBIOSIS

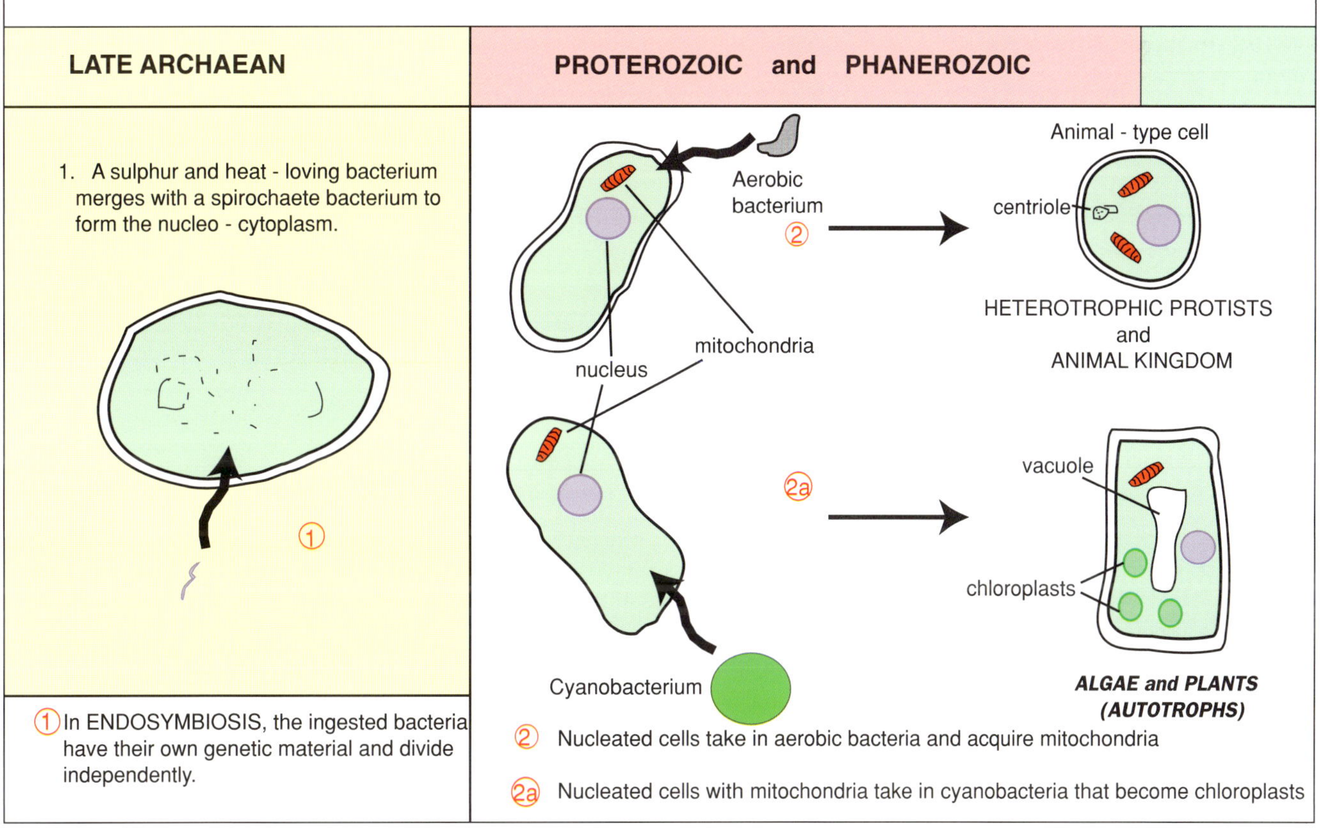

THE WHEEL OF LIFE

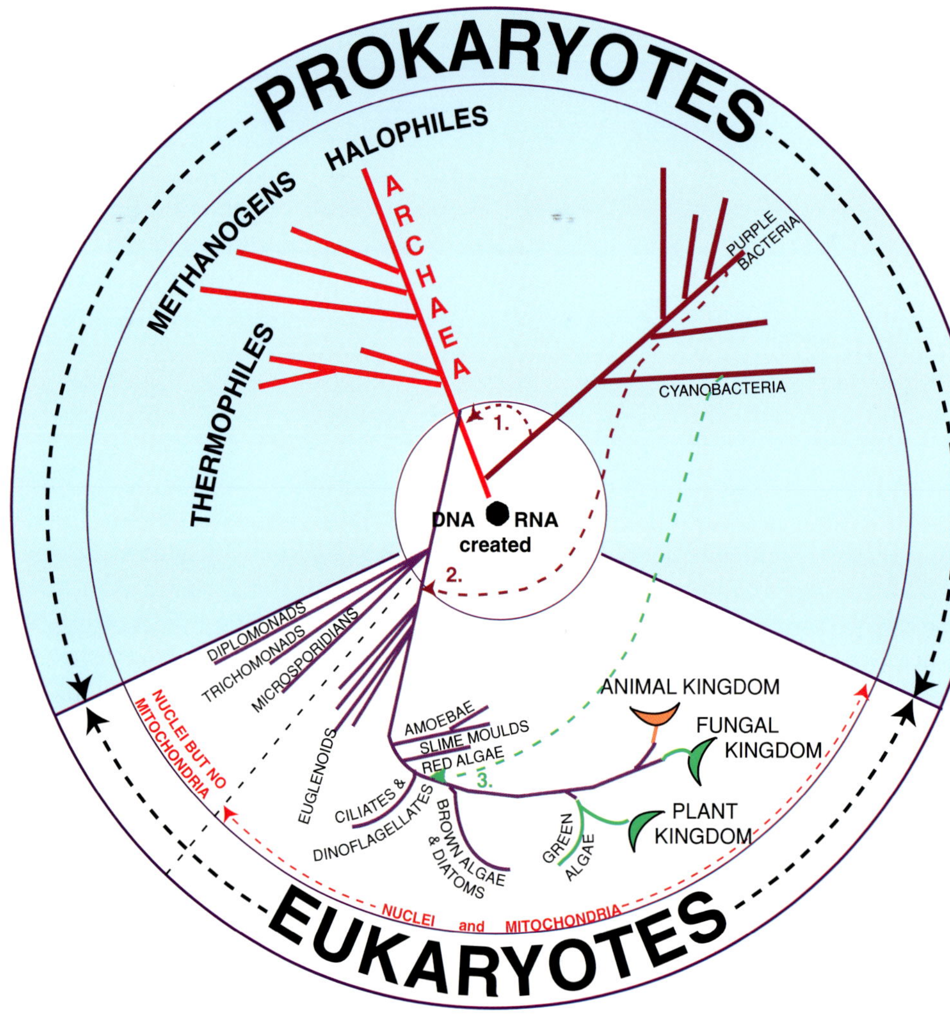

1: The first bacterial merger: A 'thermoplasma' (a prokaryotic Archaean) takes in a spirochaete to form a nucleus and thus becomes a Eukaryote.

2: A eukaryote that already had a nucleus took in as endosymbionts alpha-proteobacteria that were proficient at respiration—thereby acquiring mitochondria.

3: A eukaryote that already had a nucleus and mitochondria took in cyanobacteria as endosymbionts—and acquired chloroplasts.

The Prokaryotes, out of sight and largely out of mind, are represented here as comprising about 60 per cent of all living matter on Earth. Their contribution to running the Biosphere is only now being acknowledged.

BACTERIA THAT COUNTERACT PLANT TOXINS IN ANIMALS: THE *LEUCAENA* STORY[30]

The legume *Leucaena leucocephala* is regarded as one of the most promising yet unexploited tropical forages, potentially valuable in the rural tropics and subtropics worldwide.

It is already important in Asia and tropical America, but has been found to be toxic to stock in Australia and New Guinea. The toxicity is due to mimosine, an unusual amino acid that has no apparent function in the plant yet makes up 3 to 5 per cent of its dry weight. In susceptible animals, rumen microbes transform the mimosine into another toxic compound that affects thyroid gland function and causes goitre. The affected animals suffer hair loss, excessive salivation, poor appetite and ultimately die if not removed from the forage.

Experiments with goats in Hawaii showed that the mimosine was degraded by bacteria in the rumen. It has generally been assumed that the same suite of bacteria occupies the rumen of goats and cattle across the world. Now it is known that this is not so. However, it is a simple matter to introduce rumen bacteria from animals that are able to degrade mimosine into animals that do not have the ability, and after this inoculation they are able to graze leucaena without ill-effect.

This is an interesting example of the acquisition of an endosymbiont creating a difference between members of the same species. It gives an inkling of how the process of speciation might work over time, or at least how natural selection would favour those individuals which had a symbiont that made them suited to a particular location, thus increasing the influence of disjunct communities on the creation of diversity.

RNA)[29], which enables their origins from alpha-proteobacteria and cyanobacteria to be proven. Within the chloroplasts, the chlorophyll is located within or on the thylakoid membranes, just as it is in independent cyanobacteria.

Protist cells are huge compared with bacteria, and they display incessant internal movement as a result of their mitotic division. The 'dance of the chromosomes' seen in mitosis is widespread in protists and universal in their animal, plant and fungal descendants. The chromosomes perform a sort of micro-ballet, lining up on the equator of the cell and moving to opposite poles, following minute tubes of protein that form the lines of an ephemeral 'spindle'. When the chromosomes reach the poles, the cell divides and the spindle disappears. At the mitotic poles in animal and many protist cells are **centrioles**—discs with small perforations—that are remnants of the spirochaetes (bacteria with flagellae) that more than 2 billion years ago entered the bacterial cell and became the choreographers of the chromosome dance. The centrioles are made of micro-tubules of the protein **tubulin**, which is a component of neurones in the brains and peripheral nerves of humans. Cilia, hair-like protrusions seen on protists and in certain tissues in animals, also have this spirochaete origin—for example, the cilia that drive an egg down the Fallopian tube in humans and other mammals, and the beating tails of free-swimming gametes in algae, plants and animals, including human sperm.

Cannibalism among protists, where one ate another of its own kind but did not digest it, resulted in the **diploid** organism that had double chromosome numbers in the reproducing cell but through the process of meiosis only half that number in the gametes that were to pass the genetic material on to the next generation—thus requiring union with a haploid gamete from another reproducing cell to form another organism. Humans have diploid chromosomes, but our sex cells are haploid—so our ancestors were cannibal protists. According to Lynn Margulis[7], we are 'integrated colonies of amoeboid beings, just as amoeboid beings—protoctists—are integrated colonies of bacteria. Like it or not, we come from slime'.

The protists had meiotic cell division (and therefore sexual reproduction) long before animals appeared. But just to confuse the issue, some modern protists do not use either mitosis or meiosis. The giant freshwater amoeba *Pelomyxa palustris* reproduces by pinching in half its many nuclei when it pinches its huge cell in half; many dinoflagellates show modifications to mitosis in their forms of cell division.

THE NEXT STEP: THE ORIGIN OF MULTICELLULAR ORGANISMS

Single eukaryotic cells, basic units of Life, started to form multicellular communities roughly a billion years ago. Cooperation like this enabled specialisation to occur—some cells could concentrate on finding food for the collective, others on mobility, defence or functional organisation. As the cells became more specialised they lost the ability to survive outside the conglomerate and this led to the evolutionary sequence from multicellular protists to ever more complex life forms comprising specialised tissues. A widespread group of organisms, the **cellular slime moulds**, which still exists after surviving and thriving for an amazing billion years or more, provides clues to the origins of more highly evolved organisms. (Little is known, however, about their ecological importance, which must be considerable as they are bacterial predators and thus likely to be important links in food chains in the soil.)

The cellular slime moulds are soil inhabitants—single-cell amoebae that feed on bacteria, aggregating to form slimy 'slugs' when they run out of food and need to secure their future by producing spores that can disperse to more favourable environments.[31] They come in colours of white, yellow and purple. In Australia, as elsewhere in the world, they occur in diverse habitats, ranging from the high country in the Snowy Mountains to tropical mangrove forests like those on Lizard Island.

Single slime mould cells resemble human white blood cells—our infection fighters—and the multicellular stage displays features common to all multicellular organisms. In the collective, cells must adhere to each other, communicate with each other and control their shape. Slime moulds are useful as a simple model for how complex organisms are constructed.

Slime moulds can be cultured in the laboratory and are helping scientists to understand **morphogenesis**—the processes by which embryonic cells in plants and animals undergo specialisation to form the different sorts of cells that form the individual tissues. An average slime mould slug contains only two or three cell types but about 100 000 cells; by comparison, an adult human contains more than 200 different cell types and roughly 100 billion cells. This means that complicated processes can be seen at their simplest in the slime moulds. It has always been a mystery how cell movements are coordinated and how each cell 'knows' what kind of cell it must become. Because the slime mould slug is an aggregation of individuals, such complications of cell division and growth are avoided when studying its development.

The life cycle of the slime mould *Dictyostelium discoideum* has been extensively studied.[31] The organism grows and divides as single amoebae, dispersed in wet soil. When it runs out of food, an individual releases a pheromone-like chemical that signals other hungry individuals to aggregate. They stream together, forming a heap that falls on its side to form a slug-like creature which functions as a discrete organism despite being made of separate individuals. The slug moves through the soil for 3 or 4 days without eating, covered in a slime secreted by the cells, leaving a trail as it goes. The slug is attracted to light, so it moves to the soil surface and there it produces its fruiting body. About 70 per cent of the cells form the thick-walled fruiting body; the remaining 30 per cent make up the basal disc and stalk. The 30 per cent die, their job being to ensure the survival of the rest. Fruiting bodies can contain as few as 13 or as many as 1 million cells.

The cells that become the stalk of the mature fruiting body (pre-stalk cells) at first form the 'nose' of the slug and a type of primitive skin that envelops the creature. Pre-spore cells occupy the posterior, inner portion of the slug.

To build a multicellular organism, cells must stick together, and this they do by manufacturing adhesive molecules—**glycoproteins**—that enable them to form chains. The glycoproteins are species-specific, so different species of slime moulds stick only to their own sort. (If you mash up samples of human tissues such as liver and kidney and mix the cells together, they will stick only to their own kind—illustrating one of the reasons why the different cell types within all multicellular organisms form separate tissues.) Differential adhesion is probably just one of a variety of mechanisms that enable cells to recognise each other.

Cells must also be able to communicate with each other in order to coordinate their activities. The single amoebae aggregating into a slug send out messages to each other. The chemical used, **cyclic AMP**, is also the common messenger molecule inside cells in humans. When cyclic AMP binds to the surface of aggregating amoebae they first move towards the source of the signal; then within a minute or so they amplify the signal by releasing the chemical themselves; finally, a range of genes specific for life as a slug or fruiting body is activated in a process that takes several hours. The cyclic AMP signal is effected by its release in pulses which are controlled by the production of the enzyme phospho-diesterase that destroys it. Similar signalling in human bodies occurs where white blood cells are attracted to wounds. In plants, signalling between soil bacteria (rhizobia) and root cells triggers the formation of nodules and entry of nitrogen-fixing bacteria.[32a–e] The same sort of signalling occurs between roots and mycorrhizal fungi. (See Nitrogen fixing, pages 63–67.)

In recent years it has become clear that cell signalling is used in almost all aspects of cell behaviour and development, including hormone action, immune responses and even memory. At a biochemical level, the triggering of these processes is analogous to the events occurring in the starving slime mould when it receives the cyclic AMP signal. (Rudimentary memory and learning ability (? intelligence) are integral properties of the simplest organism. This has recently been shown by scientists at a research centre in Nagoya, Japan, where the slime mould *Physarium polycephalum* 'learned' the shortest way through a maze to a food source.)

Cell shape is critical in a cellular collective, each cell being required to maintain its shape to maintain the integrity of the whole tissue. Eukaryotic cells contain a **cytoskeleton of protein fibres**. The two structural proteins found in animal muscle (actin and myosin) have received special attention since they have been found also in non-muscle animal cells, where they are involved in cell division, movement and determining cell shape. Actin and myosin are also found in slime moulds!

Genetic engineering is relatively simple in slime moulds, facilitating experiments which have shown that cell signalling, adhesion and control of cell shape are all vital if a viable tissue with correct function is to be created. An organism has to make the right kinds of cells that can each maintain their individual right kind of shape; each cell has to 'know' *what* to become and *when* to become it. Damage to any gene, or alteration, can upset the whole process.

Glimpses of bacterial inheritance in the Animal and Plant Kingdoms

HUMAN INHERITANCE FROM BACTERIA: TRIUMPHANTLY BACTERIAL ME (AND YOU)!

1: DNA and metabolism: Life itself!

2: The Eukaryotic cell: Its structure was produced by the symbiosis of bacteria, which created the nucleus and organelles, its cytoskeleton and proteins. Humans have 30 000 genes in their genome, only twice as many as a fly or a worm. More than 200 genes can be recognised as unaltered bacterial genes.

3: The dance of the chromosomes, in both mitosis and meiosis, is orchestrated by endosymbionts (believed to have been spirochaetes).

4: Tissues were created from bacterial cells with different functions

5: The mobility of sperm, and ciliated tissues, reveal an inheritance from motile bacteria with flagellae.

6: Mitochondria, our endosymbionts that do more than release energy: Until recently, the role of mitochondria was seen as 'power plants' that generate the bulk of a cell's power needs. Their many other diverse and vital roles are now being discovered and it is established that they exercise surprisingly strong controls over important functions in reproduction, in human health and diseases, and possibly in the aging process.[36]

The mitochondria of the human cell contain 37 genes against the 30 000 present in the cell nucleus. Despite this disparity, mitochondrial genes can exercise a veto over the genome in reproduction, dertermining which egg cells will be available when a woman ovulates. In the ovaries of developing female embryos, 80 per cent of the 5 million egg cells undergo programmed cell death (a process known as atresia). While this might look like a very wasteful process, it ensures that the proteins encoded in the mitochondrial DNA (mtDNA) will interact perfectly with those encoded in the nucleus. This guarantees that there will be no incompatibility between mitochondria and nucleus in the children of the embryonic female—so a crucial reproductive decision rests with our endosymbionts! Further, mitochondria had a hand in the production of the two genders that are unable to breed with their own kind. This happened when our distant ancestor evolved a way to destroy its partner's mitochondria. Only the egg cell contributes mitochondria to the embryo, for that of sperm cells is destroyed on fertilisation—in this way the chances of passing on mutated or damaged mitochondrial genes to affect future generations, are again diminished—this time by half.

Since 1988, the relationships between hundreds of mtDNA mutations and diseases have been documented. Most of the diseases are of brain or muscles, both being big users of energy. Alzheimer's and Parkinson's disease sufferers show much higher rates of mtDNA mutations. On the other hand, it has been shown that certain other mtDNA sequences promote longevity in some people. Research on all these interesting aspects is in its infancy.

7: The chemical processes of communication between cells still use the same glycoproteins and chemicals used by bacteria for communication. The same protein, inherited from a common bacterial ancestor, is present in the lungs of both humans and lungfish (our first part-terrestrial cousin), preventing our lungs from collapsing when we breathe out.

8: The light receptor mechanisms that attract bacteria to light and enable them to photosynthesise are active in our eyes in the form of rhodopsin.

9: Our white blood cells (defenders against pathogens) were originally endosymbiont bacteria.

10: Pheromones: Our cells still use the same signalling mechanisms as their bacterial ancestors.

11: Haemoglobin: Direct linear descent of the haemoglobin gene in the Animal and Plant Kingdoms from a common, probably Archaean, ancestor is indicated.

12: Endogenous retro-viruses (ERVs) in our cells: ERVs are viruses whose genetic blueprints were trapped in the DNA of our ancestors during infections millions of years ago.[37] Most are now silently carried in the genome, but a few in some organs of the body continue to produce proteins, and even whole viral particles. This has led to speculation that ERVs have become endosymbionts, commandeered to carry out specific tasks in much the same way as, billions of years ago, invading

EGGS OR LIVE BIRTHS? THE ERVS DECIDED!

The whole endosymbiotic story is bizarre enough when one is first confronted by it. It redesigns Evolution, and the long-accepted classification of living things. Now, although the subject is still part speculation, it makes logical sense of the biggest step in Vertebrate evolution—the appearance of Mammals. It corroborates and enforces Lynn Margulis's statement about the Human vertebrate: 'We are our viruses.'[8]

Most of us do not know, and will find the information decidedly off-putting, that our unborn babies early in their genesis, and the placentas that nourish them during their incubation, are teeming with viruses.[37] These endogenous retroviruses, ERVs, that closely resemble the dreaded AIDS virus, are a normal part of every pregnancy. They are encoded in the DNA of every mammal, having invaded cells during infections millions of years ago, to become endosymbionts. Like other viruses, they cannot metabolise alone. They must insert their genes into the DNA of the cells they infect and instruct them to manufacture viral particles, which bud from the surface of the cell on the way to infecting the next one. Because they insert their genetic blueprints into the cells that produce sperm and eggs, they are passed effortlessly from one generation to the next without ever having to infect another host.

ERVs have been detected in the genome of every mammal studied. Humans alone harbour at least a thousand different ones, many of which have undoubtedly been travelling through our ancestors for multi-millions of years.

ERVs spend most of their time dormant, present only as extra segments in our DNA, but in the placenta and some foetal tissues a select number are activated, commanding the cells that harbour them to produce ERV proteins and assemble them into retroviruses. Once a baby is born they mainly return to dormancy, although a few are involved in auto-immune diseases like lupus.

The ERVs' roles in pregnancy are complex. They were critical in the emergence of the placenta and the mechanisms that protect the foetus from pathogens and from rejection by the mother. So they were critical in the transition from egg-laying to live-birth in mammals (and presumably have some role in reptiles and amphibians that have evolved from egg-laying to giving birth to live young). Under the control of ERVs, the placenta grows like a cancer, invading the uterine wall and sending out cells to grow in distant parts of the mother's body. (Fortunately their control of the process is such that after the birth, things return to normal.) The placenta has to have an extraordinary blood supply to bring nutrients and oxygen to the foetus and to remove waste products, and very sophisticated mechanisms have to be in place to prevent infection.

The ERVs are critical in preventing immune rejection of the foetus. Because half the embryo's genes are from its father, the mother's immune system would recognise it as foreign and eliminate it if ERVs did not provide protection. The protective mechanisms involved in the placenta rely on the formation of the **syncytium**—a microscopically thin membrane whose development is orchestrated by a special ERV envelope protein. This membrane is a barrier to immune system cells that would recognise and try to eliminate foreign cells.

Researchers emphasise that genome and ERV studies are in their infancy and all the processes involved are very complex, and a web of interactions between different retroviruses and chemicals made by the mother and the embryo itself is probably involved.

The fact that over the millions of years so many bits and pieces of retroviruses have accumulated in the DNA of humans, and been passed from generation to generation, is not without its worrying aspect. Although most of the retroviral sequences are inactive, it is in the nature of viruses and bacteria to swap genetic material, which means that it is possible that a new and potentially active and dangerous virus could be born within an individual. Infection of an individual with a living virus or bacterium could result in a union with one of the quiescent fragments whose genes might have suffered mutations through time, and the host would have no resistance to the new pathogen. This has happened in experiments with mice; more worryingly, current experimentation with the AIDS virus in the hope of finding a cure, runs some risk of producing unexpected surprises.

bacteria became mitochondria and chloroplasts in eukaryotic cells. Some ERVs are active during pregnancy. Our **mammalian live-birth chararacteristic** was determined initially by ERVs promoting the placenta. There is strong evidence that ERVs are involved in helping the placenta form a barrier between mother and foetus—and thus were fundamentally involved in the evolutionary changes that produced the mammalian-style of live-birth reproduction! Without viral intervention, we and all other mammals would have laid eggs as the platypus and echidna do! What would it have been like to be a monotreme and not a placental mammal?

Our immune system may also use ERVs to regulate itself. **Monocytes** are immune system cells that shed ERV proteins that have been shown in laboratory tests to have the power to change the activity of other immune cells. It is also likely that ERVs have a major defensive role of their own; the cells lining gateways to our bodies such as the lungs and placenta, and even the sebaceous glands in our skin, shed ERV proteins, creating deterrents to pathogens that would otherwise invade the cells.

For the first time, modern technologies and tools like scanning electron microscopy are revealing the truly bacterial (and viral) nature of this symbiotic biosphere in which we live and the extent to which these invisible things affect our lives and well being. For a long time bacteria have been thought of only in the context of germs or pathogens, and viruses thought of as all harmful, but new discoveries which contradict this point of view are being made at an amazing rate in this new field of knowledge. Landmark discoveries concerning the role of bacteria in stomach ulcers and heart disease, and the beneficial activities of retroviruses in pregnancy, for instance, hint at a whole sphere of knowledge yet to be revealed. Even the humble *Lactobacillus* in yoghurt is now thought to be able to protect the foetus and help it to develop a healthy immune system, decreasing the risk of asthma, when taken by mothers during pregnancy.

13: Our living cargo of bacteria: Travelling with us always are, conservatively, 80 species in the mouth; each square centimetre of the human bowel has 10 billion organisms (the number excreted each day by an adult ranges from 100 billion to 100 trillion); there are 10 million on a square centimetre of human skin (and there are approximately 2 square metres of skin per adult). Yet so minute are they that all those living on the outside of an adult would fit into a pea-sized ball; those on the wet inside would amount to 300 millilitres. About 200 species are involved. There are four times as many genes in the bacteria that live on humans than in the human.

Makes you feel like washing in disinfectant? Don't—that only increases the risk of pathological species taking the place of your beneficial resident cargo in sufficient numbers to cause problems.

PROTISTS: A MUCH UNDER-ESTIMATED KINGDOM IN THE MODERN WORLD

On land, an estimated quarter of a million species of protists live in lakes, rivers, waterfalls, warm springs, damp soils, transient puddles, dew and frost. They churn soil, and clear surfaces of bacteria. They cycle sulphur, phosphorus, silica and carbon on a global scale. They also colonise man-made structures and are a nuisance in swimming pools and in pipes.

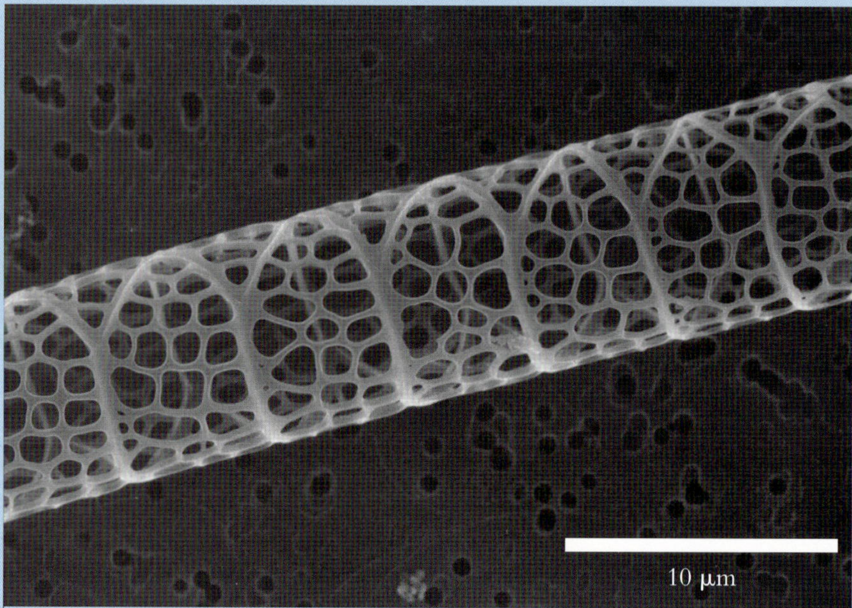

The diatom *Dactyliosolen antarctica*. *Geraldine Nash, AAD*

In the sea, protists (algae that used to be classified in the Plant Kingdom) are the major food producers and suppliers of oxygen. They, not plants, serve as the base of the marine food chain. They are the photosynthesisers (with help from cyanobacteria) performing the role carried out by plants on land. Marine algae comprise macroscopic seaweeds and microscopic algae of several different types, including coccolithophores, forams, diatoms and dinoflagellates. Vast amounts of materials flow through the bodies of coccolithophores and the other photosynthetic protists. Pelagic species synthesise nutrients for marine ecosystems far offshore. Benthic ones support the bulk of communities near shore. Gigantic blooms of coccolithophores in the oceans affect climate (see page 39) by releasing dimethyl sulphide (DMS) as a by-product of their complicated metabolic functions; on its oxidation to a sulphate in the atmosphere this provides nuclei for cloud formation.

Diatoms, radiolaria and foraminifera extract minerals from the ocean, building shelly structures of silica, chalk, glass, organic fibre and even exotic salts like strontium and barium.

SLIME

The slime that the cellular slime moulds produce, which enables them to move through the soil and also to aggregate, is a basic bacterial invention; this unglamorous material, in different forms, is still very much involved in cell behaviour in more specialised organisms. Some use it for locomotion and adhesion, with cells in certain tissues producing slimy discharges that show their derivation from ancestral bacterial slime-makers.

Slime is 96 per cent water containing a few dissolved salts, but the rest is glycoprotein, the substance that has to be present on cell surfaces in tissues of multicellular organisms to enable them to communicate with each other, and to function. (Even in *our* tissues.) The special properties of slime come from the cross-links between the glycoprotein molecules. Pressure liquefies, absence of pressure solidifies the material.[33,33a–d]

Snails use this holding and releasing property to aid their movements. The waves of contraction that ripple their muscular foot act on the slime and result in the foot being pulled forward, and stop it from slipping back. It has been found that up to 70 per cent of the energy that snails are producing from their food goes into making slime. This might seem a disproportionate use of energy—but some snails add nitrogen to the slime, allowing algae to colonise it, so that snail-trails become 'gardens' as well as routes that are followed like pathways. Limpets get back to their precise home-sites in the intertidal zone by following their own distinctive, slimy trails. Snails living in polluted water can concentrate heavy metals in their mucus, *and* decontaminate themselves in the process.

Hagfish use slime to deter predators, another use for this primaeval substance. They have evolved the most highly specialised mechanism, involving design and engineering feats that humans would find hard to equal. They keep dehydrated 'packets' of glycoproteins (mucins) in glands in their skin that remain closed until the fish feels threatened and triggers their opening. The mucins swell as soon as the seawater touches them, bursting out of the glands. The packets also contain fibres coiled up in thread cells, each of which contains a single fibre about 1μm in diameter but more than half a metre long. To pack the thread into a space just 100μm across, the fibre is coiled into a structure the shape of a rugby ball—the nucleus in one corner, the rest fibre. The expanding mucus pulls the fibres out of these cells, and the result is instant sticky goo, strengthened by the fibres.

So slime has much more to it than one might at first imagine. In fact it was a most important ingredient in the Biosphere when bacteria and protists ruled the world, through a major slice of geological time—from 4 billion to less than 1 billion years ago. (A time-slice 300 times longer than the reign of the dinosaurs.)

Sticky polysaccharides and proteins glued these ancient organisms to clay flakes and sand grains to form slimy mats. It has recently been recognised that the elephant-skin textures on rocks that were once ancient seafloors are remnants of **bacterial mats**. A picture is emerging of a curiously immobile, sticky Precambrian seafloor, before animals were there to churn it. It is also suggested now that the land may not have been the empty, sterilised-by-ultraviolet-radiation domain that has previously been envisaged. There may have been microbial mats in damp places; as time went on and more organisms became aerobic, oxygen would have been more abundant and available within the mats than elsewhere, helping evolving Life.

IMMOBILISING PREY WITH SLIME

These deep blue peripatus, *Euperipatoides rowelli*, are found in the crevices of rotting eucalypt logs in the Tallanganda State Forest in southeastern New South Wales. The texture of the skin gives peripatus an alternative common name of velvet worm. Peripatus enmesh their prey, usually insects or other arthropods, in threads of sticky slime ejected from turrets on either side of the head.

Photo Robyn Stutchbury, caption Noel Tait

When the Metazoans arrived on the scene, in the last 700 million years, dominance of the world by bacteria and protists was progressively decreased. Hot springs, very salt environments and other inhospitable places became the refuges of the once-abundant slime mats.

One of the earliest records of multicellular animal-like creatures preserved as fossils is found in rocks about 630 million years old in the Flinders Ranges of South Australia. Some of the strange creatures look like flatworms and jellyfish; others are less familiar shapes, flattened and sponge-like; the whole fauna is today regarded as an 'experimental' one and probably not on the direct line of descent to modern phyla. It is suggested that microbial mats were important in life styles of the Ediacaran fauna—they did not eat them, being filter feeders, but clung to them to stop themselves being pulled about in the action of currents and waves.

STROMATOLITES[34,34a]

A section through a stromatolite from the North Pole mine region in the Pilbara of Western Australia, age 3.5 billion years, shows the laminated form of this reef-like structure, built by some of the oldest dated living organisms on Earth. *(Specimen F63743 in the Australian Museum; photographer Jim Frazier.)*

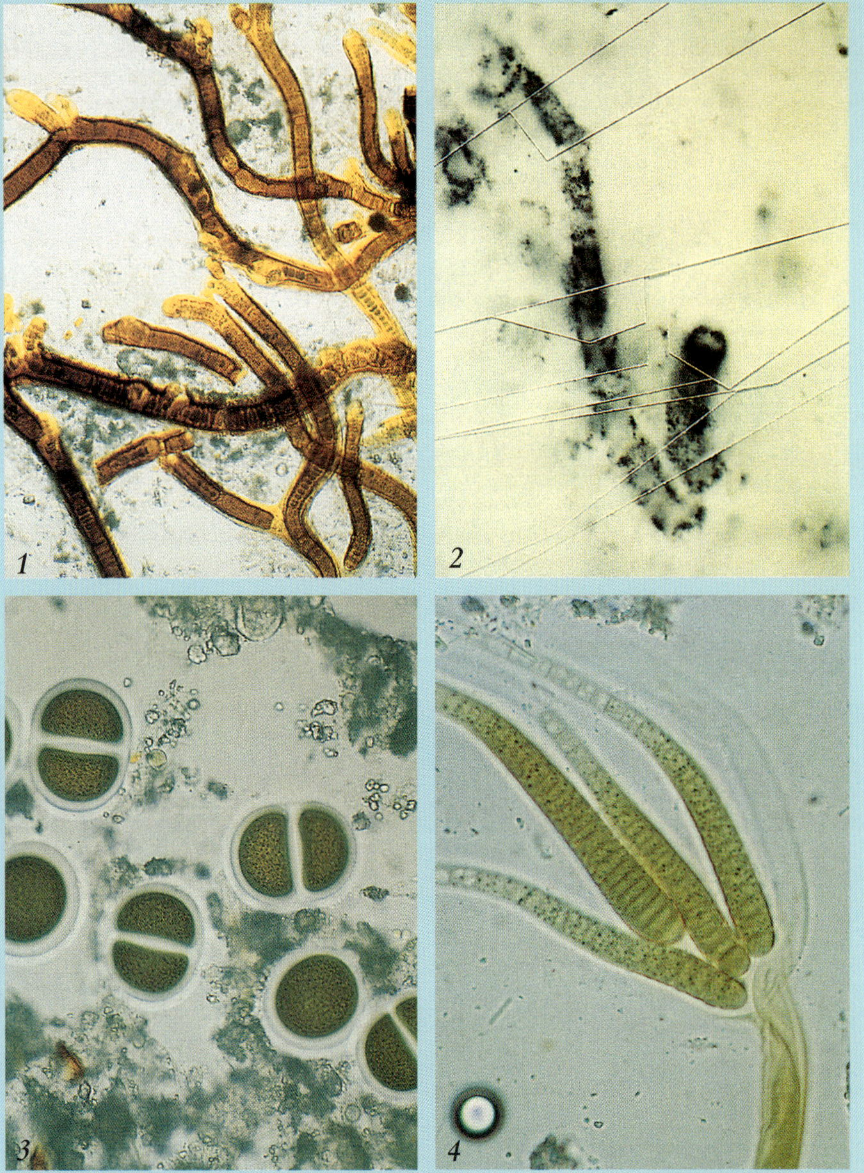

CYANOBACTERIA FROM LAKES IN WESTERN AUSTRALIA AND THEIR 3.5 BILLION-YEAR-OLD ANCESTOR FROM THE PILBARA (2)
Photos 1, 3 and 4, Linda Moore; 2, Professor Stan Awramik

Stromatolites like those preserved at the North Pole locality in Western Australia, west of Marble Bar in the Pilbara, provide some of the oldest fossil evidence of Life, dated at 3.5 billion years. Such reef-like structures constructed by microbial communities were abundant in the early seas, dominating, as far as we know, for nearly 3 billion years. Their decline started about 1 billion years ago and accelerated to the end of the Precambrian. It is likely to have been caused by increases in nutrient levels in the water as Metazoans increased in diversity and abundance in the period leading up to the 'big bang' of modern-style faunal species (including grazing marine invertebrates) in the Cambrian Period about 600 million years ago.

Today stromatolites are confined to limited habitats where specialised conditions prevail. Western Australia has a number of fossil stromatolite localities of different ages, and also a number of sites where living stromatolites are being actively constructed by organisms little changed from those which lived in the primaeval seas.

The cyanobacterium-like organism pictured, (2) from a stromatolite in the Pilbara, is morphologically like modern cyanobacteria.[34] At the early stage of Life which it represents, its metabolism would have been of the Archaean kind where carbon dioxide and hydrogen sulphide (abundant in the highly volcanic early Earth) were synthesised to form carbohydrate and sulphur was the waste product ($CO_2 + H_2S = CH_2O + S$). Modern-style oxygenic photosynthesis ($CO_2 + H_2O = CH_2O + O_2$) is believed to have originated perhaps a billion years later. The oxygen produced as a by-product first oxidised the marine environment, as evidenced by the worldwide banded ironstone deposits between 2700 and 2000 million years old.[34a] Subsequently, surplus oxygen leaked into the atmosphere, gradually preparing the world for the advanced life forms that appeared in the Cambrian Period.

Fossil stromatolites (*Conophyton*) from the Macarthur Basin, Western Australia.
M.R. Walter

The stromatolites in the Hamelin Pool at Shark Bay are now well known. When they were first recognised in the early 1950s, their construction was assumed to be typical while it was also assumed that hypersalinity was the reason for their survival. Discovery of more living stromatolites in other Western Australian locations and elsewhere in the world has since shown that these

DISTRIBUTION OF LIVING AND FOSSIL STROMATOLITES IN WESTERN AUSTRALIA

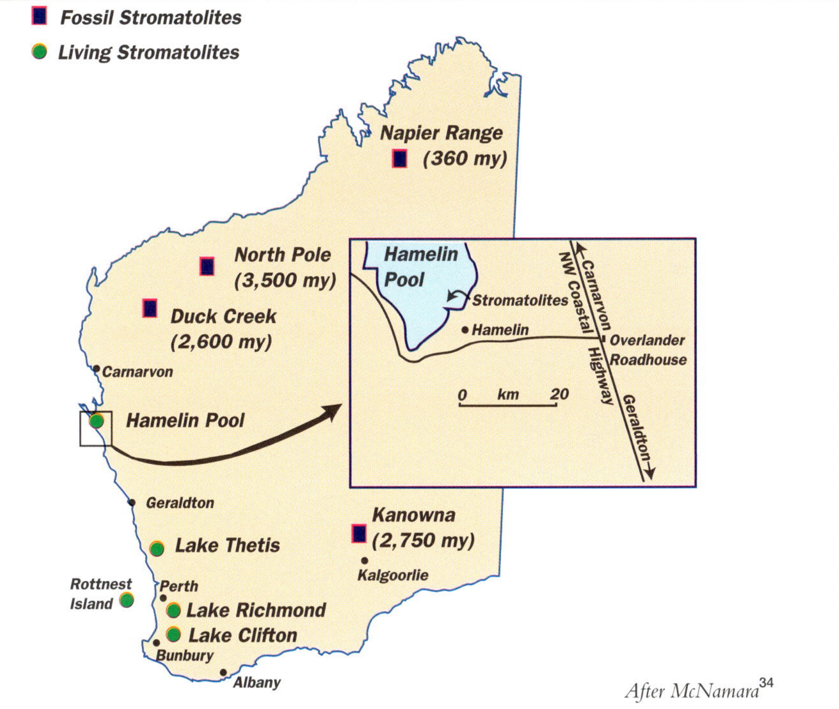

After McNamara[34]

assumptions were not justified.[34] Stromatolites have been found in such diverse and hostile locations as glacial lakes in Antarctica and volcanic lakes in Yellowstone National Park in the USA; in normal marine conditions in the Bahamas; and in lakes in Western Australia with waters varying from brackish to fresh, as we shall see. (It has been suggested[35] that an immense outcrop on Mars, 200 square kilometres in area, inside the rim of an unnamed crater in a region called Terra Sabaea near the Martian equator, might be of stromatolitic origin.)

The stromatolites at Shark Bay are constructed by the trapping of fine sediment in the sticky film of mucus that each cell secretes, the sediment grains then being bound together with calcium carbonate precipitated from the water. The cyanobacteria forming the structures are photosynthetic and have the ability to move towards the light (phototropism), thus they can keep pace with the accumulation of sediment, remaining on the outer edge of the structure where they have the sunlight they need. Their growth is now known to be exceedingly slow—less than a millimetre a year, possibly more like 0.4 millimetres—and that means that some formations in Hamelin Pool are more than 1000 years old.

The stromatolites at Hamelin Pool vary in their internal structure, depending on their position relative to the shoreline, and the different groups of micro-organisms involved in their construction at different depths. 'Pustular mat' stromatolites in the intertidal zone, created principally by *Enotophysalis*, a coccoidal cyanobacterium, have poorly defined internal structure without layering. 'Smooth mat' stromatolites occupy the region between intertidal and subtidal zones. They are constructed by *Schizothrix* filamentous cyanobacteria and have a well-defined internal texture and a smooth outer surface. Stromatolites occurring at the greatest depth, 3.5 metres, are weakly laminated, coarse-structured cauliflower-like edifices up to a metre high. They are constructed by a complex microbial community including the cyanobacteria *Microcoleus* and *Phormidium*, and some protists—mainly diatoms that secrete copious mucus which traps coarse sediment. The embayment at Hamelin Pool is hypersaline and this was originally thought to explain the presence of stromatolites by eliminating the grazing invertebrates that restrict their growth. It is now thought that the low nutrient status of the environment is responsible for the stromatolite proliferation.

In contrast to the mode of construction seen in stromatolites at Shark Bay, stromatolites in West Australian lake sites—Lake Clifton, Lake Richmond, Lake Thetis, Government House Lake on Rottnest Island and Pink Lake near Esperance—construct their framework by precipitating calcium carbonate, without incorporating sediment. This type of stromatolite is sometimes called a thrombolite. The lake waters range from saline through brackish to fresh.

The Lake Clifton stromatolites

Lake Clifton, about 100 kilometres south of Perth, is a coastal lake about 21 kilometres long and 1 kilometre wide, lying parallel to the coastline. Shallow, nowhere more than 3 metres deep, its water level rises with winter rainfall and drops in summer, when a stromatolitic reef is exposed at its northern end, 5 kilometres long and about 30 metres wide. The individual stromatolites comprising the reef are mostly less than 50 centimetres in diameter, although some are up to 1 metre across. The filamentous cyanobacterium *Scytonema* is largely responsible for building the structures. The position of the reef is determined by groundwater seepage which provides the carbonate and bicarbonate ions required for stromatolite construction.

The diverse invertebrate fauna that populates the lake includes many grazers, mainly isopods and amphipods (crustaceans) and molluscs, all of which use the stromatolites as a source of food as well as a refuge. This suggests that complex interactions promote or restrict

Stromatolites in Lake Clifton, Western Australia. *Ken McNamara*

stromatolite growth and that where conditions are right, as here where the materials needed for stromatolite regeneration are present in groundwater and the right balance exists in the ecosystem, they can survive.

A research project on the Lake Clifton stromatolites[34c] has enabled a better understanding of the whole picture regarding these ancient life forms. It has also suggested that the general name should be changed from stromatolite to **microbialite**, recognising the inherent differences between the laminated stromatolites and the non-laminated thrombolites. Thrombolites are characterised by an internally clotted texture and the absence of fine lamination; stromatolites have an internal structure of discrete laminations. (It was first though that the internal structure of thrombolites resulted from damage and alteration to the laminae caused by the metazoan grazers and borers as the structure developed. The Lake Clifton thrombolites have been demonstrated to have consistent internal morphology, and do not show disorganised laminated deposition. The principally filamentous cyanobacteria that create them are now assumed to have produced the micro-environmental chemical changes that result in the non-laminated framework, which is composed of aragonite, a crystal form of calcium carbonate.)

Most of the microbialites in Lake Clifton are thrombolites, lying on, or partially buried in, sediments of the lake basin. Radiocarbon dating has determined a maximum age of 1950 years, and net growth rates have been calculated at 10 centimetres in 100 years. The outward form of the thrombolites varies according to their situation. Small stromatolites with laminated internal structure also occur, but in small numbers.

When stromatolites declined in numbers in the ancient seas towards the end of the Proterozoic, thrombolites emerged as a still fairly abundant form, remaining throughout the Cambrian and then declining—but microbialites of either sort are rarities in the modern world.

A LIMESTONE CAVE STROMATOLITE

An active stromatolite growing in Nettle Cave in the Jenolan Caves system of eastern Australia. *R.A.L. Osborne*

EXTREMOPHILES

The early Earth, with its crust cooling and its oceans filling as water vapour condensed, was an inhospitable, over-heated, extreme environment for Life for perhaps a billion years. Its earliest inhabitants were **extremophiles**; today such organisms still exist and form thriving communities in the most unlikely places for living things to be found—on volcanic vents on the floors of deep oceans; in hot springs and geysers; in highly saline, acidic or alkaline environments. They provide us with insights into the nature and function of their ancestors and reinforce the logic of the proposition that Life's origins involved thermophilic bacteria. (The discovery of prokaryotic organisms in Archaean Era mineral deposits that are now recognised as being of deep-sea hydrothermal origin is evidence of the antiquity, and the persistence through time, of these extreme habitats. See page 45 on biogenic mineral deposits.)

Seafloor hydrothermal sites abound on the mid-ocean ridges where seafloor spreading is occurring in our modern oceans, and in associated basins. They occur, in fact, wherever submarine plate tectonic and/or volcanic activity is situated—and such sites have always been abundant throughout the geological past. At these sites, seawater penetrates into young oceanic crust, is chemically altered by interaction of heat and the surrounding rock, and rises along fractures in the Earth's crust to escape through vents in the seafloor.[23] Vents are congregated into 'fields' of different extent surrounded by an ordinary deep-sea bottom. Fields may be connected by subterranean plumbing, and many have chimneys, mounds or other structures built of the various minerals that precipitate from the heated vent fluid as it mixes with seawater. Vents do not appear to operate for very long at the same place and, although knowledge of the deep seafloor is limited and so far gained from few observations, are believed to have a life span of a few decades at the most.

The temperature of emerging effluent (vent fluid) may be as high as 400°C. A steep temperature gradient exists where the rising effluent mixes with the very cold bottom-water, whose temperature is 1° to 2°C. The vent environment has three zones with different ambient temperatures and different concentrations of dissolved and particulate substances—the vent opening zone; the near-field zone; and a peripheral zone.

Thermophilic microbes colonise the volcanic vents and their metabolic processes form the basis of food chains supporting communities that are known to involve at least 300 species of animals endemic to these deep-sea hydrothermal habitats. Fuelled by heat energy from within the planet, these deep-sea oases are the only communities on Earth whose immediate energy source is not sunlight. Sunlight is ultimately a requirement, however, because even vent bacteria must have oxygen to reduce inorganic matter to organic matter—and the oxygen in the ocean is, and always has been, the by-product of photosynthesis. Vent fluids are rich in the sulphide ions used by the autotrophic bacteria (Archaea) to convert inorganic carbon dioxide, water and nitrate into organic substances (in the same sort of processes as those by which plants make organics from the same inorganic materials in photosynthesis). The bacteria live in the subsurface plumbing; on surrounding surfaces, including the surfaces of other organisms living there; and in the effluent. Some of the microbes are symbiotic in other vent animals; others are nourished by the organic products of the vent community. In this food web, the primary producers are the autotrophic bacteria. The primary consumers are the hosts of endosymbiotic bacteria, like tube worms; grazers, like limpets; and suspension feeders. The secondary consumers are scavengers and carnivores—crabs, fish, worms and shrimps.

Vent communities have been most thoroughly studied on the East Pacific Rise and in the adjacent Galapagos Rift. Amazing and bizarre invertebrate communities of tube worms (some 1 metre long and living inside 3-metre tubes), clams and mussels dominate the vent opening zone, all housing endosymbionts and depending on them for their entire sustenance. They are physically modified, having digestive tracts that are not fully developed—in fact, the adult tube worms have neither mouth nor anus. The mussel in this group, however, has the ability to filter feed, so that when its endosymbionts do not function it is not as disadvantaged. Smaller creatures—limpets, polychaete worms, shrimps, crabs and even small fish (vertebrates)—also live in the vent zone. Nothing is yet known of their special adaptations, if any. The endosymbionts themselves can only survive in the vent zone where the water has sufficient sulphide or other reduced ions to maintain them; thus the animals that depend on them die if displaced from this zone. Filter feeders (feather duster worms, barnacles, anemones, bivalves) are attached to the seafloor around vents in the near-field zone, and opportunistic animals are found free-swimming on the periphery, depending on the vent community for their food, but avoiding materials that are toxic to them. When a vent becomes inoperative its whole community dies. Colonisation of new vents is effected by free-swimming larvae of the hydrothermal vent fauna that are always on the lookout for suitable sites to colonise.

The boundary around the vent fluctuates like a candle flame, enabling organisms that need both oxygen and sulphide to survive. Pulses of emissions, currents set up as very cold water mixes with hotter or cooler waters along steep temperature gradients, make this a very mobile zone. Some vent species are extremely heat-tolerant. The Pompeii worm (*Alvinella*) lives in tubes built directly on the vent chimney; the temperature inside its tubes has been measured at 40°C while the worm itself has been seen swimming near temperature probes that recorded water at 110°C. The temperatures tolerated by the Archaea that support the whole system are extraordinary, many living at over 100°C.

LARGE COMPLEX CHIMNEY FROM SATANIC MILLS HYDROTHERMAL SITE, PACMANUS

Picture produced using a collage of video images taken by CSIRO scientists during a manned submersible dive at 1708 metres depth. The submersible is the *Shinkai-6500* operated by JAMSTEC (a Japanese government agency). The lower flanks of this multi-spired chimney comprise many smaller chimneys that have coalesced. Most of the structure, however, is made up of a network of branching chimneys that tend to bifurcate at the top. The chimney stands on a pile of chimney talus, some of which is encrusted with snails suggesting that it is either hydrothermally active, or was recently so (fallen off?). Galatheid crabs are also visible at the base of the chimney. Note the vigorous hydrothermal activity in the background producing abundant black smoke. (Scale: the chimney is approximately 7–8 metres tall.)

© CSIRO. *Picture and caption Dr Joanna Parr*

The 'Edifice Rex' Sulphide Recovery Project

This project, a joint venture of the American Museum of Natural History and University of Washington, has concentrated on active sulphide chimneys on the Endeavour Segment of the Juan de Fuca Ridge.[24,25] Some of these chimneys rise tens of metres above the seafloor. The southernmost hydrothermal field on the Endeavour Segment was mapped and photographed using a remotely operated vehicle named *Jason*, and four chimneys in various stages of evolution were selected for recovery. Once on the research vessel they were immediately sampled for their microbial content, and extensive mineralogical, microbial and fluid analyses were later carried out ashore. (The chimneys are now on display at the American Museum of Natural History.)

Amazing imagery was obtained by *Jason*—13 large towers in one complex, one of them 23 metres high; the outer surfaces of the venting structures bathed in a buoyant, shimmering mixture of seawater and hydrothermal fluid percolating outward through the sulphide walls; and covered with dense macrofaunal communities of tube worms, limpets, snails and crabs.

A complex and interesting structure was revealed in the recovered chimneys, which weighed up to 1800 kilograms each and measured about 2 metres in length. Studies showed: vertical mineralogical zonations; complex networks of sulphide-lined flow channels; conduits lined by chalcopyrite in layers up to 10 centimetres thick; outer walls of anhydrite, zinc, iron-bearing sulphides and amorphous silica. Within the conduits were deposits of clay and

MID-OCEAN RIDGES AND ACTIVE TECTONIC STUDY SITES

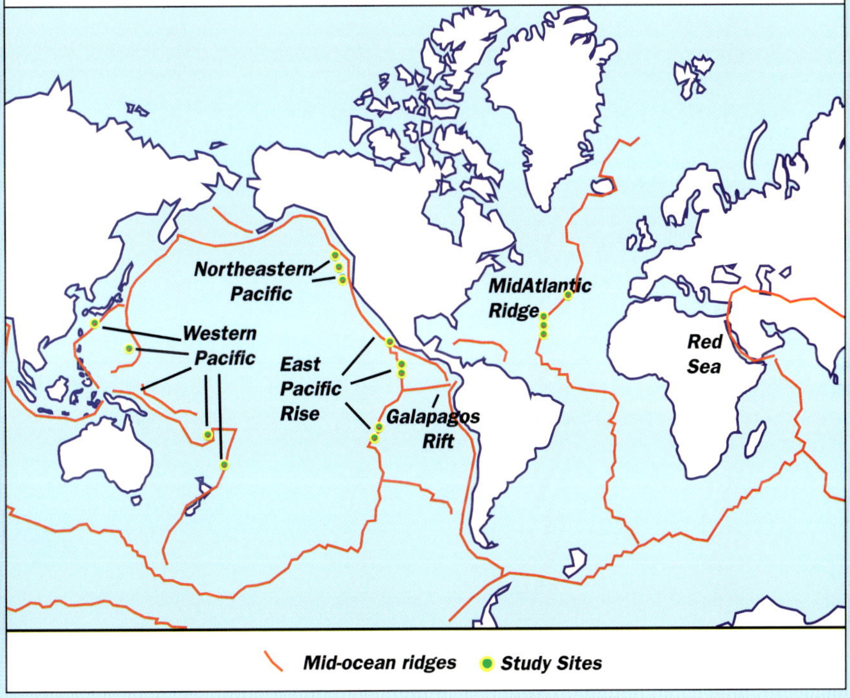

DETAIL OF MULTI-SPIRED COMPLEX CHIMNEY FROM SATANIC MILLS HYDROTHERMAL SITE, PACMANUS

Picture produced using a collage of video images taken by CSIRO scientists during a manned submersible dive at 1708 metres depth (in the *Shinkai-6500*, operated by JAMSTEC). This hydrothermally active structure comprises delicate, slender, branching chimneylets.

The temperature probe from the *Shinkai-6500* is being inserted into the base of the structure: it measured 268°C. Subsequent to the measurement, the chimney broke off and part of a black chimneylet was recovered. The chimneylet was found to consist almost entirely of copper sulphide (chalcopyrite). *Note the abundant fauna:* gastropods (*Ifremeria*), which are attached to an adjacent chimney, and blind shrimp clustering around the active chimneylet. (Scale: the temperature probe is approximately 20 centimetres long.)

CLOSE UP OF AN ACTIVE CHIMNEY

The chimney is covered in large snails (Gastropods *Alvinoconcha*, smaller hairy snails and *Ifremeria*, larger black snails) that are grazing on bacteria that colonise the chimney. The bacteria metabolise methane and hydrogen sulphide from the venting fluids at about 220°C. The water sampling tube at the left of the picture is about 7 millimetres in diameter. Picture taken by a CSIRO scientist during a *Shinkai-6500* manned submersible dive (operated by JAMSTEC).

anhydrite several centimetres thick and coated by chalcopyrite, believed to indicate an influx of partially reacted seawater into the conduit. This may be of importance in the evolution of the microbial habitat, allowing migration of micro-organisms towards the centre of the chimney.

A diversity of micro-organisms within well-defined mineral zones was discovered. Archaea comprised 40 per cent of life forms near the exterior of the structure and increased to more than 90 per cent in the intact microbial population near the hot interior. Eubacteria, in contrast, were most abundant near the outside of the edifice. Fewer cells were observed within the hot interior.

Australian research on seafloor hydrothermal sites

The most spectacular extremophile communities have recently been discovered by CSIRO scientists on the floor of the deep ocean in a subduction-related back-arc basin in the Bismarck Sea off Papua New Guinea.[26] Here the Satanic Mills vent field lies at 1700 metres depth. Hot water vented through the seafloor in 'black smoker' chimneys at up to 276°C precipitates minerals—copper, zinc, iron, gold and silver. Hydrogen sulphide and methane are also vented. The chimneys have a coating of chemosynthesising micro-organisms that oxidise hydrogen sulphide. Clustered around the hydrothermal vents, in total darkness, experiencing great pressures and a huge range of temperature over a few centimetres (2°C up to several hundred degrees) are unique ecological communities of crabs, tube worms, mussels, gastropods and fish. The food chain for the community is based on the chemosynthesis of H_2S performed by the microbes.

(Since the first research on submarine hydrothermal environments and the life that they support was reported, it has been discovered that the 'total darkness' assumed to exist on the floor of the deep sea, making photosynthesis impossible, is not total![24] Recent work on the ambient light conditions at hydrothermal vents indicates that the photon flux generated by thermal radiation of 350°C water should be sufficient to sustain low-level photosynthesis—**photosynthesis that is geothermally driven** rather than solar driven! This has important implications from the point of view of evolution of the photosynthetic habit, and for its survival in organisms that were confined to refuges at depth when deep ice covered most of the world (see Snowball Earth, page 37). The 'flashing' that occurs when water at 350°C and under great pressure is released through a vent is also the trigger for deposition of the metals that it contains, and gold, silver, lead, iron, etc. are deposited in chimney deposits.

Sulphur Springs: An Australian record of filamentous microfossils[15,27]

The Pilbara Craton is one of the ancient building blocks of Australia, and is already famous for having preserved in its rocks some of the earliest evidence of Life, in the form of the stromatolites at the North Pole locality. Recently, the first fossil evidence for microbial life in a Precambrian submarine thermal spring system was discovered, establishing that thermophilic Archaea were active here (as expected, but until then not proven) in situations like those that today occur in submarine hydrothermal zones. The volcanogenic deposit at Sulphur Springs, dated at 3 235 million years, is a massive sulphide deposit, rich in copper, zinc and lead, and even containing traces of oil. It is suggested that conditions were like those found in black-smoker chimneys at hydrothermal vents, the evidence indicating that they formed at depths of 1000 metres or more. The Archaea that occur as fine filaments twisting and turning through the deposit were anaerobic extremophiles, obtaining their energy by chemosynthesis. The oil in the deposits indicates that hydrocarbons entrained in the circulating fluids may have provided readily bioavailable energy and carbon sources to micro-organisms living around the hydrothermal systems.

Extremophiles and extreme cold

While thermophilic habitats in which extremophiles show adaptation to heat mirror the evolution of early Life when the young Earth was highly volcanic and inhospitable, adaptation to other extreme conditions has resulted in organisms capable of withstanding extreme cold.

Archaea have been found under the Antarctic ice cap, 3.6 kilometres down, above a subglacial lake—Lake Vostock—which has been isolated from the atmosphere for more than a million years. The fresh water in this lake remains liquid because of pressure of the ice above, and perhaps helped by heating from the Earth below.[12] These conditions are the closest on Earth to the frozen planetary environments of our solar system, prompting NASA to search for life under an ice-covered ocean on Europa, one of Jupiter's moons.

In Antarctica, another type of extremophile surviving extreme low temperatures and hardship is the 'endolithic' lichens that live within rock surfaces, having their photobiont (light-requiring algal or cyanobacterial) component close to the surface, and their mycobiont (fungal) component deeper inside the rock.[28]

Several hundred species of bacteria, protists and algae are known to live on the Antarctic ice pack. (See Part Four, pages 162–171.) The fish that swim beneath the ice are equipped with antifreeze in the form of special glycopeptides, proteins that bind to ice crystals and prevent them from growing inside the fish. (Trees like ash, walnut and apple suppress intracellular ice crystal formation by having antifreeze that allows supercooling to −40°C.) Extremophile adaptation of another sort in Antarctica relates to the production of sunscreen-type chemicals in the cells of organisms that live there—a response to the high ultraviolet radiation levels that are due to the hole in the ozone layer that appears over South Polar regions. The greatly increased area affected by this hole, as a result of man-made emissions that have damaged the ozone layer globally, is a source of concern if not all organisms in high latitudes have 'sunscreen' protection. The marine plankton, in particular, extraordinarily abundant in high-latitude seas, may well be adversely affected—and we now know that weather patterns are affected by breakdown of planktonic phytomass. (See page 39.)

Extremophiles tolerant of toxic, acidic, caustic and other inhospitable environments

Many different bacteria and protists are capable of living in strongly acid and toxic environments, and in intensely saline places. They are active in mines and their waste dumps and in the acid sulphate soils of coastal regions, for example. How a minutely thin cell membrane can protect the living contents of cells from such poisons is one of the great mysteries. Scientists are interested in finding extremophiles that might be used to improve mining and mineral processing—to exploit lower grade ore deposits, extract metals cheaply, and clean up waste in the environment. (See clean-up of hydrocarbons, arsenic and so on, pages 53, 99.)

Extremophile survival for extreme lengths of time

The amazing ability of extremophiles not only to survive in seemingly impossible conditions but also to remain dormant for unbelievable lengths of time, when necessary, is demonstrated by a recent discovery reported in *Nature*. Researchers at West Chester University in Pennsylvania found bacteria that were able to be reactivated after surviving for 250 million years in a state of suspended animation. These bacteria, of a type related to salt-loving species found in the Dead Sea, were enclosed in a tiny pocket of brine, within a 3.5 centimetre-wide crystal of salt, within a Late Permian deposit, deep in the lithosphere in Texas. They had been entombed, presumably in a resting spore stage, and they revived when their tomb was opened! All precautions had been taken by the researchers to ensure that no bacterial contamination occurred in the laboratory when the crystal's liquid core was opened, and their extraordinary findings have to be accepted as fact. Before this reliably dated 'resurrection' was documented, there had been reports of live bacteria being extracted from a bee preserved in amber between 25 and 40 million years ago. There had also been reports of living bacteria emerging from salt deposits hundreds of millions of years old. The possibility, or probability, of contamination of the samples in all previous cases caused doubts, but the West Chester University investigations seem to have eliminated that factor.

(Even non-extremophile species can show remarkable ability to survive in their resting phase. Eggs of copepods that had been preserved in lake-bed sediment for 300 years were able to hatch.)

Bacteria can be killed but they do not just die, according to Lynn Margulis[7], because of their ability to live on by replicating themselves; when one sees how they can persist and await their chance, one realises that nothing is impossible for them. It is not beyond the bounds of possibility that they could travel on meteorites and have 'seeded' suitable planets, including ours—is it?

ELECTROSENSORS AND ELECTRICITY GENERATION BY ANIMALS

When the processes of metabolism are broken down to the molecules and atoms that are being exchanged, they are all seen to be essentially electrical in character. The negative or positive charges carried by the element being absorbed determines whether it will enter or be excluded from a cell under whatever conditions prevail at that time. All animals generate electricity in their contracting muscles (which contract because of metabolic processes occurring inside the muscle). Each contraction is accompanied by the creation of a sizeable electric field in the animal's immediate vicinity. Some animals (and possibly many more than have been studied in this regard) have the power to detect electrical fields in their vicinity.

Knowing what we now do about bacteria, it is not too far-fetched to reason that they had this inherent capacity to detect electrical signals from the beginning. Their responses to light, their reading of chemical signals, and even the known response of some present-day soil bacteria to the Earth's magnetic field, are all evidence. (Some Southern Hemisphere soil bacteria translocated to Northern Hemisphere soils exhibit 'muddled' responses to changing conditions in their environment. When soils are drying and they should go deeper to survive, they have been recorded as going up and drying out.) Bacteria with flagellae were creating sizeable electrical exchanges in their 'proton motors' from the time they first developed this highly efficient form of mobility, and presumably they communicated with each other by recognising species-specific electrical discharges.

When bacteria with highly sensitive electrosensory ability became endosymbionts, they sometimes became the specialised cells in the tissues of some animals that are designed to pick up the often extraordinarily weak signals given out by their prey. (They also became the sensitive hairs in the balance system in our inner ears, detecting gravitational pull and direction.)

Electrosensors in vertebrates[38]

Electrosensors located in the skin of some mammals allow them to detect weak electrical fields in their surroundings. The sensors detect voltage gradients from the minute electric currents generated across their receptive surfaces. These currents lead to the generation of action potentials—pulses of electricity—that travel along nerve fibres to the brain. There they are analysed, telling the animal the size of the electric field, the direction it is coming from, and its strength. This is a form of inbuilt radar that may be giving the animal a three-dimensional view of its surroundings! It has long been known that fish and amphibians have such receptors. Because of the fishy ancestry of mammals, it is no surprise to find that some have retained them.

The platypus

It is now known that the platypus, that weird and wonderful monotreme with so many strange features that it almost looks like an evolutionary missing link, has electrosensors in its duckbill-beak.

THE DUCK-BILLED PLATYPUS
The platypus has electrosensors in its duck-like bill, enabling it to detect its prey by reading the minute electrical currents generated in their muscular movements. *Earth Sanctuaries Ltd*

The platypus's favourite food is freshwater shrimp. A single tail-flick from a shrimp was found in experiments to generate sufficient electricity to be detected 10 centimetres away by the electrosensors of a platypus. These organs are so sensitive that they can detect voltages 1000 times smaller than those very weak ones which humans can detect with their tongues.

The platypus has its sensory organs located in distinct lines on its beak; when swimming it repeatedly tilts its beak from side to side, enabling it to detect prey on either side of its pathway and register how far away it is. As it feeds at night in darkness, or in turbid waters, and as it shuts its eyes, ears and nose when it dives, the platypus must be using its radar to navigate. No primitive animal this, but one most highly evolved and adapted to its surroundings!

The electrosensors in the platypus are located within mucous glands with openings on the skin surface. Numerous nerve fibres in an expanded cup at the base of the epidermis are supplied by the 5th cranial nerve. Filamentous processes from the nerve endings radiating into the lining of the gland are believed to be the actual sensors. The electrical stimulus excites the nerve fibres directly.

The echidna

The echidna, another monotreme, has electrosensors in the tip of its elongated nose. These are fewer and far less sensitive than in the platypus, but the rest of the snout is rich in touch and temperature sensors. It is not known whether the echidna uses its electrosensors to locate food. The short-billed echidna eats ants and termites, neither of which has been shown to generate an electric field—though the difficulty of detecting such a field in the wild, where sheer numbers might raise the voltage to a readable level, is considerable. The New Guinea echidna, which lives in dense rainforest, might find an electrosensor very useful when probing deep in leaf litter. Much has yet to be learned about these animals.

Electrosensors in fish

In fish, electrosensory receptors are part of the lateral-line system that is similar in structure and function to our own inner ear and responsible for balance. Specialised hair cells sit in the bottom of minute canals that communicate with the skin surface. The electrical stimulus results in the release of a chemical that excites the nerve fibres, which are supplied by the 10th cranial nerve. The differing arrangement from that in monotremes suggests that the electrosensory function followed separate evolutionary pathways in different animal groups.

Sharks are able to detect flounder buried deep under sand by picking up their electrical signals. Electric eels generate their own electric field, strong enough to stun prey, and detect both prey and predators by the disturbance caused to their own field by the proximity of the others' fields. A small stingray, commonly known as the nummie, is well known in Australian waters for causing a numbing shock when touched by an unlucky swimmer.

Specialised electrosensory organs may not be that far removed from the specialised sensory hairs, the vibrissae, found in cats, rabbits, mice and other animals.

The animal world has much more to thank spirochaetes for than the tails on sperm! Mysteries like the ability of migrating birds to find their way, that suggests genetic imprinting with coded information about Earth magnetism, may one day be explained by an endosymbiont's ability to detect and pass on such information.

HAEMOGLOBIN IN THE PLANT KINGDOM[39]

Conservatism is a key factor in evolution. Successful molecular and genetic strategies have been preserved down the ages. The haemoglobin that characterises the blood of mammals, including us, and that occurs also in the Plant Kingdom, has an ancient lineage. An Archaean or Bacterial common ancestry (possibly as much as 3.5 billion years ago) is suggested by the discovery of the haemoglobin gene in the ancient bacterium *Vitreoscilla*. The successful molecular strategy to transport oxygen, once developed, was conserved and has been passed down ever since.

Haemoglobin has lately been found to be widespread in the Plant Kingdom. Its occurrence was originally thought to be an anomaly associated with the mechanism of nitrogen fixation in which symbiotic relationships between bacteria such as *Rhizobium* and *Frankia* and a small number of diverse plant groups enable the manufacture of nitrates. (See Nitrogen fixing, pages 63–67.) Plant haemoglobin was first found about 60 years ago in the root nodules of legumes, its role in these plants being to deliver the constant low-level concentration of oxygen needed by enzymes in the nitrogen-fixing process. Its presence was interpreted by some scientists as a case of convergent evolution; by others as a horizontal gene transfer, in which the haemoglobin gene was passed into the legume through a close association with a bacterium or an insect. The discovery of haemoglobin in other non-legume genera, including *Casuarina*, shed doubt on both explanations, and recent molecular studies have now shown conclusively that haemoglobin molecules from animals and legumes are extraordinarily similar, with identical amino acids involved in the oxygen-transporting mechanism. As a result, **linear descent of the haemoglobin gene in the Plant and Animal Kingdoms from a common ancestor is believed to be the right answer.**

CORALS ARE CONTRIBUTING TO OUR KNOWLEDGE OF METAZOAN EVOLUTION

The reef coral *Acropora* is being investigated at James Cook University, Townsville.[40] This genus belongs to a basal class of corals and has predictable, external and accessible reproduction, making it suitable for genetic studies that are providing insights into ancestral linkages and Metazoan evolution. The mass spawning events that occur regularly supply copious material for molecular biology and embryology studies. Genome studies in these primitive creatures have shown some remarkable similarities to vertebrate genomes, indicating the fundamental inheritance of gene controls from primaeval ancestors.

CORALS SHARE VERTEBRATE GENOMES—WE ALL COME FROM THE SEA!
A colony of branching *Acropora*. Big Broadhurst Reef, central Great Barrier Reef. *Peter Harrison*

CLASSIFICATION OF PLANTS BASED ON DNA SAMPLING

Although DNA sampling is only in its infancy, when it comes to revising the Linnaean classification of plants it has the potential to reveal relationships and lines of evolution that will cause major changes. Where, up to the present, gross morphology has been the determinant, many species may be found to be unrelated to others in their presently allotted genus or family, and to have acquired their determining physical characteristics as a consequence of convergent evolution. Even at this early stage when only a few genera in the different plant groups have been studied, new insights are emerging.

The conventional evolutionary tree for Plants, in which an algal ancestor gave rise to the land plants, still starts with an algal ancestor (although the Algae are now classified as Protists and the Plant Kingdom is only initiated with that evolutionary step onto the land). A single common ancestor is now indicated for the whole Kingdom—no early branch leading to clubmosses and a main branch to ferns and 'higher plants' (that is, Ferns, Gymnosperms, Angiosperms on the one line). Several separate lineages are now known to branch off and lead to dicotyledons, with primitive families with ancestral members like water lilies (Ranunculaceae) and *Amborella* in the Monimiaceae—not one evolutionary line for the broad-leaved plants; and one line to monocotyledons. When it comes to splitting up of large genera, using morphological characteristics—a pastime still much engaged in by systematic botanists—it would seem wiser to wait until sufficient DNA work is undertaken to really sort out the relationships.

DISEASE IN WILD AND DOMESTICATED ANIMALS AND PLANTS

Global trade in food, plants and animals, and the increasing mobility of people, is resulting in the spread of plant and animal pathogens all over the world. While catastrophic epidemics used to be rare, it is increasingly obvious that extinctions of wild species by the spread of diseases of domesticated animals are not beyond the bounds of possibility and may in fact be probable as a result of the introduction of viruses into wild populations that have no natural resistance to them. It is likely that future research will find that many problems, at this point attributed to chemical pollution, are in fact related to introduced pathogens.

Similarly, diseases in wildlife can have significant impacts on human health and the agricultural industry. A bat virus has been found to be the cause of infertility and deformity in pigs; wild waterfowl are believed to be the source of avian influenza in poultry; and the *Lyssavirus* in humans is of wild animal origin.

Rinderpest in Africa

The foot-and-mouth disease epidemic that afflicted sheep and cattle herds in Britain in 2001 is completely outclassed by the rinderpest pandemic that completely devastated Africa's ungulates in the nineteenth century.[41] Evidence has only recently been collated about the spread of the virus from Eritrea all the way south to South Africa (and Cecil Rhodes' property near Cape Town) within a decade of its introduction to the continent by an Italian expeditionary army which brought cattle, as was then the custom, to feed its soldiers. The virus was native to the steppes of central Asia and caused periodic outbreaks in parts of Europe, usually related to the movements of armies and their food supplies—usually the only long-distance stock movements that occurred in those times of largely local markets and livestock trade. The disease was unknown in Africa, probably because camels are not susceptible and the desert areas formed some initial barriers to its entry into the continent from the north.

Once the virus arrived, Africa, with its herding societies, was totally vulnerable to the disease. The populations of cloven-hoofed native animals (antelope, buffalo, wildebeest, etc.) were as susceptible as the cattle on which both indigenous tribes and European pastoralists depended for their survival. The epidemic spread through Tigre in Ethiopia in 1888 and followed ox-trails along the Rift Valley and westwards across Sahel herding routes via Sudan and Chad into West Africa, taking only five years to reach the Atlantic. It had arrived in South Africa on its southward rush within a decade. The mortality rate among infected cattle everywhere was 90 per cent or more.

A period of intense drought in sub-Saharan Africa between 1888 and 1892, combined with the loss of cattle, is believed to have killed millions of people. In Ethiopia alone a third of the population died. The Masai in East Africa, whose diet depended heavily on milk and meat (or blood), were also decimated as the plague spread across their drought-afflicted lands. A Masai elder is reported as telling how animal and human corpses were so thick on the ground that 'the vultures had forgotten how to fly'. The same sad story was repeated in all African countries, reducing herding nations to scattered remnant populations, the wildlife to a tenth of its original presence. (All this devastation greatly aided the 'invasion' of Africa by Europeans.)

The rinderpest pandemic had another serious consequence, the spread of the tsetse fly that carries sleeping sickness. Today this pest is still one of the greatest obstacles to Africa's development. (The trypanosome that the flies harbour is a 'protozoan'—a unicellular protist with a flagellum for mobility, and a nucleus that makes it a eukaryote, not a bacterium). It is a blood parasite in vertebrates and is spread via an insect vector.

Initially, the killing of the fly's hosts was a shock to tsetse fly populations, but as a consequence of the removal of grazing pressures that had kept vast areas as grassland, vegetation change soon provided an ideal habitat for a build-up in fly numbers. The return of woody bush vegetation provided the dense, shady conditions best liked by tsetse flies. Native animal populations rebounded more rapidly than cattle populations after the slaughter, providing food again for the flies, and they spread rapidly. The highland areas of East Africa in particular, which had previously sustained large populations of herders, changed to tsetse-infested bush and woodland; the river valleys of the Zambesi and Limpopo became fly-infested over an area eight times greater than had been affected prior to the rinderpest plague; millions of people died from the infection and large areas were abandoned to wild animals and tsetse flies. (In Uganda alone, 4 million people are estimated to have died from sleeping sickness early in the twentieth century, and about half a million people across Africa still die of the disease annually.)

Vultures in India

Griffon vultures have long been the cleaners of Indian urban and rural landscapes.[42] Materials in waste dumps and around abattoirs, and dead animals in the fields, were promptly and efficiently dealt with by these big birds, which could pick a water buffalo carcass clean in 20 minutes. Their extraordinary numbers, sometimes 5000 on the one waste dump, performed a vital service by eliminating many health risks and minimising contamination of water. A mystery virus has made the species almost extinct in India in the passage of only a few years, with packs of rabid dogs replacing the vultures at many dumps.

India's Parsee Zoroastrian community in particular has a serious problem. They have traditionally depended on the griffon vultures to pick clean the bodies of their dead—a funeral custom fundamental to their religion—because they do not believe in defiling earth or air through burial or cremation.

KOALAS UNDER THREAT OF EXTINCTION

When the immune systems of animals, including humans, is impaired by stress, individuals become susceptible to pathogens. The pressures on koalas in their remnant habitats contribute to widespread *Chlamydia* infections that affect their eyesight and their fertility. Habitat destruction and restrictions are driving them towards extinction. *Jim Frazier*

Marine creatures

A long list of marine creatures that have been seriously affected by diseases includes:

- **Seals** Over-fishing of the Atlantic drove harp seals to the North Sea in 1988. They brought with them a virus to which local harbour seals had no resistance. Three-quarters of the local animals died.
- **Pilchards** have succumbed to an unidentified disease that has seen vast numbers washed up on Australian beaches during the last decade. Abalone, prawn and lobster fisheries (both natural and aquaculture) are prone to epidemics caused by a range of pathogens. Corals of the Great Barrier Reef, like those of reefs in the Caribbean (see page 55) are killed by pathogens carried in dust or run-off from the land.
- **Starfish** along the Californian coast were killed off by a new strain of vibriobacteria in 1984. Some species recovered faster than others and maintained their advantage, and now what used to be the most common species is rare.

Birds

- The mosquito-borne West Nile virus, brought to New York from Israel a few years ago, has caused a number of human deaths and the death of at least 10 000 crows in New York City alone. It is now feared that migrating birds have spread the virus across North America.
- Songbirds in the USA are coming to terms with the bacterium *Mycoplasma gallicepticum*, which first spread from battery farms in the eastern states in 1993. Initial heavy kill has settled down to an equilibrium in which as many die as are replaced in reproduction, with overall numbers much reduced.
- British grouse populations fluctuate in response to cyclic variations in gut parasites.
- Domestic ducks harbour duck plague, a herpes virus that causes massive kills of wild ducks.
- An outbreak of poultry influenza in Hong Kong in 1996 resulted in the precautionary slaughter of a million or more birds. In May 2001 another outbreak in Hong Kong again led to massive slaughter of birds. A few human cases of the infection were recorded. It is believed that the avian virus is first modified in pigs and that humans (with their very similar DNA sequences) catch the disease from this intermediate source.

Domesticated animal diseases that have become human epidemics

When our hunter-gatherer ancestors settled down and became farmers with domesticated animals, their close association with their mammalian and avian livestock resulted in a number of diseases crossing species barriers and becoming human plagues. That this has happened is shown conclusively by DNA studies of bacteria and viruses. Very close relationships are shown to exist between, for example, measles, human tuberculosis pathogens, and disease organisms afflicting cattle; human flu viruses are related to those that cause disease in pigs, ducks and chickens; and there are probably many others whose DNA has not yet been investigated.

Vulnerability to new strains of disease to which people have no natural inbuilt immunity has frequently been shown by the decimation of populations of aboriginal peoples on the arrival of invaders or colonisers who brought measles and smallpox, for instance, to lands where they were hitherto unknown. The industrial revolution and the concentration of populations in unhygienic towns and cities led to the spread of bubonic plague, carried by the fleas which infest rats. The modern era with its globally rapid transport connections ensures that any new variant of any disease becomes a global problem.

Bacterial diseases in plants

Australia has been at pains to prevent the disease fire blight from being introduced to its apple and pear orchards from other countries. The disease is caused by the bacterium *Erwinia amylovora* that infects flowers in spring and spreads quickly through the whole plant. Control to date has been through use of copper-based sprays—that are toxic to the plants themselves—or through use of an antibiotic *Streptomycin* spray—which has been banned for agricultural use in many other countries.

Scientists at the Ruakura Research Centre in New Zealand have come up with a biological control that may prove both effective and non-damaging. The related bacterium *Erwinia herbicola* is harmless to plants, but checks the growth of the harmful bacterium by competing for space and nutrients, and also by producing a protein that is harmful to it. The problem has been to distribute the bacterium to the emerging flowers in spring. The solution uses honeybees: a powder containing the good bacterium is put just inside hives so that the bees walk through it on their way out. In recent trials, within a week more than 95 per cent of flowers harboured the good bacteria, which then multiplied and prevented infection by the harmful relative.

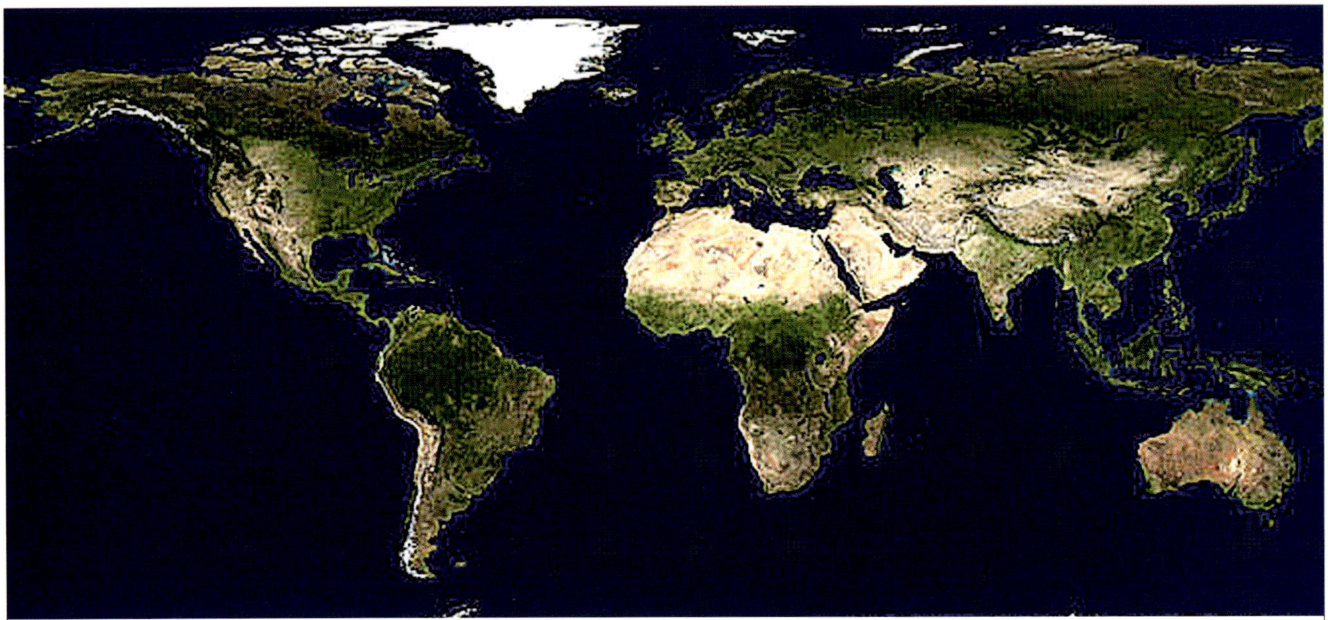

THE MAJOR GREEN BELTS OF THE WORLD
The tropical forests are Earth's lungs. Their relentless destruction has serious implications for climate change and the well-being of the Biosphere. *Dean Graetz*

Gaia—a solar-powered, living entity

The energy required for metabolism of all living things, microscopic and macroscopic, is supplied by the Sun. One of the most exciting concepts, when one first starts to think about it, is the idea that after the 4.6 billion years of evolution that has given Earth a living Biosphere (of which we are an infinitesimal part), just about all Life depends on photosynthesis. The evolution of Life-as-we-know-it would not have been possible without photosynthesis that first created and now maintains the oxygenated atmosphere, and provides carbon for the skeletons that build all living things.

PHOTOSYNTHESIS

It is quite legitimate to see 'Earth's surface as an ordered transformation of the energies of the Sun [and] living matter as a geological force transforming the Planet's surface' as described by Vernadsky in 1926[3a], because photosynthesis fuels the 'living matter' that effects the transformation. **Photosynthesis gives life to the Biosphere.**

In the 1999 botanical text *Plants in Action*[43], photosynthesis in vascular plants is described as an 'improbable process'. A weak oxidant (CO_2) must oxidise a weak reductant (H_2O) and in so doing produce a strong oxidant (O_2) and a strong reductant (carbohydrate). To achieve this 'uphill reaction' a massive and continuous input of chemical energy is required. This chemical energy is generated when sunlight strikes green plant surfaces.

Earth, Air, Life and Water—our solar-powered world. *M.E.W.*

THE RUBISCO MOLECULE

The Rubisco molecule, magnified approx x 2 million. These rotationally enhanced electron microscope images of the most abundant protein on Earth show regions of differing electron density as different colours in the upper row; the lower panels outline the regions of high electron density which reveal the contours of the subunits in the hexadecameric protein. *Cornelia Buchen-Osmond, Biosphere 2 Center, Oracle, Arizona*

Light energy has to be trapped and converted to chemically usable forms that enable plant growth and, by storing carbohydrate, provide a renewable energy source to sustain Life on Earth. Plant structure has evolved to facilitate the process and endosymbionts that are now chloroplasts and contain the light-receptive pigment were the initiators.

In photosynthesis, the energy of a photon from the sun's visible light spectrum excites an electron in the pigment molecule, chlorophyll, which transfers excess energy through a chain of reactions into two chemical forms. The first is the universal currency of energy—**adenosine triphosphate** or **ATP**—and the second is **NAD(P)H**, which provides the reducing power to convert atmospheric CO_2 into sugars. Remarkably, the energy invested in ATP within a living cell is minute at any instant in time. ATP is no energy store, rather a molecule that forms and degrades at a breathtaking rate, in turn transferring its energy to do work in the Biosphere.

Another product of photosynthesis is the oxygen freed by the splitting of water molecules; the hydrogen also liberated by the split is transferred to NAD(P)H. In this way, photosynthesis by early organisms created an oxygenated world and set in train the evolutionary sequences that have produced the modern living Biosphere.

Although much is understood about the chemical processes involved in the synthesis of vital compounds forming the basis of virtually all food chains, little is known about the physics of light capture during that first step which leads to the splitting of the water molecule. In many laboratory attempts to replicate the reaction, such violence has been required as would tear any living system apart. Yet plants do it in an orderly fashion using, simply, energy from the sun. What is arguably the most important biological reaction on Earth is also one of the most complicated reactions in the Biosphere.[44]

Chemists at Yale University have recently created an artificial version of the plant's oxygen machine and unravelled some of the mysteries, revealing a perfectly orchestrated sequence of reactions. When light is absorbed, each photon triggers a cycle of reactions in which an electron is shuttled from one molecule

Cyanobacteria from the Southern Ocean, under fluorescence microscope illumination. Magnified many thousand times. *Harvey Marchant, AAD*

NATURE'S TINIEST ROTARY MOTOR

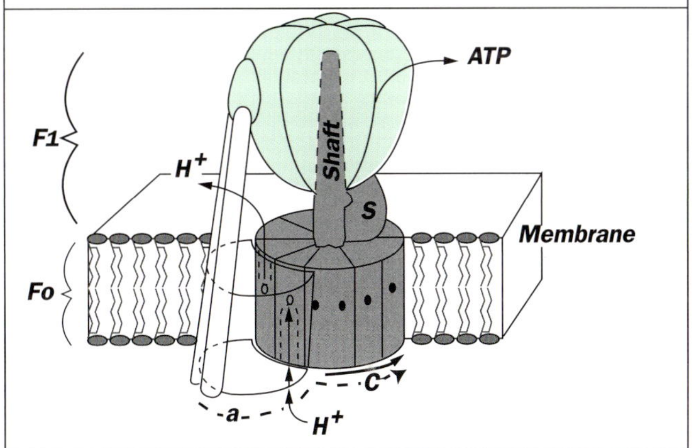

In mitochondria, an asymmetric portion **Fo** spans the inner membrane and provides a conduit for proton movement across it. This movement imparts a torque between two adjacent subunits, **a** and **c**. Torque so generated is transmitted to **F1** via the shaft and subunit **s**. The shaded portion is the rotor, the coloured portion the stator. The rotating shaft mediates an exchange of energy derived from proton flow through **Fo** to **F1** for ATP synthesis. Three ATP molecules are generated for every 12 protons that pass through this rotary motor.

Diagram modified from Brian Atwell[45]

to the next. In the first step, a photon flicks an electron out from a specially aligned pair of chlorophyll molecules, known as **P680**, and passes it to the first electron acceptor. P680 has to do this more than once if it is to capture useful amounts of energy—and in fact it does it again and again—but each time, before it can absorb another proton it must be 'reset' by another electron. This replacement electron comes from a small complex of proteins and metal ions that nestles alongside P680. This complex performs the astonishing task of pulling an electron out of the water molecule, which breaks it down into hydrogen ions (protons) and oxygen. The complex that keeps P680 supplied with electrons is known as the **oxygen evolving complex** (OEC) and is what is responsible for sustaining Life-as-we-know-it!

... A nano-electrical reaction as the foundation for a living Biosphere ...

The process of CO_2 capture has evolved variants among different plant groups and under different conditions. In the most common form of photosynthesis, known as **C3**, the first stage is the adding of CO_2 to a 5-carbon sugar phosphate to make phosphoglyceric acid (**PGA**), which is a compound containing 3 carbon atoms.[50] The enzyme that catalyses this reaction is called **Rubisco—it is present in every photosynthesising leaf or stem of Plants, in cyanobacteria and algae (protists), and is the most abundant protein in the Biosphere.** The PGA is converted into a 3-carbon sugar phosphate called phosphoglyceraldehyde. Two molecules of this **triose sugar** join to form 6-carbon sugars called hexoses; and two **hexoses** (fructose and glucose) are linked to form **sucrose** or starch. Sucrose is the form in which plants distribute most of the fixed carbon from leaves for later use in roots, stems or reproductive structures, translocated through the plant in the phloem. From sucrose, plants can also store carbon by starch synthesis in the chloroplasts. Starchy organs are important as a source of food for herbivorous animals.

Rubisco, despite its omnipresence, is an inefficient catalyst, and the photosynthetic performance of **C3** plants is restricted under prevailing conditions, with atmospheric O_2 at 20 per cent and CO_2 at only 0.037 per cent. (Rubisco evolved when the Earth's atmosphere was rich in CO_2 and virtually devoid of free oxygen.) With the advent of oxygen-producing photosynthesis the situation arose, and continues, where CO_2 and O_2 compete directly for access to the same CO_2 binding sites. Rubisco is unable to distinguish adequately between the two substrates, and in air containing 20 per cent O_2, approximately one molecule of O_2 is fixed for every three molecules of CO_2. Mechanisms to increase the concentration of CO_2 at active sites within the photosynthetic tissues have subsequently developed in some plants.

Eighty per cent of photosynthetic species are **C3** plants that use the sequence that starts with PGA.[43] Another 10 per cent have evolved a **C4** pathway that is more efficient in its use of both sunlight and water, qualities useful in the warmer and drier conditions that apply in the semi-arid regions of the world. A still more highly evolved form of photosynthesis, known as crassulacean acid metabolism (CAM), is observed in the remaining 10 per cent of plants.

C4 photosynthesis is based on CO_2 being fixed at the Rubisco sites by the much more efficient enzyme PEP-carboxylase, thereby overcoming the inefficiency of Rubisco. The initial fixation products are the 4-carbon malate and aspartate rather than PGA. These 4-carbon compounds are

STRUCTURE OF A CHLOROPLAST

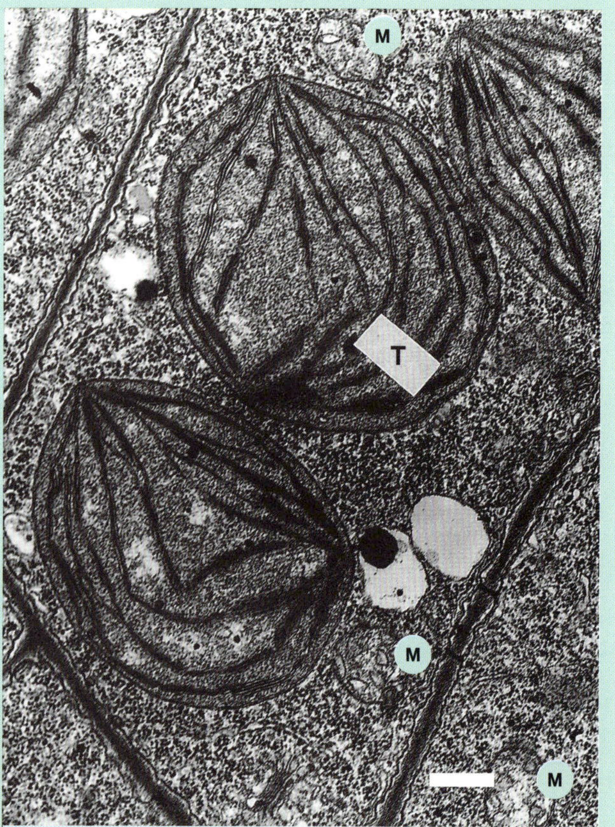

MOSS CHLOROPLASTS.
Chloroplasts in a cell of a moss, *Ditrichum* sp., from Macquarie Island. *Geraldine Nash, AAD (T=thylakoids M=mitochondria)*

Chloroplasts have a double membrane (or envelope) enclosing a gel-like material that contains all the enzymes needed for carbon fixation, many enzymes of nitrogen and sulphur metabolism, and the chloroplast's own genetic machinery. Within the gel are folded membranes (thylakoids), embedded in which are complexes that enable light harvesting and electron flow from water molecules to NADP. The complexes involve protein and pigment molecules (chlorophyll and some carotenoids) for photo-reception; molecules for transporting light energy; and ATP enzyme complexes.

'Nature's tiniest rotary motors'[45] are minute knobs found on the inside membranes of mitochondria and chloroplasts, pointing in opposite directions in the two types of organelles. (In chloroplasts during photosynthesis, the flow of protons across the membranes is into the organelle to facilitate formation of ATP from ADP and phosphate; in mitochondria, the flow is in the outward direction supplying energy for metabolism.)

pumped into bundle-sheath cells (surrounding the veins in leaves) and in effect create a biochemical pump that concentrates CO_2 at Rubisco sites.

Sugarcane, corn, sorghum and other tropical and subtropical grasses, and saltbush, use **C4** photosynthesis. **C4** plants are generally characterised by high rates of photosynthesis and growth, particularly in subtropical and tropical regions.[43] They have a competitive advantage over **C3** plants at high temperature and under strong light, using less energy in photo-respiration and fixing more of the available CO_2.

CRASSULACEAN ACID METABOLISM[46,46a]

THE CAM CYCLE—NIGHT AND DAY DIVISION OF LABOUR[46]

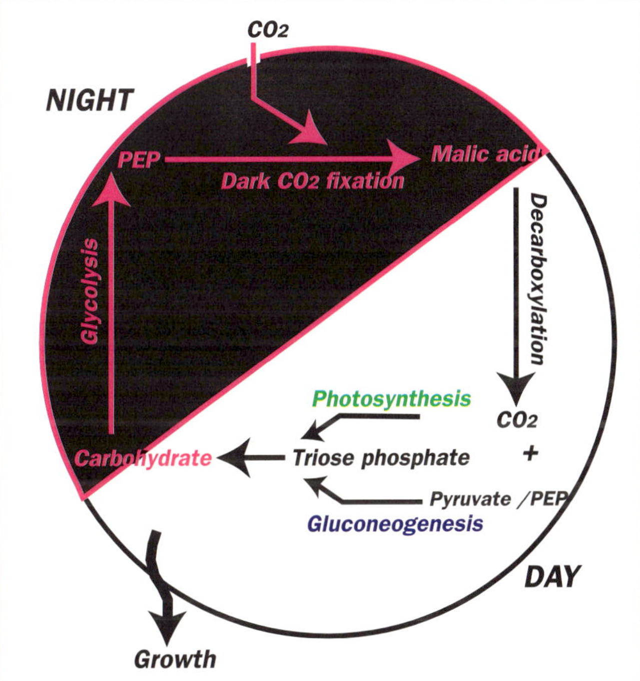

CAM is an adaptation to high temperature and, in particular, to arid conditions. CAM plants are typically succulent, with fleshy assimilatory organs—modified leaves (phyllodes) or modified stems (cladodes). The majority of cacti as well as orchids and many salt-tolerant succulents are in this category. CAM is a further modification of the **C4** system that, as we have seen, creates a CO_2 enrichment pump within photosynthetic tissues that enhances the performance of the **C3** system. Just as the anatomy of cacti and succulents is modified to cope with water-stress conditions, so also their biochemical processes have been adapted to conserve water and make the most efficient use of energy—by reversing the diurnal rhythm of CO_2 assimilation. CAM metabolism enables gas exchange at night, with photosynthetic organs taking up CO_2 and losing water during this coolest and most humid time. They store the malate assimilated in the vacuoles of their mesophyll cells for later use. During the day, with their stomata shut to conserve water, they metabolise the malate, using photo-energy and releasing CO_2. Rubisco re-fixes the CO_2 through the **C3** pathway and carbohydrate is produced, to be broken down at night to fuel the malate-making cycle that operates when the stomata are again open to allow gaseous exchange. The night-time cycle is one of acidification; during the day de-acidification takes place.

The CAM plants range from some that function only on the CAM system, to others that function by taking up CO_2 in the evening and early morning as well as at night, and others that can function all day when conditions are damp and suitable, metering their water use to atmospheric conditions.

While CAM photosynthesis has been recognised in cacti and visibly succulent plants for a long time, its use by orchids and bromeliads (where it is widespread among epiphytic species) was not investigated until more recently. Epiphytic plants are subject to severe water stress at times, even in the tropical forests where they are abundant. Plants with aerial roots attached to forest trees with little substrate need to conserve water.

At the other end of the scale from CAM plants in arid environments are the aquatic and semi-aquatic CAM plants that also need flexibility in their metabolic processes. Known as submerged aquatic macrophytes (SAMs)[46a], most of these plants occupy environments that are tidal or suffer seasonal variations in water depth and have fluctuations in CO_2 availability. Being able to switch from day to night routines as required makes their survival easier.

Succulents growing on an arid saline sandplain near Esperance in Western Australia. Exposed to heat and dryness, such plants cannot afford to lose water. Using the CAM cycle with an acid storage phase enables them to occupy niches from which most plants without this adaptation are excluded. Some of these pig-face succulents can switch from C3 photosynthesis to C4 and CAM depending on local conditions, responding to drought and flood, and to seasonality. *M.E.W.*

PHOTOSYNTHESIS IN THE HIGH-LATITUDE SOUTHERN OCEAN[20]

Phytoplankton are microscopic, floating, single-celled algae (now classified as protists, no longer as plants). More than 200 species have been identified in Antarctic waters, showing great diversity in size, shape, life style and food value to grazers. They are grouped in three classes according to size, from smallest to largest—picoplankton, nanoplankton and microplankton. (Note the degree of magnification on the micrographs.) Picoplankton and nanoplankton are the most abundant and diverse.

The cells of diatoms are encased in intricately sculptured glassy walls; other phytoplankton are covered with finely patterned scales; some can swim, others drift, encased in mucus. Phytoplankton are photosynthetic, using energy from light to convert CO_2 into sugars. Very little light is available during Antarctic winters, when the sun is either below or very low on the horizon and the sea is covered with ice and snow, further reducing light availability. In summer, the sun shines for 24 hours a day and much of the sea ice melts. Vast blooms of phytoplankton occur under summer conditions, supplying food for krill and the rest of the food web.

Concentrations of the major nutrients required for phytoplankton growth—nitrogen and phosphorus (and silica for diatoms)—are generally high in the Southern Ocean, yet concentrations of the organisms themselves are much lower than in seas with similar nutrient levels in most other parts of the world. This 'high nutrient–low chlorophyll' situation is known as the Antarctic Paradox and can be explained by a combination of physical, chemical and biological factors. The Southern Ocean is a realm of high winds and rough seas. Phytoplankton in the surface waters (the euphotic zone, where sufficient light is available for photosynthesis) can be wind-mixed to depths greater than 100 metres, where there is insufficient light for photosynthesis. The micro-nutrient iron, also necessary for phytoplankton growth, is in short supply in the ocean, and under low light conditions phytoplankton require more of it to function. Both these factors help to explain the Paradox.

(Because phytoplankon growth is limited by shortage of iron it has been suggested[47] that the addition of iron to Antarctic waters would promote phytoplankton growth, thereby increasing the removal of atmospheric CO_2 and counteracting greenhouse warming. Before any such major interference was attempted it would be necessary to ascertain whether those species that would be promoted by the presence of extra iron would transfer the carbon to deep water, achieving the desired effect, or whether its presence would promote species more likely to fuel respirational losses by the organisms of the microbial loop in surface waters, exacerbating atmospheric CO_2 accumulation.)

Some phytoplankton photosynthesise when light is available, behaving as autotrophs or primary producers, and become heterotrophs under low-light conditions, consuming other organisms. Some other planktonic heterotrophs—the zooplankton protozoa, ciliates and so on that are the grazers of phytoplankton—have developed corresponding lifestyle adaptations. They eat the photosynthetic organisms and retain their chloroplasts, which enables them to photosynthesise and increase their own nutrient levels when there is sufficient light.

The dinoflagellate *Protopteridinium latristriatum*, surrounded by chains of diatom cells, fragments of tests and bits and pieces of other organisms. *Fiona Scott, AAD*

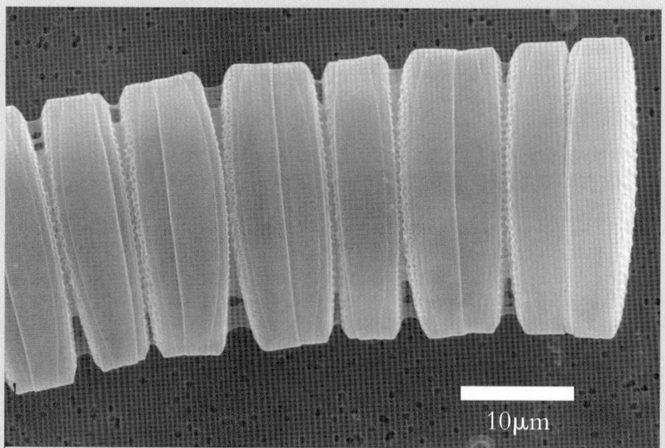

The diatom *Fragilariopsis kerguelensis*. When diatoms divide they can create short chains of cells that are easily separated. Their glassy walls accumulate as thick deposits of diatomaceous earth on the sea floor. *Geraldine Nash, AAD*

The diatom *Chaetoceros bulbosus*. *Fiona Scott, AAD*

Life's control of the atmosphere

The Atmosphere, one of the Organs of the Living Biosphere, has been modified by the activities of living organisms for more than 4 billion years. The changes made by early Life made it possible for subsequent life forms to evolve and contribute to more changes. Even today a dynamic situation exists, orchestrated almost entirely by micro-organic Life and photosynthetic organisms, but now with Technological Man making some contribution.

THE PRE-LIFE EARTH AND ATMOSPHERIC CHANGE

It is believed that in the earliest times on the newborn Earth, carbon dioxide concentrations in the primaeval atmosphere were comparable to those on Mars—three orders of magnitude higher than in Earth's present-day atmosphere. (Mars has 93 per cent CO_2; Earth 0.03 per cent.) As the Earth cooled, it was protected from the cold of Space by this greenhouse blanket, even while solar luminosity was about 30 per cent of its current value. As the Sun brightened and its energy output increased, it would have been impossible for Life to evolve until CO_2 levels decreased and a cooler world was established, suitable for the first Archaean thermophiles. The question that has to be answered is: where did the excess CO_2 go?

One possible answer, according to Vaclav Smil[50], lies in the geotectonic cycle that sequestered (and still locks up) carbon in carbonates in the Earth's crust. The process, according to this line of speculation, involves a more luminous Sun evaporating more water; water vapour and CO_2 forming carbonic acid (H_2CO_3); the acid weathering calcium silicate in exposed rock surfaces, releasing calcium and bicarbonate ions; the ions transported in run-off to the ocean; and reactions in the ocean producing sediment-forming carbonates, CO_2 and water, thus effectively locking up the carbon. (In the shallow seas of those times there would have been massive precipitation of carbonates. In modern oceans, carbonate precipitation does not occur below about 2 kilometres depth, as calcium carbonate dissolves under the pressure reached at this threshold.) Then the Plate Tectonic cycle takes over. Carbonates and other sediments are either subducted or compressed and folded and scraped off onto continental edges where plates collide. The subducted portion melts and reacts with silica, reappearing in volcanic rocks, weathering and repeating the cycle at a slow geotectonic pace that is measured in multi-millions of years. Volcanoes that erupt along the subduction zone where some of the carbonate is melting vent large volumes of CO_2, countering the locking-up process. However, the carbonate sediments that form new continental material lock up CO_2 effectively for very long periods, or practically for ever, and only slow weathering over time releases a fraction of it again. The geotectonic cycle existed before Life evolved, and the carbonate sediments were abiotic. (Life has had a major role in forming marine carbonates since. According to Dr James Kasting of Pennsylvania University, who is an expert on Mars climate matters[51], if you released all the CO_2 that is locked up in carbonate sediments today, you would get 60 atmospheres' worth.)

LIFE HAS MODIFIED GLOBAL CLIMATE FROM ITS INCEPTION

The early Earth was too hot for comfort, though it suited the thermophilic Archaea, the Primaeval Ancestors—but as its crust cooled, with its atmospheric CO_2 becoming much depleted by the very active tectonism of its original over-heated state, it was at risk of becoming too cold for Life. The Sun was a much weaker heat source when Earth was born than it is today, and is believed to have been 20 per cent weaker at 2.8 billion years ago, which implies luminosity 30 per cent reduced from the level present when Life began. It was methane-producing sulphur bacteria that helped to keep temperatures life-friendly then—100 parts per million of methane would have been sufficient.[51]

Water-molecule-splitting photosynthesisers—the cyanobacteria—followed the sulphur bacteria. Originally photosynthesis was occurring only in the seas, and the oxygen produced was used up in oxidising materials there. When saturation was reached in sea water, oxygen started leaking into the atmosphere and climate was becoming 'a machine driven by Life'. According to Dr Kasting[51], the activities of these early photosynthesisers brought on the first major ice age of 2.4 to 2.3 billion years ago. Oxygen, reacting with methane in the atmosphere, formed carbon dioxide and water vapour. The depletion of methane allowed heat to escape and started drastic cooling of the Earth. The lag between the methane greenhouse and carbon dioxide greenhouse had yet to be overcome. As the ice caps grew, temperature continued to fall, and the ice sheets themselves hastened the cooling by reflecting more heat into Space. Then, as CO_2 levels rose with volcanic activity, the Earth thawed, and increasing oxygen levels caused more oxidation, strengthening the greenhouse effect.

Then, over more than a billion years, bacteria, cyanobacteria, unicellular and filamentous algae and other protists flourished, and a green soup of photosynthesising organisms again depleted available resources of CO_2 and other essentials. A massive crash of the system about 900 million years ago saw ice age conditions return, to persist with fluctuating intensity for nearly 300 million years in the Precambrian ice age. The 'Neoproterozoic' glaciation between 750 and 580 million years ago was the most severe part of this ice age.

Paul E. Hoffman and Daniel P. Schrag of Harvard University[52] have coined the term **Snowball Earth** for this glaciation, a period of 10 million years or more when Earth was a frozen world. It was so cold that even the Equator was glaciated. (The landmasses of the world were aggregated in equatorial regions then.) Luckily for Life, and for us, remnants of the early life forms of the previous 3 billion years managed to survive, deep below the ice where water was warmed by thermal activity, and around active volcanoes. Life had by then reached the stage of filamentous algae and unicellular protists and bacteria.[52] The volcanoes eventually, slowly, built up CO_2 to such concentrations that warmth gradually returned and the Snowball melted. But the amount of CO_2 needed to thaw the globe was such that the greenhouse that followed the Snowball was equally noxious. (The planet Venus has suffered climatic swings like Snowball, and if it ever had Life, none has survived its final deep freeze.)

The **Big Bang of Species**—the coming of the multicellular animals—has been accurately dated to not more recently than 600 to 560 million years ago.[53] The Cambrian Period, whose start is determined by the first occurrence of shelly creatures as fossils, begins at 543 million years ago, and all but one phylum of the Animal Kingdom appeared in the first 10 million years. Even a chordate, *Yunnanzoon lividum*, ancestral to Vertebrates, had appeared by 525 million years ago.

It has been suggested by Paul Renne, director of the Berkeley Geochronology Center, that the sudden explosion of complex life might have been due to 'seeding' with complex organic chemicals from Space that acted as catalysts, but this 'Panspermia' theory has not received general acceptance. The explosion of forms may have a more earth-bound explanation. The fact that hot-spring communities had been widely separated geographically during the extreme Snowball conditions implies that they would have accumulated considerable genetic diversity. As conditions changed and improved, selective pressures would have been a spur to the burgeoning species diversity that we see.

(Bacteria, ever since the first major glaciation over 2 billion years ago, have been most important in maintenance of the Biosphere. According to Margulis and Sagan[7], 'Bacteria have been, and probably always will be, the most important players in maintaining the biosphere. A single photosynthetic blue-green bacterium growing under ideal conditions could, in theory, produce all the oxygen now in the atmosphere in just a few weeks.')

During the Phanerozoic Eon, from the Cambrian Period to today, characterised by the sorts of animals and plants that are familiar to us, there has fortunately been no fluctuation of climate as violent as Snowball Earth. The maturing Biosphere has had better balanced controls and feedbacks. However, some quite dramatic fluctuations may have had biological causes, while other fluctuations largely caused by the restless nature of the Planet itself in turn have affected the Biosphere.

According to Vaclav Smil[50], the Late Carboniferous to Early Permian ice age of about 300 million years ago was precipitated by the explosion of vascular land plants that took place in the Early Carboniferous, from about 350 million years ago—when the burial of vast quantities of organic matter (which created the abundant coal deposits of Northern Hemisphere lands) locked up carbon and caused a sharp decrease in atmospheric CO_2. This may be the case, but it would seem (author's personal comment, based on Gondwanan information) that the mechanisms of the Biosphere, even then, were more complex. At that time Australia was near the South Pole and half of the continent was glaciated, vegetation was limited by the high-latitude ice age situation, and no Carboniferous coals formed in Australia. As the glaciers retreated and the *Glossopteris* flora burgeoned in the whole of Gondwana, the Permian coal measures of Southern Hemisphere lands were formed, again locking up carbon (on a scale comparable with the Northern Hemisphere event) but the Late Permian in Australia was subtropical and vegetation continued to thrive, into the Triassic. This poses the question why ice age did not persist or return, if only coal formation was involved.

CO_2-greenhouse conditions rose again markedly in the Cretaceous (when subtropical conditions existed near the South Pole in the Late Cretaceous.) The tectonic activity associated with supercontinent fragmentation probably resulted in the CO_2 increase; as tectonic activity waned CO_2 levels declined.

When the whole of the carbon cycle through time is reviewed, it is obvious that it has been Life that has kept some sort of equilibrium in the carbon dioxide levels in the Biosphere.

The effect of the human technological revolution, an evolutionary advance of a different kind, on global atmospheric conditions is seen in recent damage to the ozone layer by CFCs and other gases; and in the current greenhouse warming where Life is altering biospheric conditions, and in this case too rapidly and not helping to keep the Biosphere Life-friendly.

On current estimates by the Carbon Dioxide Information Analysis Centre (CDIAC), over a period of less than 1000 years, 2–4 *billion billion* tonnes (2000–4000 gigatonnes) of carbon will be added to the atmosphere by human activity.[54] In order to visualise what that is going to mean to the global environment, scientists are looking for events in the geological past when very sudden climatic change has resulted from changes to the carbon cycle—this can be detected by anomalies in C13 to C12 carbon isotope ratios in marine sediments. One such event has been correlated with a sudden greenhouse event 55 million years ago, in the early Eocene. At this time, many extinctions of deep sea creatures occurred, and numerous mammalian orders first appeared in the fossil record, including primates.[55] This global warming, the Late Palaeocene Thermal Maximum or LPTM, is now believed to have been caused by methane bursting from the seafloor, where it had remained locked up through the ages, and not to have been caused by contemporary biological activity (although the methane itself was the product of methanogenic micro-organisms that had previously created the huge amounts that were stored). As we shall see, such an event could occur again at any time in the future!

THE ALGAE–CLOUD CONNECTION

In the Biosphere today, marine algae still affect global temperature and climatic patterns. Single-celled phytoplankton can proliferate in unbelievable numbers—6 billion tonnes is one figure suggested for their mass in the high-latitude seas when ice melt occurs in the spring. Marine algae synthesise complex sulphur compounds that balance internal ion concentrations (the salt that accumulates in their cells could destroy them without this chemical intervention). These sulphur compounds are stable in sea water[59], but are broken down enzymatically to form dimethyl sulphide (DMS), which escapes into the atmosphere. There the DMS is oxidised, forming sulphate particles that act as nuclei for cloud formation. (Rotting seaweeds in intertidal zones, and dead kelp floating offshore, make a very minor contribution to this sulphur exchange.) Clouds reflect heat back into Space and bring rain onto the land—thus exerting significant control over global climate.

The coccolithophore *Emiliania huxleyi*, covered in its decorative calcium carbonate coccolith scales. *Andrew McEldowney, AAD*

The atmospheric sulphur cycle over Antarctica is dominated by the emissions of dimethyl sulphide from marine sources.[22] Emissions from the Sea Ice Zone around Antarctica, although highly seasonal, account for at least 17 per cent of the emission of sulphur from the global ocean. The principal producer in Antarctic waters is the alga *Phaeocystis antarctica*. (See pages 68–69.)

In temperate, subtropical and tropical seas, coccolithophores—calcium-secreting, microscopic, unicellular algae covered by elaborate and decorative scales which move to allow maximisation of light input for photosynthesis—can form enormous blooms that stretch for 200 kilometres and can be seen from satellites as milky patches on the ocean.[7] (Coccolithophores only appeared in the fossil record 100 million years ago, although unicellular marine algae have a more than 2 billion year lineage.) Detached coccoliths accumulate on the seafloor in vast concentrations and are likely to be the largest single carbonate sink in marine biogeochemical cycles, contributing to carbonate in sediments.[61]

THE SOUTHERN OCEAN SULPHUR CYCLE

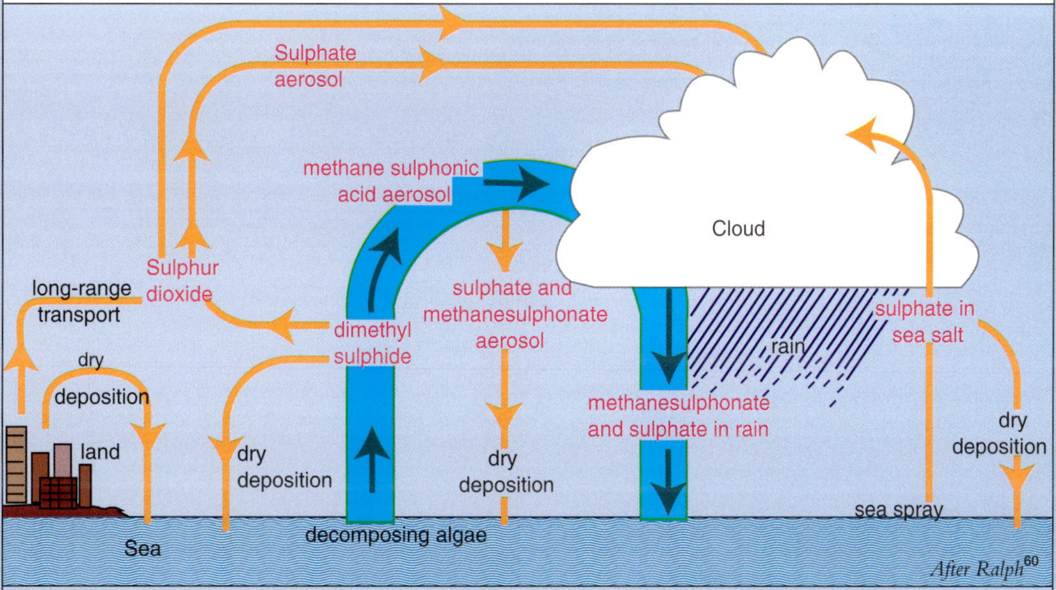

Decomposing algae produce dimethyl sulphide, a gas which on oxidation forms sulphate particles that act as nuclei for cloud formation. Some small addition to this major sulphur cycle comes via long-range transport from terrestrial sources, increased by human activities, particularly fossil fuel combustion.

THE LATE PALAEOCENE THERMAL MAXIMUM[55,56,57,57a,58]

For a brief period, about 55 million years ago, temperatures at high latitudes and in the deep oceans soared by 5° to 7°C. An abrupt drop of 2.5 to 3 per cent in the proportion of the carbon isotope C13, and an increase in C12, indicates a massive input of carbon to the ocean or atmosphere from an external reservoir greatly enriched with C12. The change in proportions occurred over about 10 000 years, and recovery to normal balance took about 140 000 years. To achieve these figures the input of carbon would have had to be at a rate comparable with the rate of addition that is occurring now from human activities. The previously held belief in a carbon cycle in which levels of carbon dioxide in the atmosphere remain relatively stable has been questioned, and it is now accepted that there must be a large carbon reservoir that in the distant past has sporadically been able to add huge quantities of C12-rich carbon to the atmosphere over geologically very short periods of time.

The best explanation is that the carbon cycle contains a large gas-hydrate capacitor, because no other reservoir could supply a large mass of carbon sufficiently enriched in C12.

Gas hydrates are ice-like solids composed of gas and water that are stable at high pressure, low temperature and high gas concentration. It is well known that continental shelf sediments contain great masses of methane (CH_4) hydrate, and presumably this has been the case through at least some hundreds of million of years of geological time. What triggered the release of 1000 gigatonnes into the ocean and atmosphere in the LPTM event is uncertain. Deep sea temperature change is definitely involved, and it has been suggested that an episode of explosive volcanism in the North Atlantic was the trigger. Warming of bottom sediments by warmed deep-ocean currents would have raised temperature in some of the traps where methane hydrate was locked up, and with the crossing of a temperature threshold, the release of methane and its oxidation to CO_2 brought on the rapidly escalating greenhouse.

There is, of course, no reason why a presently unidentified threshold will not be crossed again to allow vast quantities of methane to be released. Current greenhouse warming may mean that the world is again facing such a situation. Certainly a lot of research will be prompted by this concern. Already, unexpected large deposits of gas hydrate have been located on the Lord Howe Rise, the fragment of thinned continental crust that sank when the Tasman Sea opened along Australia's eastern margin between 80 and 60 million years ago.

UNDERGROUND COAL FIRES AND GLOBAL WARMING

The huge addition to greenhouse gases by coal seams burning underground has been little considered.[54a] Worldwide, fires are burning in thousands of underground coal seams. They are so widespread in Asia that the financial loss is forcing governments to take action to deal with them. In China's rich northern coal fields, it is estimated that more than 200 million tonnes of coal are destroyed by underground fire each year—about 20 per cent of the nation's annual production. These fires alone produce nearly as much CO_2 as is emitted each year by cars and small trucks in the USA. In India the fires are so widespread and people are so desperate for coal that mining proceeds around the burning seams. In densely populated regions, many people live on the edges of open pits that fume and flame.

Geological evidence shows that grassland and forest fires, lightning and spontaneous combustion, have all resulted in coal fires in the past. Layers of clinker in the regolith indicate where coal seams have been destroyed. Australia has an example of a burning coal deposit at Burning Mountain in the Upper Hunter region. The coal seams in the mountain have been burning for at least 15 000 years and possibly for 500 000. In Indonesia in 1997, when extreme drought resulted in major fires in rainforest, hundreds of coal fires erupted when forest fires scorched exposed seams.

The frequency of subterranean fires has increased, and human activities are to blame. Mining has exposed more deposits to surface fires whose frequency has increased because of human activities, and to the oxygen that feeds fire. Many coal fires start spontaneously when pyrite and other reactive minerals in coal are exposed to oxygen, releasing heat that builds up when there is no ventilation until the coal itself ignites. The practice of dumping rubbish into old mine shafts has also started many fires.

Satellites that map ground temperature are increasingly being used to map the extent of underground fires. Their thermal-imaging equipment is able to detect the temperature differences of only a few degrees that indicates their location.

Blair Athol Coal Mine in Queensland has the world's thickest team of black coal. *M.E.W.*

Geomicrobiology: The role of micro-organisms in geological processes

The distinction between the 'inanimate' Mineral Kingdom and the living Kingdoms becomes somewhat blurred when it is realised that many minerals are produced by Life, sometimes in crystalline form. The most common are calcium carbonate (shells) and calcium phosphate (bones). Fifty different minerals are, in fact, produced by living cells. In addition, many sedimentary rock types result from biologically aided weathering and subsequent recombinations of minerals facilitated by micro-organisms. The ferricretes (ironstone cappings and duricrusts) that are so widespread in Australia originated from weathering in the saturated watertable zone of ancient soils. Genesis of such rock continues today, particularly in Australia's Gulf country, where monsoonal rainfall, high fluctuating watertables and suitable substrate provide ideal conditions for the physical and microbial processes required. Ochre deposits, where deep chemical weathering facilitated by micro-organisms has changed hard rock into a talcum powder consistency, as at the Ochre Pits locality in South Australia, are another visible expression of microbial activity in groundwaters. Ochre deposits present evidence of great climatic changes through geological time when they occur in such arid sites as this, for their formation is known to have required high and fluctuating watertables.

In his book *The Biosphere*, the Russian geochemist Vernadsky stressed the importance of bio-geochemical reactions:

> If life were to cease the great chemical processes connected with it would disappear, both from the biosphere and probably also from the crust. All minerals in the upper crust—the free alumino-siliceous acids (clays), the carbonates (limestones and dolomites), the hydrated oxides of iron and aluminium (limonites and bauxites), as well as hundreds of others, are continuously created by the influence of life.[3a]

'INANIMATE' EARTH MATERIALS OF BIOGENIC ORIGINS

The activities of micro-organisms are involved in the formation of many rock types, including: stromatolites; some clays (produced by bacterial activity); iron ore deposits; laterites and bauxite; ochre deposits; manganese nodules (in freshwater deposits associated with cyanobacterial mats); nodules of manganese and other minerals on the floor of the deep oceans; some sedimentary carbonate deposits; chalk deposits (from foraminifera and coccoliths); the calcium carbonate matrix of coral reefs; *Globerigina* and radiolarian 'oozes' on the seafloor; some silica deposits (resulting from diatoms forming diatomaceous earths); and, believe it or not, alluvial gold deposits and opal formation.

ZONATION OF MICRO-ORGANISMS IN THE REGOLITH

The distribution of micro-organisms in the regolith is usually zoned. The uppermost subsurface zone, 10 to 50 metres deep, has active bacteria, fungi and protists. Deeper down, to between 150 or 250 metres, the organisms are mainly bacteria, with occasional others; from 250 to several thousand metres bacteria only occur. At or near the surface, microbes inhabit rock faces or the surface of mineral particles, occurring in biofilms, micro-colonies or as single cells, usually attached to the substrate by slime. Others occur in pore water. The deeper down the microbe is, the slower it is believed to metabolise. The bacteria range from aerobic to anaerobic.

The Leichhardt River, as it crosses the coastal plain to Australia's Gulf of Carpentaria, has created its own low-grade ferricrete banks and islands with the help of microbial activity at the watertable in its valley and in its saturated banks and hyporheic zone. Several major rivers in the Gulf have 'cemented' themselves into their channels in this way. *M.E.W.*

Micro-organisms and weathering

A significant role is played by micro-organisms in fundamental geological processes.[28] Bacteria, protists (including algae) and fungi operate as geochemical agents in the uppermost lithosphere and in the hydrosphere.

Some are involved in **rock weathering** through their production of organic acids or ligands which mobilise mineral constituents; others promote weathering by altering the form of iron and manganese in the rock and making their decomposition easier. Nearly all weathering agents are corrosive metabolic end-products formed by bacteria, fungi or lichens. (The collapse of concrete sewer pipes is due to the corrosive action of microbes—both inorganic and organic sulphur in sewage is turned into sulphuric acid by sulphur-oxidising bacteria. Acid mine drainage results from the oxidising and reducing activities of microbes.)

Biofilms in the weathering zone where groundwater meets the rocks of the lithosphere are concentrated microbial ecosystems where organisms range from aerobic to anaerobic.

The Ochre Pits in arid South Australia, created by deep biochemical weathering that occurred when the region was well watered. *M.E.W.*

FERRICRETE ON CAPE YORK

Brilliantly red pisolitic ferricrete gravel is used to surface roadways in the recreation area at the Jardine River in Cape York. The ferricrete layer that covers large areas of the Cape is responsible for the appalling corrugations that make travel on its roads an endurance test. The ferricrete is a biogenic deposit formed at an ancient watertable, like the bauxite at Weipa, which also occurs in pisolitic form. There the microbial activity has concentrated aluminium as well as iron in the weathering process. *M.E.W.*

CONCENTRATION OF GOLD BY MICRO-ORGANISMS

Several types of micro-organisms cause active or passive mineral formation by precipitating elements and concentrating them on or in their cells.

It has been found that microbes can promote the formation of minerals at surface to moderate temperature and pressure, where previously high temperature and pressure were thought always to have been involved in their formation in the Earth's crust. For example, *Bacillus subtilis*, a common soil bacterium, collects gold in the form of colloidal particles within its cell. These become soluble when the cell dies, and the gold crystallises and aggregates to form placer-gold particles. An example of a different process, in which accumulation of gold occurs as a result of deposition of thin coats on specific bacteria, has been described from an Alaskan locality.

In this **Alaskan alluvial gold deposit**, gold flakes occur shaped like the distinctive local *Pedomicrobium* bacterium, whose cell wall is believed to have a selective attraction for the element.[62] The process involves traces of gold in products from the weathering of gold-bearing strata being dissolved in groundwater and re-precipitated on the surface of the bacterial cells. In order for the observed build-up on the microbes to have occurred, and for the organism's form to have been maintained throughout the development of its golden skin, it must have remained alive. Gold solutions are known to be lethal to many species of bacteria, although attraction of gold to their surfaces appears to be a widespread characteristic. Presumed immunity to the toxic properties allowed the *Pedomicrobium* bacteria to remain active accumulators of gold.

The purity of the golden castings in the Alaskan deposit is extraordinary, and has led to suggestions for using bacteria to extract gold from the traces in mining waste, and from low-grade ores.

Many other bacteria have the capacity to become encased in mineral coatings. Some *Pedomicrobium* species accumulate iron and manganese oxides, and cause their build-up inside pipes in hydro-electric plants. (Selective attraction of silica to the cellulose in plant cell walls is a similar phenomenon to the attraction of gold to bacterial cell walls, in this case resulting in cell-perfect petrifactions, visible in the fossil record.)

Pisolitic ferricrete. *M.E.W.*

Microbial accumulation of gold in Venezuela[63]

Small gold specimens, ranging in size from 0.2 to 6.6 millimetres, were collected from the Quebrada Grande Mining Zone in the Orinoco River catchment. Their colour is that of 24 carat gold. Most specimens have an irregular knobby shape and finely granular outer surfaces. Scanning electron microscope photography reveals the structure of these tiny nuggets. The granules consist of numerous, variously undulating and coiled, hollow, tube-like filaments. They radiate from a subglobular, apparently dense, central region that may partly show its original composition of filaments now interconnected by interstitial gold. The distal ends of filaments may be closed and rounded, or have a small circular opening. Other pores can be seen perforating the filament walls, showing the tubes to be hollow. In some granules the filaments show thickenings like nodes and club-shaped or lobate distal ends. The specimens, as seen when cut open, are porous.

The gold-encrusted filaments comprising the Venezuelan specimens closely resemble those found in thucholite from Precambrian rocks of the Witwatersrand gold reefs in South Africa. The radial arrangement of filaments is similar to that seen in 'appendaged and/or budding bacteria' like *Metallogenium*, which is known for the manganese encrustations of its appendages. Other similar bacteria, also involved in oxidation of manganese, are *Caulococcus* and *Kusnezovia* that occur on the muddy bottoms of freshwater lakes. *Metallogenium* occurs in freshwater lakes in Europe and North America; in waterpipes and lakes in Australia, in the topmost 50 centimetres of soils that contain manganese (with a maximum occurrence at the surface), as well as in leaf litter.

The precipitation of gold onto bacterial filaments is believed to be the result of their exudation of polymers that are attractive to that metal. The presence of gold in solution in groundwater is due to the corrosive weathering of gold-bearing quartz by the organic acids produced by microbes and other organisms.

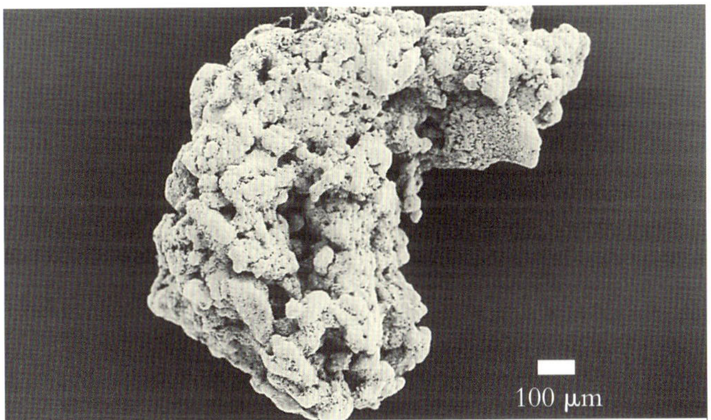

A small gold specimen. This small nugget has an irregular knobby surface. *SEM photography Gunther Bischoff*

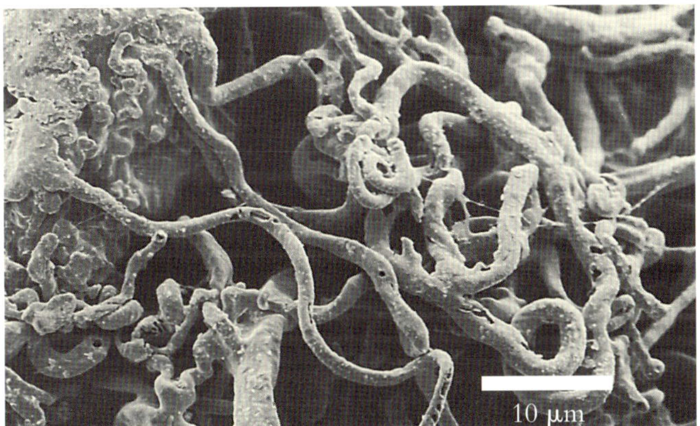

Under higher magnification the twisting and twining filaments, gold encrusted, of 'appendaged and/or budding bacteria' are more clearly seen. Very thin spider-web-like threads are seen between filaments. *SEM photography Gunther Bischoff*

EXTREMOPHILES AND HYDROTHERMAL MINERAL AND MASSIVE SULPHIDE DEPOSITS

A 3235 million-year-old massive sulphide deposit from Sulphur Springs on the Pilbara Craton of Western Australia has recently been shown to contain pyritic filaments that are believed to be fossilised thread-like micro-organisms, almost certainly prokaryotes (Archaea).[27] The deposit represents a deep-sea hydrothermal

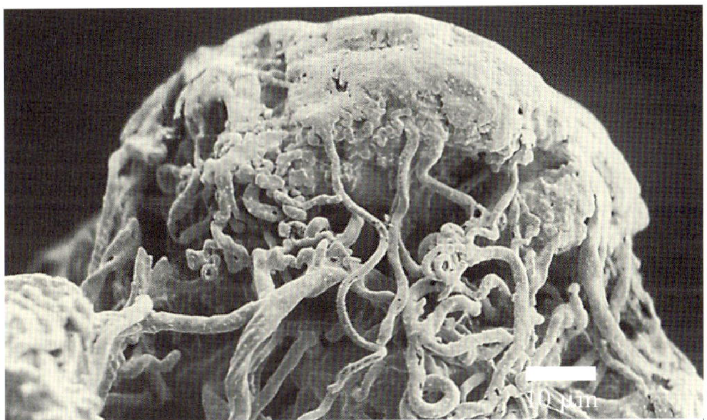

A knob, showing a top smoothed by abrasion and compaction while the specimen was being transported by water, and the unsquashed filaments in the protected part below. *SEM photography Gunther Bischoff*

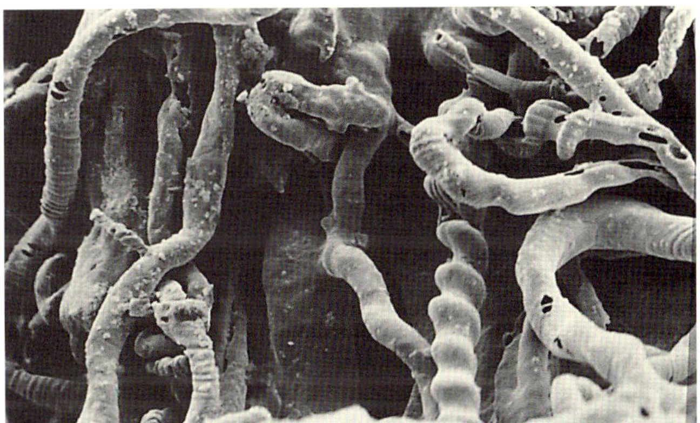

Detail can be seen on the filaments and a beautiful example of a twisted filament is revealed in this part of the specimen. In the lower right corner of the picture, in particular, note the wrinkling, the hollow nature of filaments, some terminal pores, and pores on filament surfaces, some with two segments. The purpose or function of the pores is not known. *SEM photography Gunther Bischoff*

environment where extremophiles carried out chemosynthesis, using hydrogen sulphide that came from volcanic vents and producing elemental sulphur as a by-product. The deposit is believed to have formed at more than 1000 metres depth. Parts of the deposit, believed to be the sites of chimneys, are rich in copper, zinc and lead. The composition of the ores suggests that fluid temperatures were about 300°C, similar to temperatures at black-smoker chimneys on mid-ocean ridges today.

BIODEGRADATION OF HEAVY MINERALS BY CYANOBACTERIA AND MICRO-FUNGI

Scientists at Macquarie University in Sydney, Gunther Bischoff and Robert Coenraads, made some amazing discoveries in the Northern Tablelands of New South Wales when they examined minerals in a deep lead (gravels of a palaeoriver) of Oligocene age near Inverell, Pleistocene sapphire-bearing gravels near Glen Innes, and a basalt soil near Matheson.[65] Under scanning electron microscope magnification, etched grooves, dish-like depressions and borings were seen on the surfaces of crystals of spinel, magnetite and zircon. In some of the worm-like borings, filaments have been found preserved. The minute tunnels and grooves spiral or meander into surfaces, often terminating in circular, oblique or vertical holes. The arrangement of grooves is consistent with the extension of filaments or hyphae of cyanobacteria or fungi 'searching for easy penetration into the mineral at sites of structural weakness, and etching the mineral surface in the process'.

Recently, Dr Hubert Staudigel of the Scripps Institute of Oceanography (University of California) has reported microbes leaving worm-like tracks in rocks of the lithosphere under the deep Atlantic and Pacific Oceans. In the Pacific Ocean Drilling Project, at depths of up to 6 kilometres, rock-eating microbes have been found to be present. Their numbers decrease with depth below the surface but it is claimed that they are ubiquitous below all oceans, and that their activities alter seawater chemistry, with the potential to influence interactions like carbon cycling and hence global climate. 'Deep biosphere' research has increased in the last decade, expanding horizons and emphasising the importance of 'the little things that run the world' in the most unlikely places.

ETCHING OF SPINEL CRYSTALS BY CYANOBACTERIA

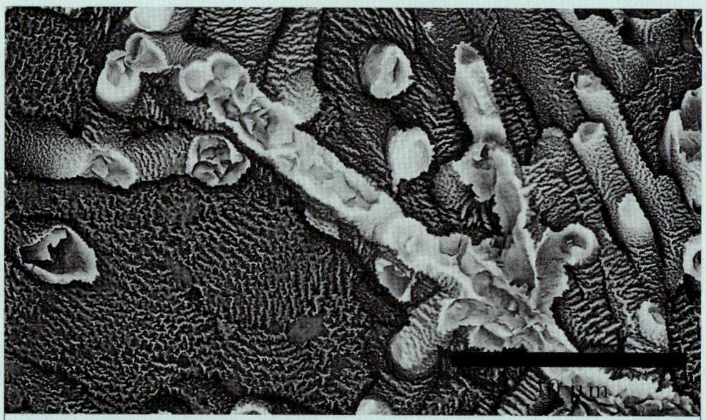

The surface of a crystal of spinel seen under SEM reveals that it is etched by shallow tunnels and grooves, and penetrated by borings. *SEM photography Gunther Bischoff*

Magnification of a small section of a crystal surface shows remains of organisms, believed to be cyanobacteria, in the etchings. The organic acids that the organisms produce are sufficiently corrosive to eat into the mineral at this microscopic scale. The delicate etching of the whole surface by bio-acids is seen. *SEM photography Gunther Bischoff*

The surface of this zircon crystal shows an etched pattern of different levels of penetration: from smooth and unetched in the lower right corner, through a curved zone of light etching, to a deeply etched region at top left of the picture. This zonation is explained as resulting from the form of the micro-lichen or fungal thallus that is believed to have been the agent of etching—a lobed thallus strongly attached in the deeper left-top corner, and lobes lightly attached (? spreading out) onto the surrounding surface. *SEM photography Gunther Bischoff*

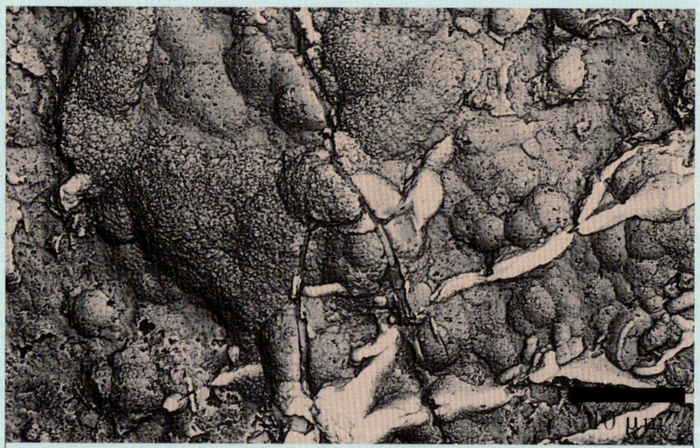

At higher magnification the etched surface and borings containing hyphae are clearly visible. *SEM photography Gunther Bischoff*

TRIBUTE TO AN AUSTRALIAN PIONEER IN GEOMICROBIOLOGY

Gunther Bischoff (1928–1999) is remembered as a Senior Lecturer in Palaeontology, for his research in micro-palaeontology, particularly in the biostratigraphy of conodonts, for his remarkable SEM work at Macquarie University, and most particularly for the enthusiasm with which he sought new scientific challenges. He was one of the first to consider the role of micro-organisms in the formation of mineral deposits like bauxite, ferricrete and silcrete, and he inspired others to research the role of microbial activity in mineral-forming processes in the Biosphere. His ideas about biological control of opal formation have since been shown to be right on target. His work on the precipitation of gold on the surface of bacteria, and the etching of the surface of mineral crystals by bacteria and fungi, was also leading edge. He was also interested in less glamorous examples of precipitation of mineral matter by bacteria—like the iron stains that afflict swimming pools. A true scientist in the classical tradition, to him everything from the ultra-microscopic to the cosmic was exciting and thought-provoking. It is the curiosity of scientists like Bischoff that has advanced the Earth Science disciplines and, equally valuably, has inspired others to follow these areas of research.

THE RELATIONSHIP BETWEEN VOLCANIC ENVIRONMENTS AND GOLD ORE BODIES

Gold is a rare element in the Earth's crust, occurring mostly as impurities in rocks.[64] Ores are rocks that contain far more gold than other rocks, either as the metal itself (native gold) or in combination with other minerals (gold tellurides, gold alloys and sulphides). Most of the world's gold has been mined from rocks more than 2.5 billion years old, in Canada, Western Australia and South Africa.

New insights into origins of gold deposits are now coming from the youngest rocks on Earth, those still forming in areas of volcanic activity. The Taupo volcanic zone in the North Island of New Zealand, where violent volcanic activity has been occurring within the last few hundred thousand years, is one such region. There, explosive eruptions have resulted in the laying down of thick deposits of volcanic ash and debris which provide excellent aquifers. Geothermal activity continues and the water is superheated. At a depth of about a kilometre the water can reach 300°C without boiling because of the pressure it is under; when intersected by bores it 'flashes' and the steam that results has been commercially harnessed to generate power. It was found, during routine maintenance of the bores and pipes which carry this high-pressure water, that scale deposits very rich in copper, silver, gold, zinc and lead build up inside them. Water reaching the well-head and discharged at atmospheric pressure, however, did not contain increased concentrations of the metals. Experiments showed that the deposition in the pipes at depth occurred when the water 'flashed' as the pressure was suddenly released.

The amount of gold precipitated was surprising considering its rarity in the parent rocks, and reasons were sought for the increase in solubility that had seen so much dissolved by the groundwater. Under normal conditions, gold scarcely dissolves in water, but its solubility increases dramatically if the water contains certain negatively charged ions and groups of ions (known as ligands) that can cluster round the gold to form 'complexes'. Ligands containing reduced sulphur, an important constituent of many geothermal fields, are particularly effective at this, and the gold sulphide complexes so formed are comparatively highly soluble. Why the gold is precipitated again as pure native gold when the water in which it is dissolved as gold sulphide 'flashes' is explained by the production of hydrogen sulphide gas from the complexes at the 'flashing' point, which sets the gold free. (See page 26 for mineral deposits in black smokers on mid-ocean ridges where 'flashing' and available sulphur occur.)

Prospecting for gold took on a new emphasis in the light of the Taupo discoveries, and the Pacific Rim, where active volcanism results from the collision of tectonic plates, is proving to be as rich a province for gold as the old Precambrian sites. The Ladolam ore body on Lihir Island, Papua New Guinea, which has formed over the past 1 million years in the collapsed caldera of a large volcano, has been estimated to contain 600 million grams of gold reserves, making it as rich as the best ore bodies known.

HYDROTHERMAL ACTIVITY IN FOSSILISATION

The role of hydrothermal activity in the formation of petrifactions of plant material has great significance in palaeobotany. Mineral-rich waters percolating through sedimentary deposits in which plant material is entombed result in cell-perfect petrifactions like those from the Lune River deposit in Tasmania. When rifting was occurring between Australia and Antarctica, in the Jurassic Period about 160 million years ago, Tasmania was in a high tension zone where crustal weakness resulted in massive outpourings of basalt that flowed down valleys, covering valley-fill deposits. Over time, cellulose and lignin in the trunks of trees and tree ferns, and in smaller pieces of wood, were degraded; the voids left by cell walls and their thickenings were filled by silicates contaminated by iron and other minerals in the mineral-rich waters. The resulting agate petrifactions are beautiful and much sought after by lapidaries who cut and polish them. Their perfect petrifaction has contributed knowledge on the structure and nature of the Jurassic plants.

Perhaps the most famous plant fossil locality in the world, because of its petrifactions of the earliest land plants (the first members of the Plant Kingdom) and a very early terrestrial ecosystem, is the Rhynie Chert deposit in Scotland. Here the combination of siliceous sinter and a precious-metal-bearing hot spring system has preserved delicate plant material, micro-organisms and invertebrates. Boiling-hot mineral-rich water, between 90° and 120°C, flooded into the vegetation on the

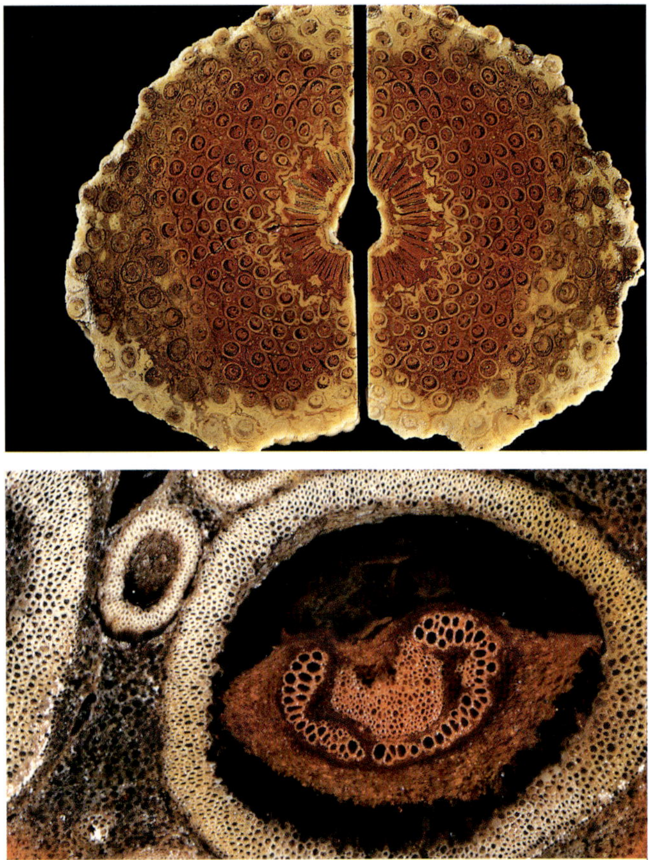

Cell-perfect petrifactions of tree-fern trunks from Lune River in Tasmania. *Jim Frazier*

edges of thermal pools, killing all the organisms and subsequently supplying the minerals to petrify and preserve their tissues.

Perfect petrifaction of the first vascular plants resulted, as well as of non-vascular plants, algae, fungi, cyanobacteria and early land-living arthropods. The presence of mycorrhizal fungi in the roots of these first land plants, visible because of this perfect preservation, is one of the most important revelations concerning Life on the land. **If it had not been for symbiotic fungi maintaining the connection between plants and soil water there would have been no land plants—and that means no Plant Kingdom (see Mycorrhizas, pages 94–95). The importance of fungal-plant symbiosis in the Beginning, about 450 million years ago, and its continuing importance in the world today, cannot be over-emphasised. Nearly all living plants today have mycorrhizal symbioses of one sort or another.**

A thin section through Rhynie Chert showing stems of early plants in cross-section. *Jim Frazier*

Transformation of organic carbon in sediments

Organic matter incorporated in sediments at the time of deposition may be incompletely mineralised by microbes. Cellulose and lignin are difficult to degrade, especially anaerobically, and tend to accumulate as humic substances. Superficially buried material may become peat. Deeply buried material may be converted into coal by heat and pressure. Some cyanobacterial and algal biomass trapped in sediments of shallow inland seas in the geological past has been gradually converted to petroleum hydrocarbons by heat and pressure.

LANDSCAPES WHERE DURICRUSTS HAVE PROTECTED FEATURES FROM EROSION

INVERTED LANDSCAPE

A mesa in a flat arid landscape. The iron-enriched capping was a biogenic deposit that formed at the watertable in an ancient landscape. What was then at the bottom of the landscape is now at the top. 'Inverted' landscape features result when erosion removes the surrounding terrain, leaving mesas standing up out of the plain protected by a capping that has prevented erosion of the layers beneath it. *M.E.W.*

SILCRETE IN A DESERT LANDSCAPE

Silcrete, both massive and scattered as gibbers on the surface in a desert landscape near Tibooburra in western New South Wales. The silcrete formed at the watertable in an ancient landscape, facilitated by bacterial activity. Erosion has worn away the softer regolith layers that covered it. *M.E.W.*

BACTERIA IN YOUR OPALS?–YES, AND HELPING TO FORM THEM![66,67]

Australia is the world's largest producer of opal. The only other country that produces a significant amount is Mexico. Australia's opals occur in sediments laid down in the Great Artesian Basin during the Early Cretaceous when the epicontinental Eromanga Sea occupied the Basin.

The Three Mile Pit at Lightning Ridge. Massive excavations in the deeply weathered sediments. *M.E.W.*

On the south-western margin of the Basin, the Coober Pedy, Andamooka and Mintabie fields produce 70 per cent of the world's opal—mainly white opal with milky background and fine colour. The opal fields in the centre of the Basin in western Queensland produce iron-hosted boulder opal that represents only 3 to 4 per cent of production; Lightning Ridge and White Cliffs in the south-eastern and south-central sector are famous for their black opal, the rarest and most sought-after type, and the very high quality of their opal of all types. (The high quality, and the occurrence of black opal, are explained by special factors that influenced their environment, as we shall see.)

Opal formation, until very recently, was believed to be a purely geological process. The account I gave in *After the Greening* summarised the process according to that thinking. It involved the deep chemical weathering of Early Cretaceous sedimentary rock about 100 million years old, laid down in the epicontinental sea in the Great Artesian Basin. The intense period of weathering was thought to have occurred about 30 million years ago, when conditions were known to have been such as to fit the scenario proposed. The chemical weathering released quantities of silica; its dissolution, forming a colloidal silica gel, and re-precipitation as the non-crystalline silicate, opal, was explained by the precise and unusual environmental conditions required for the process being current (for a considerable period) at that time.

DISTRIBUTION OF OPAL DEPOSITS

Now, however, recent research has revealed a possible biogenic origin for opal and suggested a completely different timetable and course of events.

Precious opal is the result of a unique form of deposition of a silicate from a silica-gel. Under special conditions, minute spherules (tiny balls of silica molecules) precipitate in ordered layers, resulting in an 'amorphous' gemstone with the distinct characteristics that make opal rare and valuable. The diffraction of light by the surfaces of these spheres gives the 'fire' to precious opal that distinguishes it from 'potch' or non-precious opal, which has a crystalline structure.

Because most of the world's best opals come from Lightning Ridge, including the rare black opal, the recent research that has revealed the role of microbes in opal formation has been focused on this field. Opals occur in 'nobbies' up to many centimetres in diameter, concentrated in seam-like structures and irregular patches in the Early Cretaceous host-rock, a silty claystone that in this region formed from thin layers of volcanic ash. At 100 million years ago, when fine, clay-rich sediments were being laid down both within and at the margins of the epicontinental sea throughout the Great Artesian Basin, a period of intense volcanic activity was located in the Whitsunday Volcanic Province off the Queensland coast. The explosive volcanism produced clouds of volcanic ash carried by the winds to the Lightning Ridge region where it was deposited, resulting in local volcaniclastic sediments. These sediments are very easily altered. In the Lightning Ridge area, the sediments were deposited in a coastal plain setting where large amounts of organic matter and bioturbation were common, which may help explain why this region has produced so much opal. Opalised macrofossils are commonly found in the claystone.

The involvement of microbes in weathering processes is now generally accepted, so discovery of the role of micro-organisms in opal formation as shown by current research[66] comes as no real surprise. However, what it implies in terms of the chronology of opal formation certainly is surprising.

Scanning electron microscope (SEM) examination of host-rock and gemstones at Lightning Ridge has revealed microbe communities of multi-millions of organisms per cubic centimetre. The most common microbes are aerobic bacteria—actinomycetes (*Streptomyces*, *Nocardia* and *Micromonospora*) and myxobacteria. Thirteen other species were identified, along with some fungi. The microbes occur as mycels, mats, biofilms, globular colonies, networks, swarms and single individuals. All the organisms are the same age as the opal.

The microbes are of the kind that occur in soil, in organic muds deposited under still conditions, or in a surface-fouling biomass. They require a nutrient-rich (rich in cellulose and chitin), near-surface aerobic environment with a temperature range of 23° to 35°C, and neutral pH. The specific environment required by these microbes indicates the conditions under which opal was produced and enables the determination of a new timetable for their production involving a few months, not the hundreds of thousands of years originally envisaged.

The formation of opal occurred in muds at the edges of the contracting epicontinental sea. Populations of bacteria grew along grain boundaries of the smectite clays that were as yet uncompacted, excreting acids and enzymes that resulted in biochemical weathering of clay minerals and feldspar (see photos 1 and 2). The colloidal particles produced were stored in the compaction water of the mud, particularly in voids where gas bubbles had been replaced by pore water—sites where sizeable nobbies could develop. The silica gel itself was then a suitable substrate for bacterial films to grow on, like the agar plates that are used in laboratories; the bacteria found among the spherules were determining the specific pH that was resulting in their continuing production. The size of the spherules increased with increasing pH and when they reached a critical size they settled out in horizontal layers. The host-rock was being silicified with silica freed by bacterial weathering at the same time as the spherulites were forming and settling in organised layers in the voids. (When melanin was added, by bacterial activity interacting with additives in the volcanic ash, the black opal fire that is so much more dramatic than the fire of common opal resulted.)

The opal is thus 'syn-diagenic' in character—that is, it formed at the same time as the surrounding sediments were consolidating into the claystone host-rock in which they are found today. Much of the opal in the Great Artesian Basin may therefore be the same age as the sediment that enclosed it, somewhere around 100 million years, forming in the aerated sediments exposed as the epicontinental sea dried up.

In the very iron-rich muds in the centre of the Basin the opal was precipitated from gels in voids inside ironstone concretions. On being split open, the concretions sometimes reveal an opal in the centre. Considerable thickness of opal has sometimes also formed along cracks.

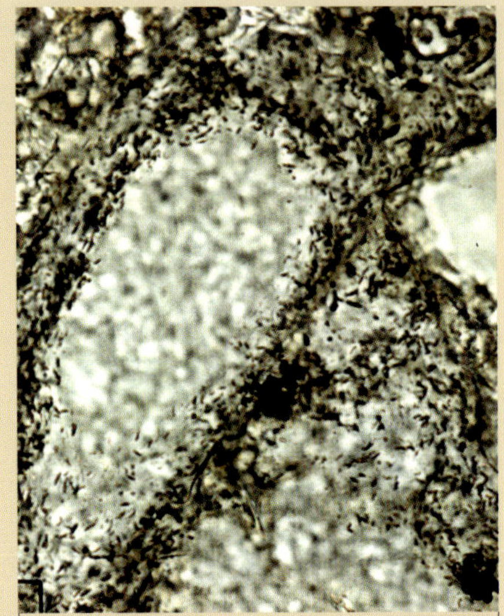

1: Claystone host-rock adjacent to a nobby. The central grain is an opalised clay grain (probably a former feldspar grain) lying between two unaltered grains of volcanic quartz. The matrix of the claystone, and the clay grain, have been invaded and corroded by bacteria and are now opalised (potch). The bacteria have produced acids and enzymes that have chemically attacked the clay grain and the matrix to produce silica that has then formed opal. Specimen from Lightning Ridge. *SEM photography Dr John Watkins*

2: Bacteria *Cytophaga* in the contact between potch and siltstone. The density of bacteria increases approaching the siltstone (a microbe front) where they release acids and enzymes that attack and chemically alter the siltstone to produce silica for opal (potch). Specimen from Lightning Ridge. *SEM photography Dr John Watkins*

Boulder opal from Yowah Opal Field. *Jim Frazier*

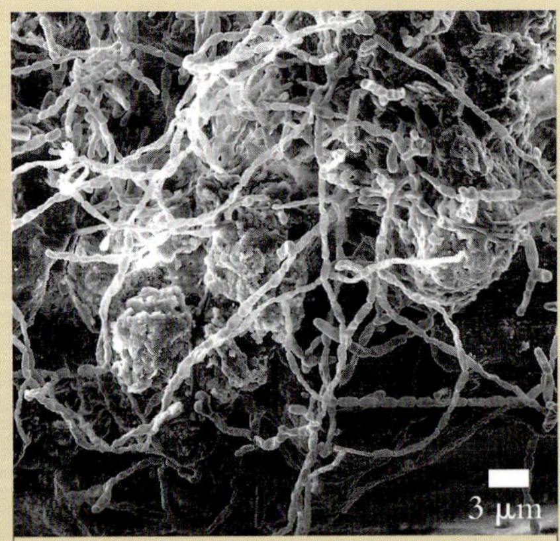

3: *Streptomyces* on black opal in a specimen from Lightning Ridge. *SEM photography Professor H.J. Behr.*

CASE STUDY: THE BIOGENIC ORIGINS OF BAUXITE

One of many types of mineral deposits whose creation is assisted by microbial activity is bauxite. Bauxitisation can occur on many rock types, provided they have high aluminium content[69], including quartz-rich rocks, previously thought to be unfavourable, and in cool climates as well as tropical.[(69a)] A strong geomorphological control is evident, with sites of aggressive leaching predisposed to bauxitisation. A great length of geological time was also believed to be a requirement, as old land surfaces were usually involved. It has been recognised over recent years that much of the formation of bauxite is an ongoing process, and deposits are much younger than the old land surfaces on which they occur, commencing on incision of the drainage patterns in the distant past and continuing up to the present. It has long been recognised that more than geochemical leaching is involved and that there is a biological control of the processes, involving micro-organisms that concentrate the aluminium and lock up the iron. The accumulation of residues so rich in iron and aluminium after dissolution of kaolin (the fine clay fragment when the lithospheric material is hydrolysed) cannot be explained by leaching processes alone. The mobility of elements in a manner not constrained by pure geochemistry requires explanation, and biological intervention is the only one that fulfils all the requirements. Bauxitic profiles are known to produce CO_2, evidence of the biogenic origins of deposits.

The huge deposits at Weipa on Australia's Cape York are currently being mined by Comalco Aluminium.[68] The deposit is spread over 2500 square kilometres in a low plateau that extends 30 kilometres inland from Weipa and about 100 kilometres to both north and south. A section of the deposit shows a poor topsoil layer about 60 centimetres deep; bauxite up to 10 metres deep; a mottled zone with hard nodules (ironstone, sandy clay, sand); and kaolin to various depths below.

The bauxite is mined after first removing the surface overburden (which contains seeds, humus and other organic

The mining at Weipa involves complete removal of the bauxite layer, down to the ironstone layer. *M.E.W.*

The pisolitic bauxite is scooped up by huge front-end loaders. *M.E.W.*

SECTION OF THE WEIPA DEPOSIT

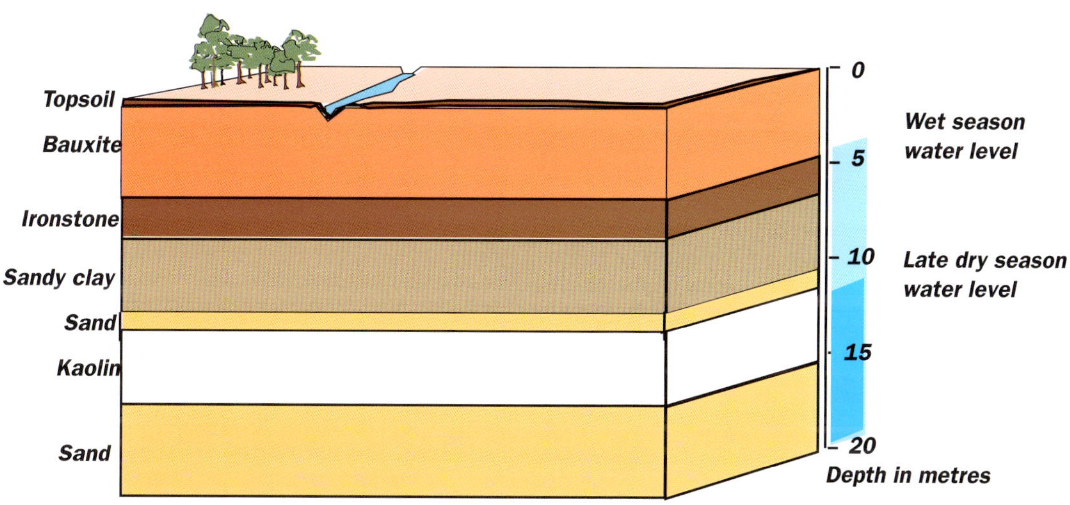

When the ore has been removed it is washed and graded in a beneficiation plant using the abundant water in the shallow aquifer (the same watertable that helped create the deposit and the leached clay zone below it). The plant uses more than 4 million litres of water an hour, much of which is recycled and returned to the aquifer.

matter); this material is kept for regeneration purposes once the underlying deposit has been removed. The deposit comprises an estimated 3.7 billion tonnes of bauxite. Nine million tonnes are mined annually and shipped elsewhere for refining and smelting, yielding 2.5 million tonnes of aluminium and revenue of $240 million.

The deposit is mined down to the ironstone zone. Underneath the bauxite, which has a pisolitic structure—rounded pea-like bits cemented together—pure clay occurs in lenses that are also mined for use as kaolin in the finish on printing paper. The clay and bauxite layers have resulted from leaching caused by the downward percolation of surface water and the effect over millions of years of the rising and falling of the watertable, where bacteria have been active agents in the saturated zone. The watertable is recharged by annual monsoonal rainfall of nearly 2 metres a year.

The soil profile at Weipa. The topsoil is removed when mining starts, and is stored for future return to the site when the underlying bauxite layer has been removed. *M.E.W.*

The bauxitic layer on top of the white clay zone can be seen in the sea cliff near Pera Heads, south of Weipa, making a dramatic coloured backdrop to the beach. *Leeanne Rosel, Weipa*

MICRO-ORGANISMS AND OIL RECOVERY

With current technology, only a proportion of the oil in an oil well is recoverable. In particular, the heavy crudes which are characteristic of many oil fields in the USA have been difficult to extract, and large reserves are still available if extraction could be economic.

It has been realised that surfactant-producing bacteria might be used to mobilise the inaccessible portion. Surfactants are substances that reduce the surface tension of the oil (bio-detergents) and make it more mobile. Research is being carried out by oil companies to find the right mix of extremophiles to introduce into the wells to alter the viscous material into something more manageable. Many of the crudes contain contaminants like sulphur compounds and heavy metals that can be broken down into soluble forms by bacterial activity; these would then dissolve in the water that is always present in wells and be more readily separated from the oil. The removal of sulphur by specific bacteria would be of particular benefit in countering the formation of acid rain and other environmental problems.

Australian research into **microbially enhanced oil recovery** (MEOR) was carried out during the 1970s and 1980s at the Baas Becking Geobiological Laboratory of the Bureau of Mineral Resources in Canberra.[60] Many suitable microbes were isolated and cultured, mainly from the genus *Pseudomonas*. The MEOR approach was to inject bacteria in a nutrient solution under low pressure into the well. Supported by the nutrients, the bacteria would seek out and proliferate at the water-oil interface. The surfactants they produced would alter the physico-chemical properties of the water-oil mix and thus the oil would be more easily flushed out. Research along these lines is still being carried on by oil companies, but they do not publicise their methods or results.

THE USE OF BACTERIA AND ARCHAEA IN MINERAL PROCESSING[70]

The mining industry in Australia, which annually generates $37 billion in mineral exports, is becoming increasingly sensitive to environmental issues, but at the same time has to reduce operational costs if it is to compete on global markets. Extraction of metals from low-grade ores cheaply and efficiently has thus become a challenge, and there is increasing interest in the use of biological agents in the many processes that mining and ore concentration involve. The breakdown of undesirable by-products, such as the cyanide from gold extraction, is carried out by bacteria, and they also can be used to extract minerals directly from ore.

Bacteria facilitate the oxidation of sulphur and iron through the process of bacterial leaching or bio-oxidation. The bacteria use the energy released for cell maintenance and growth, and metals are freed from the ores in the process. By using bio-oxidation techniques, it becomes economic to treat marginal mineral deposits that would not be viable under conventional processing. The gold industry, in particular, has been using such techniques in recent years and two commercial plants are operating in Western Australia.

Major ore deposits of gold, copper, nickel, cobalt, uranium and other minerals have been formed through geological time by the activities of archaean and bacterial extremophiles on mid-ocean ridges and in other submarine environments associated with tectonism and volcanism (see pages 24–27), resulting in mineral sulphide deposits. Current major technologies for metal recovery from mineral sulphides are smelting, bio-oxidation and pressure leaching. Smelting has serious environmental consequences; pressure leaching requires expensive plant and can only be economically viable when ores are high quality.

Presently, only a few species of Bacteria and Archaea are used in biomineral processing. Most are acid-tolerating chemolithotrophs that obtain their energy by oxidising inorganic substrates in mineral sulphides. *Acidithiobacillus ferrooxidans* is efficient in the bio-oxidation of refractory gold ores and concentrates but has proved to be less efficient as the bioleaching agent for metal recovery from the more refractory sulphidic ores such as chalcopyrite and molybdenite. Moderate thermophiles, that tolerate temperatures between 35° and 55°C, are now considered to be the dominant leaching microbes in mineral processing operations, giving increased bioleaching rates in chalcopyrite. This has prompted research into organisms that can function in the thermophilic temperature range (60° to 80°C), and an expedition to retrieve such organisms from black-smoker volcanic vents at the bottom of the Bismarck Sea brought back material from which it was hoped to culture strains that could be used by the mining industry. Research is ongoing.

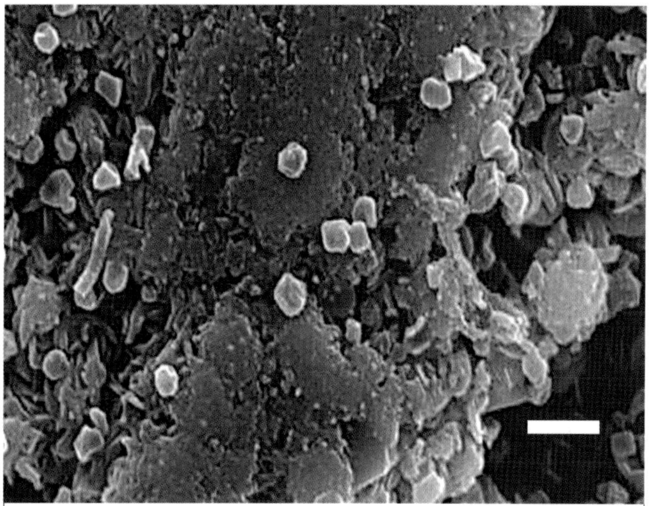

Archaeal cells isolated by Jason Plumb from material brought back from the Bismarck Sea *by Peter Franzmann and Luke Zappia. Luke Zappia, CSIRO Land & Water*

Making toxic mine wastewater fit for recycling

More than 70 per cent of Australian gold production occurs in Western Australia. From 15 to 30 per cent of the resource is present in arsenopyrite deposits and is not readily extractable without prior treatment.[71] Roasting of the ore in smelters in the Kalgoorlie region required massive stocks of firewood, altering landscapes in the past, and the release of toxic sulphur and arsenic oxides to the environment as by-products of smelting is undesirable. Youanmi, Harbour Lights and Wiluna mines used bacterial oxidation as a pre-treatment for refractory gold ores prior to the extraction of gold by cyanidation. One reason why the practice has not been more widespread is the need for a plentiful supply of good water, and water is usually a limiting factor in Western Australia.

Research aimed at developing ways of reducing toxic substances, mainly cyanide, in tailings dams, and making wastewater fit for recycling, has had priority and has involved cooperation between the CSIRO Centres for Groundwater Studies and Minerals, the A.J. Parker Centre for Hydrometallurgy and BacTech (Australia) Pty Ltd. Two cultured strains of thiocyanate-degrading bacteria isolated from Youanmi wastewater in laboratory experiments were used in rotating biological contactors, where they formed biofilm that converted the toxins to harmless products. The removal of up to 99 per cent of available thiocyanate from mining wastewater in fixed film, laboratory and full-scale bioreactors has now been demonstrated, with only phosphate having to be added.

The Millstream Falls, North Queensland. Water is almost exclusively the carrier of the products of weathering and recycling. *M.E.W.*

Part Two

CYCLING OF MATERIAL IN THE BIOSPHERE

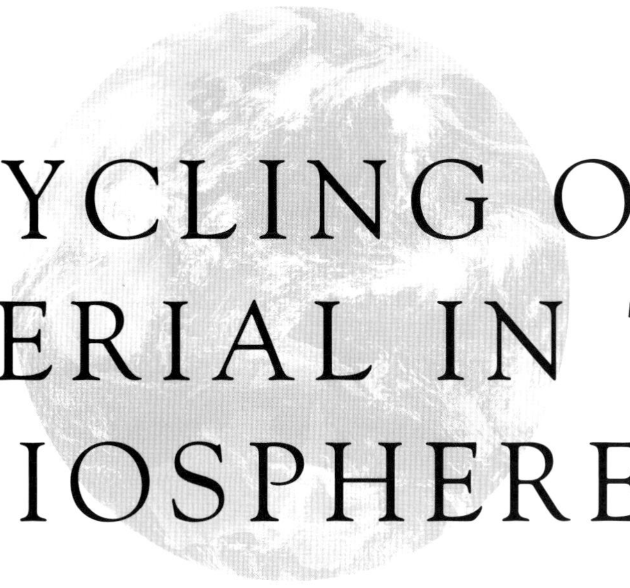

The Biosphere is a closed ecosystem, in which cycling and recycling of elements has proceeded since Earth's genesis. During 4 billion years of evolution, Earth's finite available resources, the critical components of living matter, have been used again and again in endless cycles of creating and maintaining new generations, the decomposition of dead tissue or organisms, and returning of elements to inorganic stores for ultimate re-use. Six of the most important elements in sustaining Life are carbon, hydrogen, oxygen, nitrogen, phosphorus and sulphur.

Energy recycling is another matter. The Biosphere is solar powered and the Earth is an **open energy system** that reflects about a third of incoming radiation back into Space, allowing the rest to be absorbed by the atmosphere, land, waters, and living organisms, and returned to Space as heat.

A number of different kinds of processes and of major cycles are involved in the cycling and recycling of materials in the Biosphere.

- **Bio-geochemical processes** are those in which Life affects or effects the chemical processes that are involved in building up or breaking down matter, facilitating a continual migration of atoms from inert matter to living matter and back again. Such processes are a main focus in this book as they are basic to our symbiotic planet.
- The **natural sedimentary cycle** facilitates biospheric flows of minerals (including phosphorus, potassium, silicon and so on, essential for plant growth) without the intervention of living things.
- The **water cycle**, involving hydrogen and oxygen, is the most rapid circulation system within the Biosphere. It is a straightforward physical process involving evaporation, transportation through the atmosphere, precipitation, and water-flow, to the oceans (mainly). With more than 97 per cent of the Earth's water held in the oceans, 90 per cent of all water evaporated comes from the surface of the sea. About half of the solar energy absorbed at the surface of the Earth is transferred to the atmosphere through evaporation and condensation of water. This is because evaporating water absorbs heat as it changes from a liquid to a gas, then releases heat into the atmosphere when it condenses into droplets.

Modern Earth does not lose significant amounts of water to Space—an important consideration when Life and water are inseparable. Today, because the top of the troposphere is cold, water falls back to the Earth's surface[223] and the stratosphere is extremely dry.

Water is almost exclusively the carrier of the products of weathering and recycling. (Wind transport is responsible for only a very small proportion.) Water is fundamental to Life. Living organisms typically are 60 to 80 per cent water. 'Animated water' would be a good description for Life, according to Lynn Margulis.[8] Where there is liquid water on Earth, no matter what the physical and chemical conditions, there is Life, with extremophiles able to occupy niches that one would have thought presented insurmountable problems, as we have seen. This realisation has fuelled the controversy about whether life exists elsewhere in the Universe.

Water is the carrier of nutrients and metabolic products in living things; it is also the mobiliser and carrier of nutrients and all the other substances that make soil a living organ of the Biosphere. It creates microscopic watery highways for the transit of soil organisms between particles in the regolith. The concept of the **hydrosphere** can be expanded to include the whole Biosphere as water is the facilitator, and soil processes, basic to plant-based food-chains on which biospheric life depends, are water-related.

- The **cycling of oxygen** is more complex. Sixty per cent of the generation and consumption of oxygen is due to photosynthesis and respiration, two components of the carbon cycle. Oxygen's presence in the atmosphere is the result of photosynthetic life through the ages, but in this symbiotic world another interaction exists, with specialised bacteria involved in the **sulphur cycle**, and leads to processes in a **sedimentary-tectonic cycle** that maintains oxygen in the atmosphere at a more or less constant level.
- The **phosphorus cycle**: Phosphorus is vital for the maintenance of living organisms, acting as an agent for energy transfer through the generation of ATP and ADP (adenosine triphosphate and adenosine diphosphate) from inorganic orthophosphate.
- The **carbon cycle** and the **nitrogen cycle** are dealt with in detail below, as is the **sulphur cycle**, whose role in global weather patterns along with the role of early sulphur bacteria have been discussed elsewhere in this book (page 8). Carbon, nitrogen and sulphur cycling occur as part of the natural biospheric cycles, but their ability to form stable gaseous compounds that can be cycled through the atmosphere renders them 'doubly mobile'.

HOW OXYGEN IS KEPT AT CONSTANT LEVELS IN THE ATMOSPHERE[50]

Oxygen in the atmosphere (O_2) is in the form of molecules made of two atoms. The bulk of atmospheric oxygen originated in the photosynthetic activities of ancient bacteria and protists, as we have seen, and cyanobacteria, algae and plants continue to play essential roles in the production and consumption of oxygen in today's Biosphere, accounting for about 60 per cent of the total. The amounts produced and consumed in this biogenic cycle more or less balance each other. Because a very small proportion of new phytomass (approximately 0.3 per cent) is buried in sediments and cannot be oxidised, a small surplus of oxygen is produced by Life—but the oxygen content of the air is not rising because this fraction is used up in the oxidation of volcanic gases and the weathering of sedimentary and volcanic rocks.

Volcanic gases are often sulphur-rich. Maintaining and balancing the surplus oxygen from the biogenic cycle is dependent on functions within the sulphur cycle. Specialised bacteria transform sulphate ions to hydrogen sulphide gas, releasing oxygen. The hydrogen sulphide reacts with iron minerals to produce pyrite. The burial of pyrite in sediments removes sulphur to where it cannot be oxidised. Only when the sediments are eventually exposed to weathering will oxygen be consumed.

The more or less constant level of oxygen in the atmosphere is thus maintained primarily by a sedimentary-tectonic cycle where burial of biomass is counterbalanced by mineral weathering and the oxidation of volcanic gases.

The natural sedimentary cycle

Weathering of rocks, here creating a beautiful landscape feature in a Hawkesbury Sandstone cliff, is proceeding all the time. The rock becomes sand that washes into the sea to become part of a sedimentary deposit in the course of geological time. *M.E.W.*

The natural sedimentary cycle involves the cycling of minerals, including such essential-for-life ones as phosphorus, potassium and silicon, without the critical involvement of living organisms. Weathering strips about 0.1 millimetre, or less, from the Earth's crust each year, in the process liberating minerals from the rocks. Transported by water, deposited on ocean floors or in large water bodies in sinking geological basins, they again become the substance of rocks. (Only a small fraction of water-borne materials becomes incorporated into living tissues, a process which delays its ultimate reduction to elements and long-term storage.) In the distant geological future, geotectonic movements will see these inert stores becoming part of the continental crust, and weathering (now aided by living organisms) will once again see the minerals altered and their dispersed elements becoming part of the cycling of materials in the Biosphere.

The Hawkesbury Sandstone, a gigantic deposit largely of

quartz sand that was poured into Australia's Sydney Basin by rivers that carried the products of weathering of high mountains in Antarctica in the Triassic Period (between about 220 to 200 million years ago) is an example of such cycling. The sediments became rock, storing the minerals; uplift and erosion saw the creation of the modern Hawkesbury Sandstone plateaux and landscapes; the current re-mobilising of materials on the floodplains and in the shallow, infertile soils on the plateaux is again a bio-geochemical process.

On an infinitely greater, Earth-building scale, going back to the Beginning, carbonates formed by chemical weathering of rocks in the initially CO_2-rich atmosphere (before there was Life) locked up many atmospheres' worth of carbon dioxide[50], as we have seen. Fortunately, though levels were dramatically reduced in this hyper-active sink process, enough CO_2 remained to keep the Earth warm enough for Life. The much cooler Planet today still cycles carbonates via subduction zones and by adding carbonate rocks onto continental edges in its plate tectonic cycles, but at a very slow rate compared with that in the beginning, and it is fortunate that Earth's volume is such that its inner fires are still large enough to keep it alive, for it is core heat that drives plate tectonics. (Mars, by contrast, was not as fortunate. According to NASA scientists it was too small to sustain the processes that have seen Earth become a living Entity—its core cooled quickly, its plate tectonic cycles ceased, and it became the dead, frozen world it is today.)

WIND AS AN AGENT OF DISPERSION

Wind is an important agent for the long-distance dispersion of living and inanimate matter in the Biosphere. Wind can carry fine particles of loose sediments, weathered rocks, topsoil with its organic matter, and living organisms, over very long distances. Any crustal or oceanic compound, solid or liquid, can become passively airborne. The cosmopolitan nature of fern, lycopod and moss floras is a result of the easy distribution of minute spores in wind currents, even in the jet streams at high altitudes. Spiderlings, temporarily frozen and with web fragments attached, circle the globe in jet streams, occasionally landing and thawing out alive. On newly formed volcanic islands they can be the first invertebrates to appear.

The transport of salt in rain is perhaps the most widespread wind-borne phenomenon. With cloud formation predominantly occurring over the oceans that occupy more than two-thirds of the Earth's surface, salt in minute doses that add up to be sizeable amounts over time is deposited on the continents in rainfall. In well-drained regions this salt is largely dissolved in run-off water and makes its way back to the sea. In old, flat, badly drained or inwardly draining continents like Australia, a large proportion remains on the land. Where there is sufficient rain, over time the salt accumulates at watertables at varying depths below the surface, to be maintained there as long as the hydrological balance of a region remains reasonably constant. The problems with salinisation of soils and waters in Australia due to rising watertables are well known and extensively documented.

Dust and sand from arid regions are often carried amazing distances to distant places with quite different climates. Dust from the Sahara is regularly detected on Europe's Alpine glaciers and on Caribbean beaches; red dust from Australia frequently colours the snow on glaciers in New Zealand. On 10 March 1991, about 50 000 tonnes of dust, mixed with pollen, was transported from Algeria to northern Scandinavia, some 7000 kilometres away.[50]

Prolonged drought in the Sahel of sub-Saharan Africa since the 1970s has increased atmospheric dust fivefold. Dust contains bacteria, viruses and fungi[72] that can kill coral. It also contains iron that promotes the growth of algae that smother coral reefs. The Caribbean is particularly badly affected—a billion tonnes of African dust settles there each year, brought by the trade winds. Outbreaks of white band and black band diseases of coral, and 'coral plague', have coincided with the years when dust load was heaviest. The soil fungus *Aspergillus* appeared there in 1983, a year of bad dust, and has since killed 90 per cent of sea fans (soft corals). Sea urchins have also declined and the algae they graze on have multiplied. Iron in the dust might trigger unusual algal growth in remote reefs, upsetting the ecological balance and facilitating invasion by organisms harmful to coral.

(The North Atlantic Oscillation which caused the Sahel drought from 1976 to 1999 eventually shifted into its wet phase, and in 2000 there was good rain in North Africa. It will be interesting to see how this affects dust problems and reef health.)

In the Hawaiian Islands, fine fragmental angular quartz in the surface horizon of soils has been shown to have arrived as dust.[73] The same material has been found in deep-sea cores taken from north and south of the islands. With modern technology it has been possible to date the fine-grained potassium micas found with the quartz in the surface layers of the deposits. They prove to have a 200 million year age when dated by potassium argon; as the island of Oahu, from which the tested soils come, is only 3.5 million years old, this proves their foreign origin.

Similar long-distance transportation of dust and its incorporation into soils is seen in Australia, where a large dust component has been recognised in many soils, often attributable to the extraordinarily dry and windy last glacial stage of the Pleistocene ice age. Lichens growing on rocks on the Great Divide of eastern Australia have been found to have trapped small pockets of soil with components that indicate a distant origin.

The ultimate in dust distribution occurs in volcanic eruptions, when even global climate has been affected in the past—for millions of years during times of intense volcanism, and for months in more recent times, as when Krakatoa erupted in 1883. Large amounts of mineral elements can be dispersed over long distances in such events.

Bushfire smoke has been shown to transport minerals and nitrogen; strong winds can result in their distribution to distant places as smoke drifts across continents and oceans.

THE LIFE-BRINGING PROPERTIES OF WATER

Water has many properties that make it the bringer of Life to the Biosphere.[50,74] Water can absorb a relatively large amount of heat before its temperature rises, having five times the capacity of common rocks in this regard. Thus while the Earth's solid surfaces heat and cool rapidly, producing a wide range of temperature extremes, water bodies gain and lose heat more slowly, making them excellent temperature regulators and preventing large temperature swings.

Water vaporises only at a high temperature and must absorb a great deal of heat energy before it transforms to the vapour state. It is thus a good medium for transporting latent heat in vapour and for energising wind-flow and other atmospheric processes. Because water is a good conductor of heat, its role in determining ocean eddies and currents, and thus influencing global climate, is paramount.

No other liquid would be as good a moderator, keeping conditions optimal for life in true Gaian fashion. Water's heat-conducting ability is important in living cells.

Fresh water and dilute seawater have their maximum densities at temperatures above the freezing point. (Pure water freezes at 4°C; at decreasing temperatures with increasing salinity.) This feature is important in controlling temperature distribution and vertical circulation in water bodies.

Water has the highest surface tension of all liquids except mercury. This is important in the physiology of the cell, controlling surface phenomena, drop formation and capillarity. The dissolving power of water is greater than that of any other liquid—more substances, in greater quantities, are soluble in water—which is important in both physical and biological processes.

Because water is transparent, light can penetrate a considerable distance into water bodies, allowing photosynthesis. The ability of water to absorb radiant energy, which is large in the infra-red and ultraviolet bands, results in water vapour being responsible for about 75 per cent of the natural greenhouse warming that protects Earth from the cold of Space.

Water in the solid state (ice) is less dense than as a liquid, and this is important in partly frozen lakes and seas, and also in weathering. The 'polar' structure of water's molecule, rare in liquids, is important in many reactions, and in rock weathering. In the water molecule, the two positively charged hydrogen atoms are arranged at one end, and the negatively charged oxygen atom at the other, resulting in a **polar liquid**. The positively charged end is attracted to the negatively charged surface of a clay or other material, and other water molecules line up similarly, joining negative and positive ends like so many tiny bar magnets. This forms a layer of 'ordered water' with increased molecular pressure. Alternating wetting and drying of some rocks, particularly shales and mudstones, causes them to fragment into blocks and disintegrate. It is believed that the repeated wetting and drying allows the water molecules to become increasingly ordered, assuming a quasi-crystalline nature and exerting expansive force that thrusts against the confining walls. (Many fossil collectors will have discovered the effects of ordered water when they left their shale specimens out in the weather and returned to find piles of fragments). A small amount of surface disintegration results in less susceptible rocks over time when they are alternately wet and dried.

Crater pits in the rock platform at Durras, New South Wales. *Cliff Ollier*

This 'slaking' form of weathering is responsible for another weathering effect—**water-layer weathering**—seen on coastal rock platforms and in other locations where the surface is alternately saturated and dried and becomes a continuous broad flat surface. It is not uncommon for such surfaces to be pock-marked with 'miniature volcano' structures with raised rings and central craters, often limonite-lined. These form when initially small pits are filled by tide or rainfall, keeping the immediate area uniformly wet. The surrounding platform is alternately dried and wetted, being lowered by surface slaking until eventually the iron-enriched crater rims protrude above the flat surface, enclosing enlarged craters.

LARGE-SCALE VEGETATION CHANGE: ITS EFFECT ON CLIMATE

Vegetation does not only influence climate by its consumption of the greenhouse gas carbon dioxide. **Evapotranspiration** is a global source of water vapour, the water that rises through the plants' systems and evaporates through the stomata on the leaves affecting ground temperatures.

This becomes a vital factor in forests and other areas of very dense vegetation. The latent heat of evaporation is locally cooling—it has been estimated that if all the forests disappeared from Amazonia, to be replaced by grassland, temperatures there could be expected to be between 2° and 5°C higher.[50] Change from forest cover to grassland involves a local decrease in evapotranspiration of about 20 per cent, with considerable decrease in cloud cover. Dry seasons become longer, and with increased dryness come more fires and more erosion. Air circulation and wind strength change with the decrease in surface roughness associated with change from tree canopies to grass, and as ground temperatures rise (particularly over eroded land) updrafts result in less rain falling locally, often where it is needed most. (The changed rainfall patterns in better-watered agricultural regions of Western Australia that grade into arid zones, where it is claimed that there has been no 'decent' rain since 1974, can be attributed to the massive clearing that has locally altered rainfall patterns.)

If all the Asian forests were to disappear (and they are well on the way), possible gross alteration to the patterns and amounts of monsoonal rains could be expected.

The desertification of arid lands, a global phenomenon and a serious problem in Australia, is another aspect of vegetation change affecting climate. Loss of plant cover grossly affects water infiltration, ground temperature and erosion. It has been shown that degraded land actually receives less of the local rainfall than adjacent areas that have retained good vegetation cover. (The 60 million trees cleared in Queensland in 1999, and the continuing clearance since, should be causing alarm bells to ring, not only about salinity risks but also in this context.)

CHERRAPUNJI, THE WET DESERT

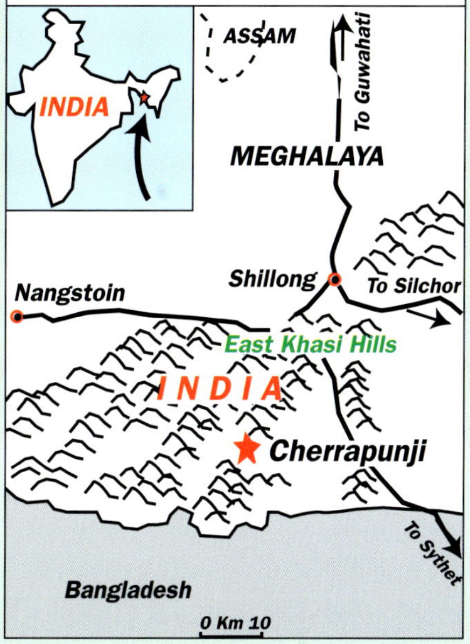

An amazing example of the desertification that can follow forest clearing in an excessively wet climatic region has been documented in the Cherrapunji region of Megahalaya in the Indian state of Assam, on the northern border of Bangladesh, perched about 4000 feet above sea level near the southern edge of a limestone plateau that drops abruptly into Bangladesh. Years of record rainfall, averaging 12.7 metres (500 inches) a year, make this reputedly the wettest spot on Earth.[75] Decades ago, oak forests clothed the region of the East Khasi Hills, but shifting agriculture eventually resulted in their complete removal. Erosion of the soil followed, at a calculated rate of 300 tonnes per hectare a year. Now in places the soil is so thin that even grass will not survive, and elsewhere barely 30 centimetres covers the limestone. Water now goes straight through to the limestone and pours out in devastating floods into Bangladesh, with waterfalls thundering out of the cliffs at the plateau edge. An intense shortage of river water results on the plateau because all run-off water dives into the porous and tunnelled limestone—the rivers only flow when rain is actually falling. Only re-establishing the forests would fix the problem, but vegetation needs soil—and so a **wet desert** has resulted!

CLOUD STRIPPING ('OCCULT PRECIPITATION') IN TROPICAL RAINFORESTS

The rainforests of tropical North Queensland play a significant role, as yet little understood, in the water cycle of that region.[76] Scientists from CSIRO Land & Water are attempting to develop methods for accurately measuring the amount of water that forests strip from low-level clouds and fogs, estimated to be as much as 30 per cent of the total water flowing into high-altitude sites. The water stripped from the clouds as they brush against trees becomes 'throughfall', that falls directly through the canopy, and 'stemfall', that runs down the tree trunks. The trees use little of the water that they strip, and the rest is released to the system by streams and rivers. When accurate measurements can be made of the amounts of water involved at a range of different sites, the information will assist policy makers concerned with sustainable water allocation. As 'ecosystem service providers' the forests have an additional intrinsic economic value because of their contribution to the catchment water cycle.

The benefits of conserving existing forest and of replanting marginal land in upland catchments are already visible at this stage of the research. The increased water supply that would result from replanting degraded catchments would be worth more than the returns from poor agricultural land.

Cloud contributes to the water that sustains remnant forest and feeds the Wollemombi Falls in upland northern New South Wales. *M.E.W.*

THE PHOSPHORUS CYCLE

Healthy plant growth depends on the availability of adequate concentrations of phosphorus in solution in the soil. It is the rate of supply of phosphorus to the plant (its bioavailability), rather than the amount in the soil, which governs the extent to which phosphorus affects plant growth. (In the oceans, 1 atom of phosphorus is needed for every 106 carbon atoms in phytoplanktonic biomass so phosphorus is again the limiting factor, restricting photosynthetic activity.)

On a world scale of soil fertility, Australian soils rate extremely poorly, due to their low phosphorus content.[77] This, more than levels of any other single nutrient, governs soil fertility. Although some profiles have increased phosphorus in surface soil, phosphorus is relatively immobile in the landscape and soil fertility in general is largely related to inheritance from parent material. The other major nutrients (nitrogen, sulphur, magnesium, calcium and potassium) are more mobile. Phosphorus in soils ranges from 0.02 to 0.5 per cent, averaging 0.05 per cent. When phosphorus is added in fertiliser, only some is used—the rest remains fixed in the soil. The mechanisms involved are not fully understood, but clay minerals and iron oxides react with phosphorus and bind it to soil particles. Around roots, the organic matter that they produce displaces phosphorus from iron oxide surfaces, and increased bacterial activity makes more of the nutrient available.

A wide range of fungi and bacteria are able to solubilise inorganic phosphorus. Mycorrhizal associations with fungi improve phosphorus nutrition; the fungal hyphae increase the total root surface area; more area means more production by roots of substances that dissolve phosphorus. Increasingly, inoculation with micro-organisms is being used to increase soil fertility. The soil fungus *Penicillium bilaii* is produced commercially for inoculation; it colonises roots easily and promotes plant growth by increasing available phosphorus.

A wheat crop requires only several parts per million of phosphorus in the top 15 centimetres of soil, which is usually a small fraction of the amount present even in low-phosphorus soils, indicating that only a small amount of soil phosphorus is available for plant growth.[78,131k] The wheat plants can only use the phosphate anions present in solution in the soil.

In natural ecosystems, the supply of phosphorus from the soil to the biota is extremely slow, and plant communities have evolved many strategies to retain and recycle phosphorus that has entered the biological cycle. Much of Australia's native vegetation has adapted to the low phosphorus status of the continent's soils through:

- extensive mycorrhizal infection of roots (see pages 94–95);
- evolution of proteoid roots (see pages146–147);
- developing lower internal phosphorus concentrations;
- using mechanisms that release phosphorus from sources with low solubility;
- retaining phosphorus by re-distribution within the plant before senescence;
- changing root:shoot ratios;
- reducing root metabolism;
- having more carbohydrate available for the mycorrhizal fungi, symbiotic bacteria and non-symbiotic microflora associated with roots.

Exports of phosphorus from natural ecosystems are usually insignificant (unlike the situation in agro-ecosystems where significant quantities of phosphorus are removed from the system annually).

Natural inputs of phosphorus into the Australian continent are from dry deposition and precipitation from the atmosphere, but very little is known about this aspect. Biomass burning results in volatile losses of phosphorus, but a large proportion is returned in rain or particulates and the phosphorus budget of the soil, over time, is probably little affected.

Human intervention in the phosphorus budget has been enormous. Inputs are predominantly from imported phosphatic fertilisers, or from raw materials used for their manufacture within Australia. Rock phosphate deposits have been exploited globally since the late nineteenth century; Australia's first major imports were in 1905 when 16 580 tonnes were brought in for single superphosphate manufacture. Up to 1990, more than 100 million tonnes of rock phosphate were imported, providing more than 15 million tonnes of phosphorus as single, double and triple superphosphates and compound phosphatic fertilisers for improving soil fertility. Rates of importation of rock phosphate slowed and by 1990 production of phosphoric acid for fertiliser manufacture at Kwinana, Newcastle, Brisbane and Geelong had ceased. Instead, greater quantities of triple superphosphate, di-ammonium phosphate and mono-ammonium phosphate have been imported since.

Early imports of rock phosphate from Nauru, Christmas and Ocean Islands had high concentrations of cadmium; for many years market gardeners in the Sydney region were producing vegetables with higher cadmium levels than recommended by the World Health Organization.

The doubly mobile elements and their cycles

Carbon, **Nitrogen** and **Sulphur** are doubly mobile in that they occur in soluble water-borne compounds and also in gases.[50] Their cycling is therefore more rapid than that of mineral elements. They combine with other elements in the course of their recycling, either losing electrons when the process is one of oxidation, or gaining them in the process of reduction. The chemical transformations involving all three elements are largely biological, with plants, protists and bacteria usually dominating as agents. Invertebrates make a smaller contribution, although in arid lands termites are directly significant in carbon cycling and indirectly in nitrogen cycling.

THE CARBON CYCLE

Grassland fires, like this one in the Northern Territory of Australia, add both methane and carbon dioxide to the atmosphere, and cause losses of nitrogen and other nutrients. *M.E.W.*

A great deal of carbon cycling and recycling took place in the newly formed Earth of more than 4 billion years ago, according to Vaclav Smil[50], whose work on natural cycles has been my major source for the subject. He believes that in the very beginning Earth had an atmosphere with CO_2 concentrations three orders of magnitude greater than today's—similar to that of Mars, which has 93 per cent CO_2 by volume. This blanket of gas, and Earth's not-yet-cooled crust, would have resulted in initial air temperatures above 200°C, even though the Sun's luminosity was 30 per cent less than it is now. The CO_2 reacting with water vapour produced carbonic acid, which weathered calcium-silicate rocks in the crust, releasing calcium and bicarbonate ions which on entering the primordial seas created carbonates, CO_2 and water. The seafloor sediments, over the course of hundreds of millions of years, were either subducted into the mantle as tectonic plates moved, to re-emerge in volcanic activity and to be weathered and eroded again, or were compacted, folded and smeared onto the edges of continental plates—some to weather and erode again later. Only when greenhouse gas levels fell far enough to have produced an environment with temperatures suitable for Life (in spite of a cooler Sun) did conditions become suitable for the first living organisms to emerge.

Where forests are drowned when valleys are flooded to create dams, methane produced by the anoxic decomposition of biomass disperses into the atmosphere for as long as it takes for all the wood to disappear. *M.E.W.*

This earliest history is of course highly speculative, but the story from the time when Life was born and started to produce the Webs of Life that became the living Biosphere is easier to imagine and re-create. (See page 37 on the influence of Life on global climate.)

The biospheric carbon cycle is today dominated by CO_2 flows, but the atmosphere contains other carbon compound gases. Methane (CH_4) and carbon monoxide (CO) are always present in trace quantities. Natural methane, the most abundant organic molecule in the atmosphere, is generated largely by anaerobic decomposition that takes place in lake bottoms and waterlogged swamps, and increasingly in flooded paddy fields where rice is grown. The agents of decomposition are methane-producing bacteria, strictly anaerobic. It is estimated that between 100 and 200 million tonnes of methane are released into the atmosphere each year, mainly from the boreal peat bogs and temperate swamps of the Northern Hemisphere.

The 1.5 billion cattle introduced into the Biosphere by humans produce 40 to 60 million tonnes of methane annually. Methane is a greenhouse gas that contributes to global warming. *M.E.W.*

Metabolic processes in the gut of herbivores, particularly ruminants like cattle, are another source of atmospheric methane—one that is becoming a more serious element in global warming as more people in the developing world include larger amounts of meat in their diets, and intensified feed-lot production becomes more widespread. Termites, which also have methanogenic bacteria in their guts to enable them to digest cellulose, are another source of methane, particularly in the tropical savannas. (See page 140.) Incomplete combustion in forest and grassland fires is another source of methane.

The cycling of continental carbon

Photosynthesis is overwhelmingly responsible for the exchange of carbon between the atmosphere and the continental realm. (Weathering, for instance, involves less than 0.5 per cent of the photosynthetic carbon uptake of the land.) Plants and soil carbon are constantly involved in exchanges with the atmosphere, so **changes in the extent and type of vegetation have profound and long-lasting effects on the levels of atmospheric CO_2.**

About 300 000 species of plants, and an unknown number of micro-organisms, are involved in terrestrial photosynthesis today. It has been calculated that in pre-industrial times, before human modification of the environment occurred on a large scale, terrestrial photosynthesisers took up annually about four times as much carbon per square metre as is used by ocean organisms.[50]

While terrestrial photosynthetic organisms produce only twice as much new biomass per year as is produced by phytoplankton, land ecosystems are estimated to store nearly 200 times as much phytomass. The large and diverse store of organic compounds in this terrestrial phytomass provides food for the innumerable species of living organisms that are unable to synthesise their own (heterotrophs, which includes humans). (See Photosynthesis, pages 32–33.)

The accepted view has been, until recently, that the bulk of the Earth's living matter (all but a small fraction of the carbon in organisms) is stored in the tissues of plants—**the phytomass**. The growing recognition of the significance of the invisible world of prokaryotic life challenges that assumption.[220] It can now be confidently stated that **the inclusion of prokaryotic carbon in global models will almost double the estimates of the amount of carbon stored in living organisms.** While it is extremely difficult to calculate the astronomical abundance of the prokaryotic organisms, their carbon content is estimated to be between 60 and 100 per cent of that in the phytomass. (Equally significantly, prokaryotic life contains about ten times more organic nitrogen and phosphorus than is stored in plants!)

Globally, photosynthetic plants, cyanobacteria and protists use over 100 billion tonnes of carbon, in the form of carbon dioxide taken from the atmosphere, each year. Carbon dioxide is returned to the atmosphere in respiration and through the decomposition of organic matter. Carbon must be recycled rapidly because it is the essential element for Life and is a rare element outside living matter. It represents only 0.05 per cent of the Earth's crust (silicon represents 28 per cent). Even in the atmosphere, in the form of carbon dioxide, it amounts to only about 0.01 per cent of the total mass (two orders of magnitude less than the rare gas argon). Yet carbon extracted from the atmosphere, where it is equivalent to a 'trace element', makes up half of all dry biomass. United with other elements, it forms a great variety of biospheric compounds.

Planetary stores of organic carbon, accumulated over billions of years of biological activity, are 99 per cent kerogens (preserved in oil shales), whose aggregate mass is three orders of magnitude greater than the mass of all the Earth's coals and four orders of magnitude greater than the biomass of all living organisms.

Soil carbon

Organic matter in soils is estimated to contain between 1200 and 1800 gigatonnes of carbon—at least twice as much as in all the plants growing on the landmasses of the world. Over 90 per cent of soil carbon is in dead biomass derived through activities of micro-organisms and soil invertebrates decomposing leafy and woody tissue and roots.[50] (Once again, the prokaryotic contribution is not sufficiently included in this calculation.) The other 10 per cent is found in living roots, bacteria, fungi, protists and soil invertebrates. Soil organic matter is a complex mixture of humic, amino and fatty acids, carbohydrates and aromatic carbon compounds.

Decomposition of some of the organic soil compounds happens easily; others persist for centuries. Temperature and availability of water control the rate of decomposition and thus the amount of CO_2 that is returned to the atmosphere.

It has been estimated that 10 per cent of pre-industrial atmospheric CO_2 was cycled via soils every year, with an average in-soil residence time for carbon of about 30 years. Much variation exists between rates in different ecosystems, however, tropical grasslands having a 10-year residence rate and peat swamps and tundra 500 years. In the anoxic conditions that prevail in swamps, mires and bogs, decomposition is excessively slow and peats result—in some cases to be converted into coal over millions of years and locked up as almost pure carbon. (Globally, peat is estimated to contain 400 gigatonnes of carbon. Accumulations of locked-up carbon through geological time involve mind-boggling figures—10 million gigatonnes of carbon in kerogens; 6000 gigatonnes in fossil fuels; incalculable amounts of the same order in coal.)

Roots, living and dead, comprise much of the huge amount of carbon that is stored in the soil. This picture shows canola roots growing within a subsoil biopore. The previous crop was wheat, and the two brown roots in the same pore are residual dead wheat roots. Note numerous root hairs reaching from the canola roots into the periphery of the pore. *Margaret McCully*

The impact of fire on soil organic matter (carbon) in Australia[79]

Termite castles in burnt savanna on Cape York where termites are major cyclers of carbon. Serious damage to the trunks of trees decreases their ability to withstand the next fire. *M.E.W.*

The Australian continent spans 35 degrees of latitude, from temperate southern Tasmania at 45°S to tropical Cape York at 10°S. Fire is experienced across the whole continent, but at different characteristic intervals that reflect current land use and climate—every 1 or 2 years in the tropics; 10 to 20 years in the arid core; and 50 to 100 years in the temperate south (according to official figures, but these figures would bear close scrutiny, particularly for the 'temperate' category). Historical and archaeological evidence reveals that while the frequency of fire has changed over the millennia in step with cultural and climatic changes, it has always been present. Satellite observations today enable estimates of the extent of burning, allowing calculations to be made of the amounts of carbon stored or lost as a result of biomass burning.

The average area affected by fire in Australia is an astounding 30 million hectares a year, 90 per cent of it in the tropical savannas of Queensland, the Northern Territory and Western Australia. Most of this burning is initiated by people. The extra-tropical 10 per cent comprises hazard-reduction burning in forests, involving 1 million hectares a year, and wildfires. In all, it is estimated that 150 million tonnes of carbon are burned—this represents 145 million tonnes volatilised as CO_2 and 5 million tonnes carbonised to charcoal. By comparison, national fossil fuel consumption is approximately 85 million tonnes a year. **Fire on the scale practised deliberately in Australia is a major contributor to global warming.**

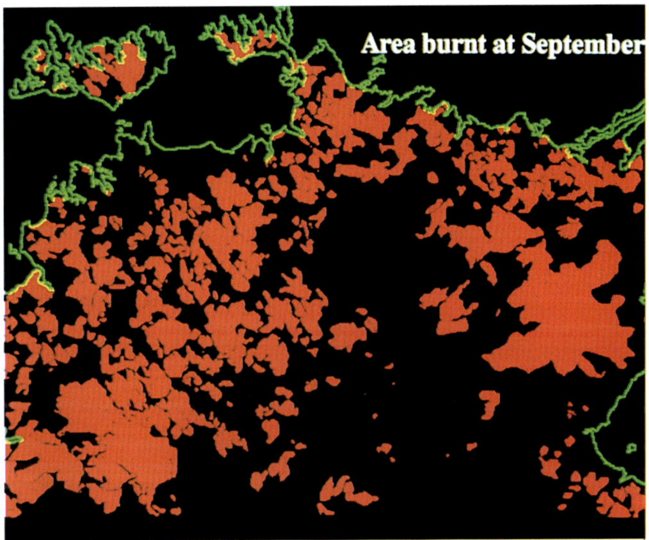

A satellite image of the Top End of the Northern Territory (Melville Island top left corner). The red colour shows the area burnt in September 1990. Burning is mostly annual, sometimes biennial, and largely started by humans, in this part of the tropical savannas, and has devastating long-term effects on biodiversity. Up to 1.7 million tonnes of nitrogen are lost annually in these fires. *Dean Graetz*

Fire has a number of direct impacts on soils. Heating at the soil surface (300° to 425°C) can cause mineralogical transformations, such as conversion of iron oxides to maghemite in the presence of organic matter. Such changes only occur within the top few centimetres of the soil. Even at much lower temperatures (60° to 120°C), heating sterilises soils and causes water repellance. Between 200° and 450°C, organic materials, living or dead, become charred, leading to the formation of smoke, soot, charcoal and ash. Smoke and soot can travel great distances and have significant climatic impact. Charcoal and ash remain mostly where they are formed but may be transported later in erosional processes.

Significant amounts of organic nitrogen can be lost through gases formed in biomass burning—sometimes as much as 50 per cent. It has been estimated that tropical savanna fires in Australia annually lead to a loss of 1.7 million tonnes of nitrogen. At still higher temperatures (more than 600°C) phosphorus may also be lost. (Imagine the nutrient losses involved in the disastrous Christmas 2001 bushfires in New South Wales alone.)

The products of incomplete combustion form a continuum, from partially charred plant materials to fully graphitised 'black carbon' particles that are highly resistant to microbial decomposition and persist in soils and sediments for long periods. This inert fragment may be 50 per cent of the organic carbon, and it contains nitrogen (in the same carbon-to-nitrogen ratio as that of the plant material from which it was made). So both carbon and nitrogen are effectively locked up for decades or even centuries. It has been estimated that 10 to 26 million tonnes of charcoal (black carbon) are formed annually, worldwide, from savanna fires. Theoretically, without a mechanism for decomposition of charcoal, the Earth's surface carbon pools would be converted into charcoal within 100 000 years. The mechanism for decomposition of charcoal is not clear, but fortunately it does exist, probably as a combination of microbial and chemical oxidation processes. Very finely divided charcoal particles cause dark colouration of soils; the charcoal is mobile, moving like clay particles and accumulating in low points in landscapes.

Carbon in the lithosphere

Carbon in the lithosphere is stored in carbonate rocks. Limestones, comprising calcite and aragonite, are most common, followed by dolomite. Weathering of carbonates withdraws CO_2 from atmosphere and soils; rivers carry the products of weathering mostly as dissolved and particulate inorganic carbon. Only about half of all carbon entering streams from terrestrial ecosystems reaches the sea, however—the rest is used up in the respiration of aquatic organisms or deposited on floodplains.

Plants have accelerated the recycling of lithospheric carbon since the birth of their Kingdom with the origin of land plants.[50] Plants have been instrumental in the weathering of silicate materials, resulting in the formation of bicarbonates that are water-soluble and carried via rivers to the sea. Roots liberate CO_2 through their respiration, and carbonic acid is formed in the soil moisture; plant litter decomposes, releasing more CO_2, and plant organic matter is the source of humic acids. All these acids react with mineral matter, facilitating weathering.

Carbon in the ocean

A huge reservoir of carbon is held in the oceans—calculated to be 37 000 gigatonnes of inorganic carbon dissolved in the vast volume of seawater.[50] By contrast, pre-industrial levels of CO_2 in the atmosphere have been calculated at 600 gigatonnes. Carbon dioxide is continually being exchanged across the interface between sea and air and at pre-industrial levels an equilibrium existed, with fluctuations of about 75 gigatonnes in either direction. (Fluctuations are responses to variations in temperature, pressure, photosynthetic productivity, respiration and decay rates.) Less than 1 per cent of the carbon stored in the ocean exists as CO_2, with inorganic carbonates (particularly calcium carbonate), and carbonate and bicarbonate ions accounting for most of it.

Marine micro-organisms play a vital role in the global carbon cycle. (See also Introduction, page 12 and Part Four, pages 164–171.) Primary production by marine phytoplankton can be regarded as having two components—recycled production, depending on nutrients recycled in the surface layers of the ocean; and new or export production that is supported by nutrients supplied to the photic zone from the land, the atmosphere, or upwelled from the deep ocean.[22] The process by which export production is transported from the well-lit (photic) zone to the deep (dark) ocean where, over long time periods, the carbon is remineralised to inorganic forms, is referred to as the **biological carbon pump**. The pump reduces CO_2 concentrations in surface waters of the ocean and the atmosphere and increases the amount of carbon at depth.

HOW MARINE LIFE BALANCES THE BIOSPHERE'S CARBON BUDGET

Marine life acts as a living pump to remove carbon and nutrients from the system and balance the Biosphere's carbon budget. About 98 per cent of the water in the ocean is beyond the reach of solar radiation, and the water temperature remains fairly constant, between 2° and 4°C. The dense, stable waters at depth are capped by the lighter, warmer waters of the mixed layer, which lies from 50 to 100 metres below the biologically richer surface layers. Carbon dioxide is fairly readily and speedily exchanged between the surface and mixed layers and the atmosphere, but very little exchange occurs in the deeper waters. Most of the interchange between the mixed and deep layers is driven by the circulating currents that connect the world's oceans.

These vast conveyor-belt-like currents are determined by salinity and temperature differences in zones at different depths. Only in areas where the deep waters upwell through the surface layers are they fertile and carbon dioxide-rich—becoming sites for massive phytoplankton blooms. Such upwelling sites in the central and eastern equatorial Pacific alone support up to half of all oceanic plankton production, and are the largest marine source of carbon dioxide.

Marine phytoplankton, concentrated in near-surface layers, uses up huge amounts of carbon dioxide in photosynthesis. Much of the manufactured carbon remains in circulation in the mixed layer, cycling among the organisms in the food chains. A small amount is transferred into deeper waters through carbonate precipitation and detrital sinking, but at least 90 per cent is returned as carbon dioxide gas through respiration of the organisms that eat the phytoplankton, and it either escapes into the atmosphere or is re-used by phytoplankton. A fraction of the carbon in the dead bodies of micro-organisms and other marine life that feeds on the phytoplankton falls to the ocean floor at great depth, thus removing it from circulation. The rapidity of the life cycles of phytoplankton, some of which double their mass in a day or two, means that in spite of their small individual size and the fact that their total biomass is about 0.5 per cent of the continental biomass, they can withdraw at least half as much carbon from the mixed ocean layer as land plants take up every year from the atmosphere. The high latitude circumpolar oceans, in particular, where in summer photosynthesis by phytoplankton occurs for 18 or more hours a day, act as carbon sinks.

The bulk of marine carbon is effectively isolated from the continental part of the Biosphere, although the total mass of carbon in marine sediments is about 50 per cent larger than the amount in terrestrial plants and soils.

If the oceans were lifeless, levels of atmospheric CO_2 would be far higher than they are today. Vaclav Smil[50] has estimated that if marine life stopped balancing the carbon budget, within a thousand years there would be more than three times as much CO_2 in the atmosphere than is present now (and what a greenhouse!). Marine life prevents large shifts in the equilibrium by constantly removing CO_2 from the near-surface layers.

THE NITROGEN CYCLE

Nitrogen is an essential component of Life—a component of DNA and RNA, the Stuff of Life—whose self-maintenance is a complex and many-faceted metabolic process that gives Life to the Stuff because only when it is metabolising is it alive.

Fundamental processes of converting atmospheric nitrogen into forms suitable for absorption by plants are *the* most basic, underpinning the photosynthesis-based food chains that sustain all higher life forms in the Biosphere that makes this Earth a Living Planet.

More than 50 per cent of global nitrogen cycling is performed by living organisms.[32] With the exception of nitrate assimilation, all reactions of the biological nitrogen cycle are performed by prokaryotes. Microbial processes result in the nitrification of ammonia following the ammonification of organic matter or of ammonia-producing fertilisers like urea.

Nitrate assimilation occurs in leaves in the photosynthetic process, and recently has been shown also to be a function of mycorrhizal fungi in root tissues, whose contribution to the nitrogen supply of their hosts is now being recognised. **Under stress conditions, more than 80 per cent of plants are colonised by VAM (vesicular arbuscular mycorriza) fungi that take over their phosphate and nitrogen nutrition.** Advances in plant genetic technologies have recently made it possible to show that genes involved in nitrate assimilation are differently expressed when plants are colonised by mycorrhizal fungi, and that soil bacteria are endosymbionts in the fungi, assisting in the nitrifying and denitrifying processes. Once again the interconnectedness of organisms working together in the biosphere is emphasised and 'living soil' is seen as a reality.

Nitrogen is commonly the limiting nutrient in terrestrial ecosystems. It is particularly scarce in many semi-arid grasslands. (And how many original grasslands are now croplands, particularly in Australia's limited agricultural regions?)

Nitrogen is an even rarer component of the Earth's crust than carbon, averaging about 25 parts per million, and in ocean waters only 80 parts per hundred billion.[50] Almost all is in the form of unreactive N_2 in the atmosphere, a form that cannot be used by plants or animals. Reactive nitrogen is a minor constituent of the Biosphere, but essential for Life. It is present in molecules of DNA and RNA that store and transfer all the genetic information that is required for the building of proteins and is a key compound of proteins, the 'stuff of life'.

Lightning and falling meteorites fix a small fraction of atmospheric nitrogen without biological intervention. The creation of DNA in the Beginning probably occurred through such natural processes; its synthesis in laboratories using lightning-like electrical discharges demonstrates how it occurred on the early Earth. Possibly about 2.5 kilograms per hectare per year of usable nitrogen is deposited by rain in thunderstorms.

More than 90 per cent of the nitrogen in soils is bound in living and dead organic matter. One square metre of soil in a temperate grassland in the USA has been estimated to contain up to 10 billion bacteria, billions of fungi, millions of algae and hundreds of millions of other protists, 10 million nematodes and hundreds of earthworms. Such populations represent several tonnes of biological matter per hectare. Yet the amount of stored soil nitrogen that they release by breaking organic macro-molecules down into amino acids, nucleic acids and amino sugars is only about 1 per cent of the total nitrogen stored in the soil—representing about 20 to 60 kilograms per hectare. *Pseudomonas*, *Bacillus* and *Clostridium* bacteria; actinomycetes like *Streptomyces* (that produces a useful antibiotic); and fungi like *Penicillium* (the source of the first antibiotic) are the main organisms involved in the breakdown of mobile nitrogen. They release their organic nitrogen as NH_3.

Nitrogen fixation

Living organisms get their nitrogen, almost exclusively, from the atmosphere's vast stores of unreactive N_2. Before they can use it, it has to be incorporated in reactive compounds by the process known as **nitrogen fixation**, carried out by soil micro-organisms. Nitrogen-fixing organisms are known as **diazotrophs**. Some live free in the soil, others enter into symbiotic

NODULE FORMATION IN A LEGUME ROOT

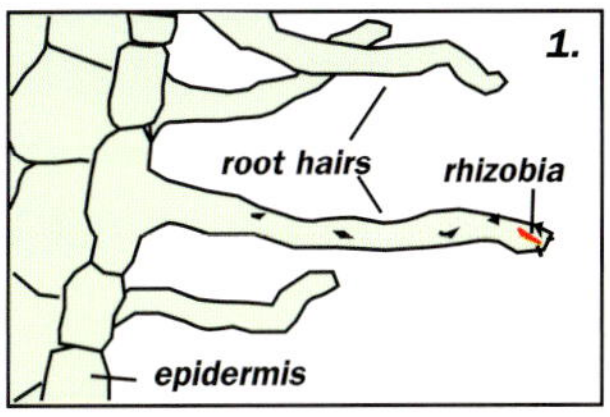

1. Rhizobia from the rhizosphere adhere to the root hair.

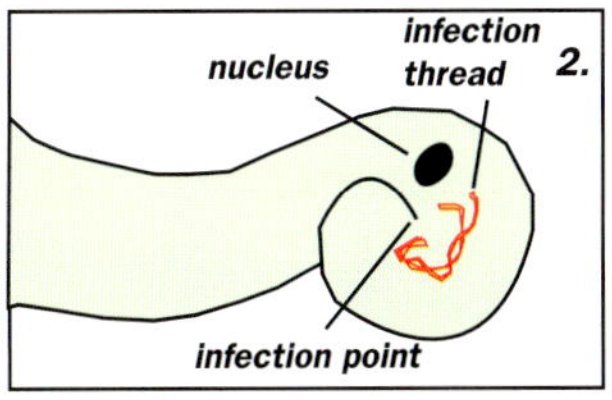

2. Nod-factors induce root hair curling. Bacteria enter at infection point and infection thread starts to develop.

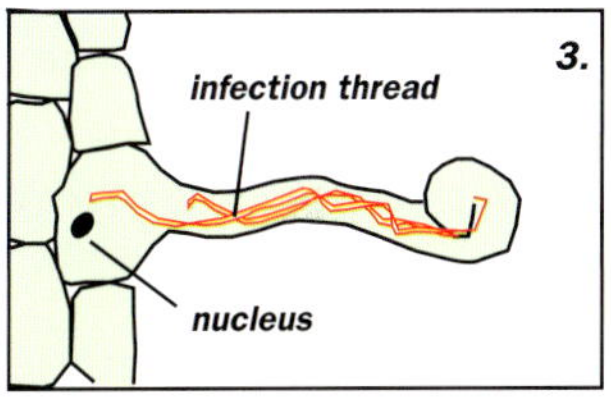

3. Infection thread follows nucleus down the root hair into the epidermis.

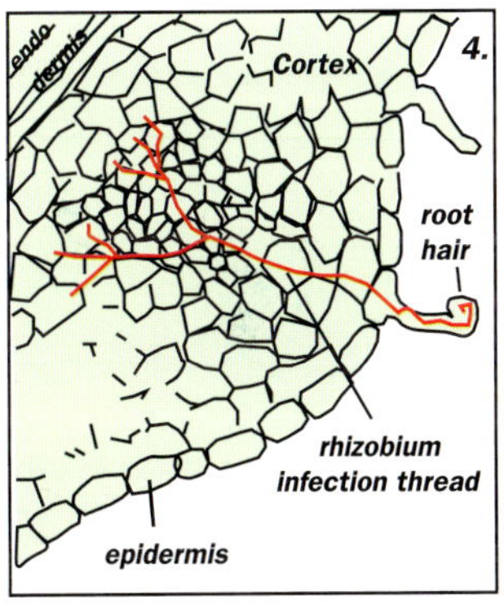

4. Infection thread ramifies near the dividing cells of the cortex that are forming a nodule.

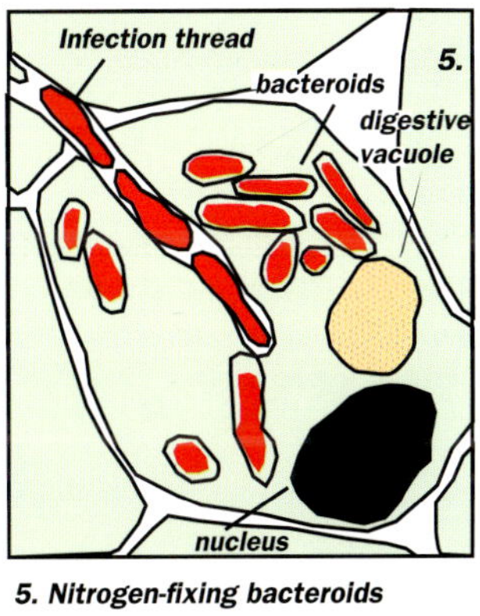

5. Nitrogen-fixing bacteroids released by the infection thread into nodule cell.

relationships, mainly with legumes but also with some grasses, and with trees of the genus *Casuarina*. In the process of nitrogen fixation, di-nitrogen's strong bonds must be broken and the atoms incorporated into ammonia, in which form they can be absorbed by plant roots and converted into amino acids. Much is lost through leaching of the highly soluble nitrates and volatilisation of ammonia; some is removed from the system by microbes that use nitrogen instead of oxygen in their respiration; some clays bind nitrogen ions to their particles, removing them from circulation into storage; and some of the processes involved in decay of vegetable matter that produce humic acids lock up nitrogen in the soil for century-long periods. Generally, decomposers of vegetable matter in soils release nitrogen only slowly. Thus there is little natural accumulation of soil nitrogen.

Nitrates can accumulate in groundwater as a result of leaching of the products of nitrogen fixation. Nitrate concentrations are significant in much of the groundwater in the arid zone of Central Australia, often exceeding World Health Organization recommended levels. The nitrates are the product of micro-organisms, mainly cyanobacteria, which have been found to be widely distributed, even in the hot red dunes of Australia's central deserts.

Nitrogen-fixing cyanobacteria were studied near Yulara (Uluru/Ayers Rock) and in the Ti-Tree Basin north of Alice Springs.[19] (See box on Cyanobacteria, page 9.) Coccoid and filamentous cyanobacteria were found to be randomly distributed in soil samples. A diversity of species was found. Among them, *Scytonema* and *Calothrix* are both capable of living in symbiosis in lichens or free-living in the soil. All the organisms in such an environment must have the ability to survive the extraordinarily dry and hot conditions that prevail most of the time, and be able to revive with rain. The discovery of a water-stress protein in the cyanobacterium *Nostoc commune* explains the ability of this particular species to endure desert conditions.

Technological Man has learned to synthesise nitrogenous fertilisers and today an ever-increasing volume of global food production for a grossly excessive and ever-expanding human population depends on their use. By this means, cereal production, for example, has doubled without requiring more land for crops. But the problems that follow prolonged and intensive fertiliser use are legion. They include soil acidification; loss of, or alteration to, balances in microbiological content, with increasing pathological consequences for crops; and build-up of contaminants.

While industrialised nations were originally the major users of synthetic fertilisers, today the developing countries annually apply 40 million metric tons of nitrogenous fertiliser to the soil, half of all that is used in agriculture worldwide. By 2025 consumption may have increased by 60 to 90 per cent, with two-thirds of the increase occurring in poor countries.

The Green Revolution thinking after the Second World War was that artificial fertilisers, plant and animal breeding programmes, mechanisation, irrigation, and technological fixes generally, would solve all the world's problems. It is becoming increasingly apparent that a far deeper understanding of the natural processes that underpin sustainability is necessary, and that includes the much-neglected realm of soil microbiology. The use of legumes in crop rotations has been an accepted and more natural way of boosting soil nitrogen—but it is not without its own problems, often causing acidification.

A great deal has yet to be learned of how to achieve the symbiotic balances that result in sustainable agricultural practices. To this end, research is ongoing, worldwide, to achieve more biological nitrogen fixation and plant growth promotion by various associative and endophytic systems (within plant cells, in root nodules), through the promotion of naturally occurring soil bacteria, some of which have extraordinarily widespread, virtually global, distribution. Such bacteria have been part of the symbiotic network since the Beginning.

The bacteria that have the ability to fix nitrogen all rely on the production of **nitrogenases**, enzymes that comprise two protein molecules—the larger containing atoms of molybdenum and iron; the smaller only iron.[50] Nitrogenase tolerates no more than fleeting exposure to oxygen, so the bacteria form an oxygen barrier in the nodule tissue, carrying the oxygen needed for the respiration of the nitrogen-fixing organelles in molecules of leghemoglobin, a protein whose high affinity with oxygen prevents the gas from reaching and inactivating the nitrogenase. (The versatility of bacterial symbionts and the lengths of geological time that were required for the perfection of such relationships are a never-ending source of amazement.)

The amount of nitrogenase in the Biosphere may not be more than a dozen kilograms, according to Vaclav Smil[50], being present only in micro-organisms, and then only in a limited number of genera. (Professor Ivan Kennedy doubts this calculation, arguing that root nodules in legume plants comprise about 2 per cent of the fresh weight of the plants and are 20 per cent bacteria. As nitrogenase amounts to 2.5 per cent of the protein in *Rhizobium* bacteria, and as 25 tonnes per hectare of soybeans is easily achieved, a figure of about 2.5 kilograms of nitrogenase per hectare is implied. Therefore, he says, 'there must be tonnes of the stuff!')

The question has to be asked why no 'higher plants' (the Plant Kingdom) have even a rudimentary capacity to fix nitrogen without the help of micro-organisms. The answer, as I see it, would appear to be that the Plant Kingdom was a latecomer that appeared only when Life moved onto the land about 450 million years ago, when the soil was already inhabited by all sorts of bacteria, fungi and protists. The harsh conditions that prevailed then would have required symbioses of all sorts with micro-organisms in order to survive in the first instance, and then to evolve over time. With such symbioses through time creating the new species that were subject to natural selection as environmental conditions changed, we see the working of evolution according to the Margulis model (SET). Looked at this way, it is no longer surprising that plants left nitrogen fixation to their attendant micro-organisms. Why do it themselves when there was always some organism around doing it anyway, particularly when the process requires a lot of energy? Nitrogen fixation has such a high energy requirement that micro-organisms have to consume 3 to 5 grams of carbon for each gram of usable nitrogen reduced from atmospheric N_2.

Bacterial nitrogen fixation is influenced by many environmental conditions. The trace elements the fixers need—molybdenum and iron—are often limiting factors; and they need more phosphorus than non-fixing species do. Both acid soils and high temperatures inhibit the process; the host plant's photosynthetic rate must be sufficient for the rhizobia's carbon-based food to be guaranteed; too much inorganic nitrogen in the soil suppresses fixation; the energy the micro-organisms need may not be available where light is inadequate, as in understorey environments or in turbid or deep water.

The amount of nitrogen fixed in different ecosystems thus differs greatly. In some arid grasslands it has been shown to amount to only a few kilograms per hectare annually, while *Acacia* savannas like brigalow can receive 100 kilograms of nitrogen per hectare from their nitrogen-fixing symbionts. The seemingly large productivity of brigalow soils after clearing only persists, however, until all the stored nitrogen has been used up in cropping, which mines the residues. After a few seasons, the natural infertility of these soils becomes apparent.

Although the general rule applies that most members of the Plant Kingdom require their nitrogen to be in soluble broken-down form, there are some exceptions. In some infertile soils, some plants have evolved a way of absorbing their nitrogen in dissolved organic compounds without involving micro-organisms. Others produce chemicals (polyphenols, mainly tannins) in their ageing leaves which, on leaf fall, bond to soil nitrogen and make it available for absorption by their roots. Many perennial plants move nitrogen from their leaves before leaf fall, storing it in stems and roots so that it is available to supply up to a third of the nitrogen needed for new growth in spring.

Nitrogen fixing by legumes

Nitrogen-fixing bacteria of the genera *Azorhizobium*, *Bradyrhizobium*, *Mesorhizobium*, *Rhizobium* and *Sinorhizobium*, collectively known as rhizobia, are responsible for most of the nitrogen fixed biologically. Rhizobia enter the roots of legumes and induce the formation of nodules. Inside the nodules, rhizobia become endo-symbionts (or bacteroids) in which di-nitrogen is reduced to ammonia. Such bacteroids are responsible for a large proportion of the fixed nitrogen available for plant growth.

Nodule development is controlled by the exchange of signals between the roots of the legume plant and rhizobia. These signals are of three types:

- Flavonoids emanating from the plant interact with rhizobial **NodD** proteins that serve as sensors and activators of transcription.

Rhizobium nodules on roots of medic (*Medicago* sp.). *CSIRO Land & Water, Adelaide*

- A second set of signals is synthesised when the NodD-flavonoid complexes activate transcription from ***Nod*-boxes** to produce **Nod-factors** (lipo-chito-oligosaccharides, molecules of chitin to which fatty acids are attached). Nod-factors provoke deformation of root hairs and allow the rhizobia to enter the root through infection threads. In effect they act like hormones, inducing the empty nodules.
- A third set of signals is necessary for continued invasion of the nodules. Amongst these are extra-cellular polysaccharides (ECP) and related compounds, as well as proteins exported by the type-three-secretion system.

Root hairs have only a small window of opportunity when they can be infected by rhizobia. Infection by VAM—vesicular arbuscular mycorrhiza—is subject to the same limitation and is also triggered by the flavonoid signals. (VAM fungi have been symbionts ever since the Plant Kingdom originated when Life moved out of the waters and onto the land about 450 million years ago. *Rhynia*, one of the earliest vascular plants whose cellular anatomy is preserved, has VAM in its root tissues.) Inoculating soil with flavonoids can enhance uptake of rhizobia and fungal symbionts.

Nitrogen fixing by non-legumes

Reseach into nitrogen fixation in members of the grass family (Graminae) using asymbiotic bacteria may prove significant and valuable for agriculture. With world populations heavily dependent on cereals, facilitating and promoting such symbioses in farming practice may represent the way of the future.

Asymbiotic bacterium *Azospirillum* and its relatives [32,32a–c]

Bacteria of the genus *Azospirillum* may promote improved plant growth and increased yields of forage and grain grasses under a variety of environmental and soil conditions. Originally the enhancement was thought to be a result of microbial nitrogen fixation, but it is now known that most of the positive effects derive from changes to the form, size and function (morphological and physiological changes) of the inoculated plant root systems, leading to an enhancement of water and mineral uptake. Promotion of root hairs and generally improved root development are beneficial to the plant, enabling seedlings to become established as plants with increased biomass, more efficient water use and improved phosphorus absorption.

In most of the plant species studied, it has been found that *Azospirillum* colonises the elongation zone of roots, both increasing the density and length of root hairs and promoting the elongation and proliferation of lateral roots, thus increasing total root surface area. The secretion of growth-promoting substances by the bacteria is at least partially responsible for these beneficial changes.

Where co-inoculation with *Rhizobium* and *Azospirillum* has been used, biomass, nitrogen content and crop yield have all shown significant increase. The positive effects are attributed to increased nodulation, enhanced nitrogen fixation rates and the general improvement of root development—a symbiotic response where the activities of the organisms bring mutually favourable results and the plant is the beneficiary. Increased root efficiency can replace up to 35 per cent of artificial nitrogen fertiliser use under sub-optimum fertiliser and water conditions. The value of increased crop yields is higher than inoculation costs, making this good business practice.

Strains of rhizosphere bacteria that assist in promoting nitrogen fixation and nutrient uptake are the **'biofertilisers'** of today. Research into their use and function is at the cutting edge of agricultural science and vital to the concept of **sustainable intensification** that has to be the goal for food production in a degraded world. (The more sustainable and less degraded regions will have to be used in conjunction with every available kind of scientific and technological assistance to ensure the highest possible levels of productivity and prevent their degradation as far as is possible. Marginal and degraded lands currently used for cropping must be used in different ways, and hopefully be restored to health and some sort of economic use.)

A team at the Rutherglen Research Institute[32] has been working to overcome two factors that are increasingly threatening sustainability of farming practices in south-eastern Australia—soil acidification and poor water use. The former results from the use of artificial fertiliser and legumes in rotations; rising watertables result where crops fail to use all the water available because their times of maximum water use are out of step with rainfall patterns (unlike the deep-rooted perennial vegetation that they replaced). One strategy for combating these problems involves planting perennial grasses like phalaris and cocksfoot; breeding programmes have attempted to produce strains that are both winter active (to use the winter rainfall that otherwise becomes an addition to, and nutrifier of, groundwater) and acid tolerant. So far breeding trials alone have not been successful, as poor seedling vigour has resulted in weeds being able to out-compete the grasses at the time of establishment, and subsequently. When a range of rhizosphere bacteria from long-established phalaris and cocksfoot pastures are added to acid soils when attempts are being made to establish new remedial pastures, it has been found that they significantly enhance root growth and performance, with greater water and nutrient uptake, and improved health and strength of the plants. Different microbial communities are involved in making aluminium, iron and di-calcium phosphates soluble, thus improving osmosis, nitrogen-fixing ability and cellulolytic activity. Inoculation by a mixture of all types gives maximum beneficial response.

Frankia-based nitrogen fixing on *Casuarina*

Frankia is an actinomycete, the active agent of nitrogen fixation in the Casuarinaceae, the she-oak family which is endemic to Australia and nearby South-East Asian and Pacific islands. It is a family closely related to the Betulaceae, in the order Fagales. I give these relationships because *Frankia* is also the nitrogen-fixing agent in red alders (*Alnus*, Betulaceae) in the Pacific Northwest, a fact that I find interesting because these ancient, primitive families are among the earliest families of flowering plants, whose origins seemed to be in the Permian Glossopterids I used to study. Their symbiotic relationship with a specific microbe different from the usual bacterial rhizobia is probably a very old one. Their nitrogen-fixing nodules inhabited by micro-organisms of very different sorts suggests that nodule-forming symbiotic relationships may have evolved separately several times during the evolution of Angiosperms.

Actinomycetes are filamentous bacteria that occur as stringy masses in waters and are among the most plentiful of soil organism. They secrete enzymes and break down tissues, freeing nutrients which they absorb along the surface of their filaments. They are responsible for the heavy, heady smell that is so noticeable when rain starts to fall on dry earth.

Nitrogen fixing by cyanobacteria

Where rhizobia are the major nitrogen fixers on the land, cyanobacteria are the primary nitrogen fixers in the ocean. Many are free floaters, whose blooms can be so concentrated that they colour patches of the sea, often far from land. Others are symbionts living inside diatoms. They also form dense mats on coral reefs and abound in saltwater marshes. The largest source of fixed nitrogen in the Arctic is cyanobacteria living in symbiosis with fungi in lichens.

Only cyanobacteria among the free-living nitrogen fixers provide substantial amounts of reactive nitrogen. About 60 genera are known, with *Nostoc*, *Anabaena* and *Calothrix* the most

Rhizobium nodules on soybeans. Image publication *CRC Soil & Land Management, Adelaide*

important. They produce heterocysts (special enlarged sections like buds along their filaments) with strengthened membranes of many layers to prevent gas diffusion and to protect their nitrogenase from oxygen damage.

Human interference in natural nitrogen cycles: Synthetic nitrogenous fertilisers

Human interference in the flows of reactive nitrogen in the Biosphere was minimal until agriculture became an established practice, and then slowly increased as intensive agriculture developed. It was the technological breakthrough enabling the synthesising of ammonia, and thus the production of nitrogenous fertilisers, that raised human intervention in the nitrogen cycle to a new and disturbing level. (The Haber-Bosch synthesis has a dark and dreadful history, the chemical research that led to it being inspired by the search for ways of manufacturing poison gases and explosives for war.)

The first nitrogenous fertilisers based on ammonia became available in the early 1920s. The difficult economic times of the 1930s and the war years kept production to below 5 million tonnes until the 1950s, when improved techniques enabled production of affordable fertilisers on an ever-increasing scale. By the end of the 1950s, 10 million tonnes a year were being used. By the 1990s, synthetic fertilisers annually contributed about 80 million tonnes of nitrogen to the soil, compared to at least 30 million tonnes fixed by legume crops. Added to this, gas emissions from combustion of fossil fuels contributed over 20 million tonnes of reactive nitrogen a year (and created acid rain).

The total of reactive nitrogen that humans add to the Biosphere each year now surpasses natural flows. While the world worries about increased carbon dioxide and Greenhouse, the dangers inherent in the disturbance of other balances, and the active part that nitrous oxide, for instance, plays in destruction of the ozone layer, go relatively unnoticed.

A real dilemma exists—our protein-requiring, grossly over-populated world is now only capable of feeding the multitudes because of the high yields that are obtained from its finite and limited agricultural soils through the use of synthetic fertilisers. It has been said that 2 billion of the world's 6 billion population are only alive because of the availability of artificial fertilisers. The situation is never as simple as statistics make out, however, for famines are endemic in sub-Saharan lands and in many other parts of the world. Some problems could be cured if wars and social conflicts in those lands could be eliminated and the crop surpluses of affluent countries could be humanely distributed. If meat-hungry Western societies were willing to redirect their agricultural enterprises from meat-producing activities to plant-based systems, and if the developing countries would forgo their adoption of Western eating habits, enough food could still be produced to feed the numbers living now. The energy budgets of high-input agricultural systems are often negative, however, using more kilojoules than they produce—so they are clearly not long-term systems.

If global population is set to double in the next 20 years, as has been predicted, massive change and reduction in living standards in the developed nations is just around the corner and, sadly, massive famine and misery in the less-developed populous nations. Technological fixes are not available to solve future shortages, and the damage already done and being done to the Biosphere by human interference in natural biospheric processes renders it ever less able to support Life.

CASE STUDY: BIOFERTILISER RESEARCH—TOWARDS SUSTAINABLE AGRICULTURE

(This case study on current research programmes at the SUNFix Centre at the University of Sydney was prepared by Professor Ivan Kennedy.)

Azospirillium brasilense cells growing on the surface of wheat roots. The individual cells shown in the magnified image are stained using a reporter gene, lacZ. *Ivan Kennedy*

At the SUNFix Centre at the University of Sydney, a concerted effort is being made at the dawning of the twenty-first century to reduce humanity's reliance on industrially produced chemical fertilisers and to substitute more sustainable biological sources using sunlight to provide the energy needed. To improve the extent of use of atmospheric nitrogen as a biofertiliser, the means of improving the survival of *Rhizobium* cells once they are inoculated onto legume seed are being researched by enthusiastic young scientists like Rosalind Deaker. Exposed to hot, dry conditions in soil or on legume seed, bacteria will die at a very rapid rate as a result of the loss of their cell water. With such a large surface to volume ratio because of their minute size, micro-organisms often suffer a huge penalty when they are subjected to adverse conditions for even a few minutes. A million cells applied to lupin or clover seed can nearly all be killed by disruptive forces during the drying process because of damage to their DNA molecules and proteins beyond their capacity to repair. With so few living rhizobial cells, the process of nodulation described earlier may no longer be possible and the ability of the legume to access nitrogen from the air is lost!

Australian companies like Bio-Care Technology, located near Gosford, who are also very supportive of this research, take its benefits to farmers by providing them with inexpensive peat inoculants carrying up to 10 billion *Rhizobium* cells per gram to inoculate their legume seed like clover, medics, lucerne, lupins, soybeans and many other grain legumes or lentils. Because of the very low nitrogen status of its soils, Australia's scientists in universities and CSIRO by necessity have led the world in the development of *Rhizobium* technology; heroic individuals like Jim Vincent at the University of Sydney and Lex Parker in Perth at the University of Western Australia, both of whom died in the last two years, developed outstanding national and international reputations and were in high demand everywhere because of their experience and knowledge of the methods needed to bring this organic form of nitrogen to a point where it can easily be used by farmers.

Remarkably, like all true solutions that rely on Nature's methods, the cost of using such inoculants is extremely small at only two or three dollars per hectare. By contrast, the 100–500 kilograms of nitrogen they typically add to crops and the soil each year would cost more than $100 per hectare if purchased as chemical fertiliser, and the development of the *Rhizobium* technology has brought a benefit to agriculture in the country as a whole to at least $50 billion for Australian produce such as wool grown in rotation with legumes and lupins. The huge benefit from symbiosis is now taken for granted.

In the past ten years, SUNFix has turned its attention to bringing the same benefits as *Rhizobium* can bring to the growth of non-legumes such as grasses and cereals like rice, wheat, maize and sugarcane. *Rhizobium* cannot nodulate these non-legumes but a range of other soil bacteria, many of them diazotrophs, can live in mutually beneficial associations with plants. These micro-organisms, including species such as *Azospirillum*, *Azotobacter*, *Klebsiella*, *Acetobacter*, *Pseudomonas*, *Citrobacter* and many others too numerous to mention, stimulate the growth rate of plants, increasing the grain yield possible.

In work in Vietnam, SUNFix has shown that rice farmers using inoculant biofertilisers containing a group of three of these microbes can reduce their need for chemical fertilisers like urea to half and still obtain an increase in rice production on their farms amounting to 120 per cent of their maximum production with chemical fertilisers. Apparently, these biofertilisers produce phytohormones that affect the architecture of plant roots, making it possible for them to extract nutrients like nitrogen, phosphorus, and metals like calcium, magnesium and potassium from soil much more efficiently. The microbes specifically make nutrients available to their companion plants, increasing their action state by mobilising and energising them. Given that the soil often contains huge reserves of insoluble phosphorus built up over many years of applying chemical superphosphate, as well as most of its nitrogen being locked up in unavailable forms as mentioned earlier, the potential for future application of biofertilisers is also very large.

Studies in Brazil have shown that this beneficial suite of micro-organisms in sugarcane that mobilise nutrients for the plant is lost when chemical fertilisers like urea are applied, leading to dependence by farmers on chemical fertilisers from then on. It seems that the more organic approach to growing crops available when biofertilisers are used may be Nature's way, which may be lost when large quantities of chemicals are used. The claim of organic farmers that a delay of several years may be necessary before maximum yields can be obtained after changing the farming system, is consistent with the time needed to re-establish a full suite of micro-organisms that were lost under the synthetic-chemical farming system.

In collaborative international research funded by the Australian

Centre for International Agricultural Research and AUSAid, SUNFix is now concentrating its efforts on developing the quality control and technology necessary for effective application of such inoculant biofertilisers by farmers. Local village production of biofertilisers may bring social and economic benefits to villagers and farmers as a result of their greater self-reliance. But there is need to make sure that the biofertilisers contain sufficient numbers (up to 1 billion cells per gram) of the most beneficial microbes, and simple test measures that will allow farmers to check this themselves wherever possible are being developed. There is also a need to make sure that the micro-organisms can survive in soil and associate effectively with crop plants and that the best microbes for the particular crop are used.

This is an important area of research for the twenty-first century (see also *Biofertilisers in Action*—in press, RIRDC, Canberra) that promises a more sustainable future for Australia's agricultural and natural ecosystems, provided the needs of these simple symbiotic microbes are catered for in our land management. There is the promise of a new and true Green Revolution if cultivars and crops are selected that can benefit the most from such mutual associations. The full benefits possible from symbiosis, pointed to by Margulis as the true goal of evolution, which have been largely excluded by the widespread and unsustainable use of chemical fertilisers that often double as environmental contaminants, would then be available. This approach also needs to be integrated into the solutions now being sought for problems such as salinisation and SUNFix's action programme seeks to do this.

Disposal of nitrogenous wastes from animals and humans

The disposal of nitrogenous wastes from animals and humans is an ever-increasing problem. Half the world's 1.5 billion cattle and 1 billion pig populations alone are raised in confinement, where their half of the 80 million tonnes of nitrogenous waste produced by just these two species must be disposed of into soil, streams or ponds.

Disposal of human wastes has become critical, with some still going straight into soil or rivers in undeveloped nations. In the industrialised world, most goes into the sea and into rivers, treated or untreated. Little is being recycled, and increasing amounts of precious fresh water are being lost in sewerage systems. The increased nutrient levels of all waters, including groundwaters, as a result of animal waste (including human) and nitrogenous fertiliser run-off from agriculture, is leading to toxic cyanobacterial blooms, the proliferation of harmful bacteria and degeneration in the quality of water bodies.

High nitrate levels in drinking water are deleterious to health, and groundwaters are often found to have been contaminated by quite remote sources.

The Great Barrier Reef, Australia's most valuable and unique natural ecosystem, is suffering from increased levels of nitrogenous nutrients. On the landward side of the complex, along great lengths, proliferation of algae and the cyanobacterium *Oscillatoria* are smothering the coral; while the crown of thorns starfish reaches plague proportions in the nutrient-enriched waters.

Earth's carrying capacity before artificial fertilisers were available

Historical records of the amount of food produced in traditional intensive agriculture are few and far between. Vaclav Smil[50] attempted to assess nitrogen balances for pre-industrial ecosystems from a number of sources, ranging from Stamford Raffles' descriptions of farming in Java in the early 1800s to intensive surveys of cropping in China during the 1920s and 1930s as recorded by John L. Buck. Their figures on crop yields, animal husbandry and manure application enabled Smil to reconstruct nitrogen flows and carrying capacities in what were the world's most productive traditional farming systems.

The oldest and simplest systems were those in which a single crop was produced in a field each year, in conjunction with extensive fallowing. In the most complex of the traditional systems a rotation of about six crops, three a year, was used in each field, a level of use which required intensive recycling of organic wastes.

In early permanent systems, the nitrogen supplied by farmers through the use of organic waste amounted to less than 20 per cent (and even as low as 10 per cent) of the nitrogen available to the crop. Such systems could rarely support more than 2 people per hectare, even in regions where the soils were fertile and there was abundant water. In ancient Egypt, productivity along the Nile was limited to only 1.5–2 people per arable hectare; in northern China, less than 2 per hectare applied until the Tang dynasty (618–906 AD); in the fertile Mexico basin, prior to the Spanish conquest in 1519, the figure was 4 people per hectare.

By the early 1800s, Dutch farmers were supplying up to 75 per cent of the nitrogen used by their crops, and the same proportion was being applied by peasants in South China's most productive regions, sourced from recycled human and animal wastes, green manures and food legumes. In rice-growing regions, nitrogen fixation by the *Anabaena* bacterium–water fern *Azolla* symbiosis was additional. All these inputs added about 200 kilograms of nitrogen to each hectare of arable land, enabling the production of 200–250 kilograms of food protein.

Protein output is what determined the carrying capacity in the very largely vegetarian societies of the pre-industrial world. In the most fertile areas, where peasants could grow crops year round, 11–15 people could be supported per hectare (with 90 per cent of their food energy coming from staple grains). The Aztecs, with their terraced fields raised above the shallow lake waters in the basin of Mexico, achieved this level of productivity.

Less fertile regions and those with seasonal restrictions on production obviously were less productive. Where only two crops a year were harvested, only about 5 or 6 people could be sustained per hectare, even when meat and dairy products were added to the diet. Much of pre-industrial Europe fell into this category. While more and more land could be brought into production by clearing forest and natural grasslands, and by draining wetlands, intensification of production was not an urgent priority. The increasing use of nitrogen-fixing legume crops in rotations, adding more manure, ploughing-in crop residues—the improved systems of the 1800s—delayed the arrival of the moment when the demands of an expanding population came up against the nitrogen barrier.

By the middle of the twentieth century, synthetic nitrogen fertilisers were making a marginal difference to global food production. Their expansion over the next 50 years to a level where half the nitrogen used by the world's crops is now synthetic has enabled global population to reach 6 billion. China, for example, applies about 200 kilograms of synthetic nitrogenous fertiliser per hectare in its most intensively cultivated provinces (where 350 million people live), and sustains 20 people per hectare on a high-protein diet, and more than 30 people per hectare where the diet is largely vegetarian. (An amazing example of 'sustainable intensification', if such application of fertiliser was indeed sustainable over the long term.)

THE SULPHUR CYCLE

Sulphur, which is plentiful in emissions from volcanoes, was abundant in the early Earth with its high level of volcanic activity, where the sulphur bacteria that use hydrogen sulphide in their metabolism evolved. In the world of today, volcanic activity is limited, but sulphur bacteria still function in anoxic places and are a source of methane. Human interference in the natural sulphur cycle has been considerable. With the coming of industrialisation, the combination of fossil fuel combustion and smelting processes produced sulphur-bearing emissions whose volume now surpasses that from natural sources. These man-made emissions cause acidification of soils and waters and corrosion of structures.

Inputs of sulphur from the atmosphere in the Southern Hemisphere are low by comparison with the highly industrialised Northern Hemisphere. A major component comes from sea spray. As happens with salt carried by wind and rain, the huge area of operation of Australia's continental coastline results in a significant acquisition over time, and atmospheric input to the continent has been estimated as 1152 kilotonnes per year. Human inputs to agricultural lands are less than atmospheric inputs on a continental scale, but as fertilisers are used on only about 4 per cent of the continent, where they are used the local sulphur budget is very much altered. (Single superphosphate contains 10–12 per cent sulphur.)

The huge plume of sulphur-bearing smoke from the smelter at Mt Isa in Queensland used to carry its pollution hundreds of kilometres downwind. Grazing properties below its path suffered vegetation damage and other ills; several were abandoned. The introduction of filters to trap the particulate matter and cure the pollution problems of the township and the surrounding countryside enabled the establishment of a sulphuric acid industry. An unbelievable 3000 tonnes per day was regularly produced. This gives some idea of the volume of material that can be carried in a plume of smoke and its potential as a source of substances toxic to living things, including us. *M.E.W.*

Sulphur is not rare in sea water, where it exists mainly as the sulphate anion. It is uncommon in the Earth's crust—three orders of magnitude rarer than carbon—but is vital for Life. No proteins can be synthesised without it. Its rarity is explained by the fact that the bulk of Earth's biomass exists as woody tissue of trees and shrubs that contains no proteins, while the biomass of animals, although protein rich, represents less than 1 per cent of all living matter.

Coccolithophores, which start their life as small golden planktonic algae and form wonderfully decorative resting forms that have overlapping 'coccoliths' made of calcite, are responsible for transferring considerable amounts of sulphur from the store dissolved in seawater to the atmosphere. Some large brown seaweeds have the same ability. The dimethyl sulphide (DMS) they produce is an important weather moderator. On oxidation in the atmosphere, sulphate particles from DMS act as nuclei for cloud formation over the oceans. (See page 39.)

Anaerobic sulphur bacteria (Archaea) were among the first living organisms on the early, cooling Earth. The atmosphere then was without the oxygen that would later begin to accumulate when photosynthesising organisms split the water molecule in their reactions. Using the H_2S (hydrogen sulphide or rotten egg gas) that was abundant in that highly volcanic world, the green and purple sulphur bacteria used electrons from the gas to replace those lost by pigments excited by solar radiation, and produced sulphur as a by-product. The same bacteria today, probably unchanged in 4 billion years, still carry out this sort of photosynthesis in anoxic muds, in coastal bacterial mats and in volcanic hot springs. (Only when water-molecule-splitting early organisms developed 'normal' photosynthesis could they move away from the sulphurous and saline places where life began, and start to colonise the rest of the environment.)

The sulphur bacterium *Beggiatoa* today lives in sulphur-rich sands, oxidising elemental salt and H_2S just as its primaeval ancestors did. In seawater the sulphur it produces can combine with iron to form pyrite. The acid sulphate soils that occur in coastal regions, and today are such a major problem in parts of eastern Australia, are the product of anaerobic sulphur bacteria. Some anaerobic 'thiopneute' sulphur bacteria in the ocean also release sulphur in a form that can combine with iron to form pyrite. Some other sulphur bacteria oxidise ammonia to nitrite, and nitrite to nitrate; some oxidise methane to methanol.

With the eyes of the world focusing on increasing levels of carbon dioxide in the atmosphere as the culprit in greenhouse warming, little is ever said about increasing methane levels. The Asian rice paddies that feed a huge proportion of the Earth's people are a major producer of the gas. In spite of the fact that agricultural improvements have resulted in the doubling of rice harvests since 1950, paddy fields have also expanded, now covering 40 per cent more of the Asian agricultural landscape. It has recently been found that draining the paddies briefly in the middle of the growing cycle effects a great improvement in methane control, so possibly changed agricultural management

Gypsum crystals in Nelly Creek, Lake Eyre. *M.E.W.*

is an answer to this problem. All the paddy-field methane, added to that produced by cattle, whose numbers have increased globally over the same period to the astounding figure of 1.5 billion, must be changing sensitive balances. (Symbiotic bacteria in the rumen, helping with digestion of cellulose, are the source of methane produced by cattle.)

Some sulphur bacteria can tolerate amazingly acidic conditions; and some, like *Sulfolobus*, that lives in nearly boiling-hot springs, can only survive at above 50°C. Recent studies of chimney-like structures on mid-ocean ridges in the deeps, where darkness is total, have shown amazing communities of organisms whose food chains are based on synthesis of sulphur by micro-organisms (see Extremophiles, pages 24–27).

Oceanic sulphur can be sequestered without bacterial interaction. As seawater evaporates in shallow basins in arid lands, carbonates are precipitated first; sulphate compounds, mostly **gypsum**, form when water volume has been reduced by 80 per cent. Huge evaporite deposits, predominantly anhydrite, have been formed in the geological past by the drying-up of seas. The stranding and drying of the Mediterranean Sea 6 million years ago, in the Terminal Miocene Event, resulted in the formation of such anhydrite deposits deep in the Mediterranean Basin. So much salt evaporated then that the salinity of seawater globally decreased by 6 per cent, and presumably much sulphur was locked up at the same time. Today no major anhydrite deposits are being formed.

Salt lakes are also sites for gypsum formation as a result of evaporation. Gypsum crystals today lie scattered over desert surfaces around Lake Eyre in Central Australia, glinting in the sunlight, and high gypsum dunes lie on the downwind side of the lake, a legacy of the extraordinarily dry glacial stages of the Pleistocene ice age. At the time the saline waters of the lake, and of all the other lakes in the dry and windy centre of the continent, dried, and salt and gypsum were deposited, as in marine evaporites. The winds blew across the dry lakes, sweeping the lake bottom deposits into dunes on the far side. On Lake Frome, gypsum dunes form prominent features 20 metres high on the south-eastern shoreline; on Lake Callabonna's north shore lies the distinctive 'Cobbler Dune'.

Sulphur in Australian agricultural soils

It has long been established that large areas of Australia are sulphur deficient.[77] This is limiting for agriculture, since 1 kilogram of sulphur is required for fixation of 14 kilograms of atmospheric nitrogen; legumes are used for nitrogen fixation in Australian improved pastures.

Acid sulphate soils

Acid sulphate soils (ASS) are formed when sea or brackish water mixes with sediment containing organic matter. Although they are part of the Biosphere's natural sulphur cycle, their presence is currently causing great concern in Australia.

Recent acid sulphate soils (formed in the last 10 000 years) are found in the coastal zone, and are presenting serious problems for agriculture, urban and rural expansion and infrastructure, in eastern locations from the Victorian border to Cape York, across the tropical north and in isolated areas in Western Australia, South Australia and Victoria. They occur in low-lying areas, coastal embayments and estuarine floodplains, and are still being formed today in mangrove and salt marshes, estuaries and tidal lakes. Their inland extent beyond present-day tidal limits is related to major sea-level rises of the past.

In their formation, sulphate ions in seawater reacted with iron oxides in sediments in the presence of organic matter from mangroves and other coastal vegetation that was being broken down by micro-organisms. Iron sulphides, mostly in the form of pyrite (FeS_2), were produced in quantity by the chemical reactions. These remain inert in anoxic conditions. In the inert form these soils are known as **potential acid sulphate soils** (PASS). When disturbed by drainage or cultivation, the sulphates are exposed to the air, and oxidise to form sulphuric acid (H_2SO_4), acidifying the soil, soil water and groundwater, and eventually drainage waters and streams. In moving through the soil, the acid strips iron, aluminium and sometimes manganese from the soil, creating an acidic brew which is toxic to plants, to organisms in rivers and estuarine environments, and is highly corrosive to man-made structures. The leaching of acid from the deposits is progressive, and can continue for hundreds of years.

This dam, excavated in an area affected by acid sulphate soil, is surrounded by black sulphidic sediments, and the water is red with iron pollution. *Rob Fitzpatrick*

An island of black acid sulphate soil in an eroded scald is a legacy of rising saline sulphate-rich watertables. *Rob Fitzpatrick*

The size of the actual and potential problem in coastal regions of Australia is enormous—2 million hectares, containing 1 billion tonnes of pyrite. One tonne of pyrite can produce 1.5 tonnes of sulphuric acid!

The impact of these soils on farming, fishing, aquaculture, urban development and engineering works has been extensively studied and risk maps have been made to assist in their identification. A real risk of damage to coral structures has been identified for the Great Barrier Reef from acid released by cane farming and other land uses.

Only recently has it been realised[81] that PASS also have widespread distribution in inland Australia. Dryland salinity, the consequence of land-cover change and rising saline watertables, has the potential to produce acid sulphate soils and soil acidification. The combination of waterlogging and agriculture in areas where the geology involves rocks rich in iron, sulphur and heavy metals, leads to formation of acid sulphate soils. Anaerobic sulphur bacteria use the sulphate to help degrade organic matter, producing the iron sulphur compound pyrite that is inert until exposed to air. The tell-tale signs of acid sulphate patches are black bogs, slimy red ooze and scalded, infertile areas. The patches are reported from the West Australian wheatbelt, South Australia, south-west Victoria, the Yass valley in New South Wales, and south-east Queensland—their full extent is yet unknown.

Farmers need to be able to identify, prevent and manage such areas on their land, otherwise erosion, acid and heavy metal run-off, reduced water quality and reduced crop yield are inevitable. The CSIRO, through the CRC for Soil and Land Management, has produced manuals for management in South Australia's Dundas Tablelands and Mount Lofty Ranges, where the problems were first identified and studied. Even where relatively small areas are affected, the impact on water quality is significant, as the soils constantly discharge acid and toxins which rain carries into watercourses. Planting trees in groundwater recharge areas helps to lower watertables; affected areas have to be fenced off so that no added disturbance can further expose the affected soils to the air and cause acid leakage. Acid-tolerant plants such as phalaris, fescue, strawberry clover and tall wheat grass planted in the seep areas help to control the problems; the control of erosion and run-off in adjacent areas helps to protect waters.

The only good that can theoretically be said to come of the situation is that by analysing the seep liquids it is possible to locate ore deposits in the underlying rock.

NUTRIENT CYCLING IN THE SOIL

The decomposition process

Living organisms are largely constructed from **biopolymers**. These are complex organic molecules consisting of simpler sub-units of the same, or similar, kind linked together in chains. These constituents are the simple sugars that make up complex carbohydrates; the fatty acids in lipids, the molecules that comprise fat; and amino acids in proteins.

The decomposition of biomass takes place in stages, facilitated by different bacteria.

- Bacteria of the genera *Bacillus*, *Clostridium*, *Acromonas* and others that occur just about everywhere on this bacterial Earth are the first decomposers. They are omnipresent in air, soils and waters, and attack both soft and tough biomass tissues (and are responsible for a lot of food spoilage). They have evolved 600 or more enzymes as their agents of external digestion. They excrete these specialised proteins into their surroundings to break down the biopolymers into simpler compounds which they can absorb and metabolise. Their activities start long before tissues die—millions may be present on a leaf that has just unfurled from a bud. Some, like clostridia, thrive only in the absence of oxygen, multiplying inside tissues and degrading complex sugars to butyric and acetic acids, releasing hydrogen sulphide and ammonia gases. They create the stench of decomposing materials. The dangerous botulism bacteria are of this sort, and have the added ability to excrete powerful neurotoxins. These anaerobic bacteria have a lineage that extends back to the beginning of Life in an anaerobic world, and today they live in deep aquifers, flooded soils and marine muds, in plant tissues and in animal guts.
- The next stage of biomass decomposition is dominated by the highly diversified genera *Azotobacter*, *Enterobacter* and *Pseudomonas*, commonly present in soils and waters. During this stage, most of the organic carbon is returned to the atmosphere as CO_2. The residue is incorporated into soil humus or sediments.
- Such final decomposition as occurs is effected by bacteria suited to operating in a nutrient-poor environment.

Cellulose and lignin

The decomposition of cellulose and lignin is a vital recycling service, in view of the fact that a massive 90 per cent of the Biosphere's above-ground plant tissues are in the form of living and dead woody biomass, composed of these strongly bonded (hard to break down) biopolymers. Some of the bacteria involved, including *Cellulomonas* and *Clostridium*, have evolved mechanisms to keep their cellulose-breaking enzymes in large complexes attached to the cell surface, not released into the surroundings, thus maximising the availability of the soluble break down compounds for their own use. Particulate plant organic matter, however, persists in soils because of the inability of micro-organisms to degrade it. Much of this residual material is cellulose that is protected by a coating of lignin or other organic material and has such a small surface area that it does not provide anchorage for decomposers.

Lignin is particularly hard to break down. It is a three-dimensional biopolymer comprising highly cross-linked alcohols, and is resistant to bacterial enzymes. Decomposition by bacteria is slow and woody material is primarily degraded by fungi, often with considerable help from boring invertebrates.

Termites have cellulose-decomposing *Bacillus* species living symbiotically in their hindgut, enabling them to feed on woody materials, but they frequently depend on white-rot fungi to start the decomposition. Fungi produce powerful enzymes that breakdown organic matter.

Charcoal, the product of burning of woody tissues, is not degraded by micro-organisms to any extent and much of the fine particulate matter in some soils is this resistant form of pure carbon.

The age of organic matter in soils increases with depth.[82] At a depth of 1 metre, the mean residence time for organic matter in a range of soils has been found to be several thousand years.

THE CRUST OF THE EARTH

The chemical composition of the Earth's crust, surprisingly to non-chemists like me, is 93.77 per cent oxygen by volume, or 46.60 per cent by weight, and oxygen represents 62.55 per cent of the atoms forming its components.[73] The next most abundant element is silicon: 27.72 per cent by weight, only 0.86 per cent by volume, and 21.22 per cent of the atoms. Aluminium, iron, calcium, magnesium, sodium and potassium, in decreasing order of abundance, make up most of the rest, with other trace elements accounting for only 1.5 per cent by weight. This means that

> the crust can be regarded simply as a packing together of oxygen atoms, with all the other elements accommodated within the packing voids. However, such an accumulation is only made possible by the presence of the other elements, for they are the positively charged cations, which hold together the negatively charged oxygen anions. Because the most abundant cation is silicon, it follows that silicon–oxygen combinations, or silicates, are the most abundant minerals in the Earth's crust[73]

and therefore the rocks mainly to be considered in the rock cycle and soil formation.

The **lithosphere** that was created when Earth coalesced from a fiery mass and started to cool comprised igneous rock, but as soon as water arrived, followed by first Life, surface rock began to be worn away and sediment started to accumulate. Washed into water bodies, this material consolidated eventually into sedimentary rocks; and they in their turn were eroded away into sediment. The turnover and recycling of lithospheric materials had begun.

Regolith is the term used for the whole sequence from topsoil through lower soil horizons, subsoil and layers of altered rock (**saprolite**) to the solid, fresh rock of the lithosphere. Constant cycles of creating regolith, changing and removing and remaking it, have been going on since Earth's beginning.

The 'soil' that accumulated on early land surfaces was largely sediment only, transient except in swampy areas. There soil bacteria would have been operative—anaerobic Archaea at first in the saturated muds and at depths wherever there was pore water; later, under aerobic conditions, bacterial mats formed in suitably damp places. It seems likely that some time in the Precambrian, terrestrial ecosystems comprising cyanobacteria, green algae, fungi, lichens and perhaps, later, mosses and liverworts, formed biological soil crusts, like those that are found today. (In Antarctica, for instance, such crusts are the only terrestrial 'vegetation', and the lichen soil crusts of Australia's arid lands still play an important colonising role under harsh conditions of heat and dryness.)

When the ancestral Plants emerged (from green algal ancestors, not more than 450 million years ago) and started the major colonising of the land world, there was vegetation to hold the soil and stabilise surfaces, and it gradually spread into increasingly drier places. A terrestrial invertebrate fauna could then emerge from the water. The terrestrial realm of the Biosphere became one of its fully active organs with the proliferation of plant biomass and a corresponding burgeoning of the soil micro-organisms.

THE COMMONER CHEMICAL ELEMENTS OF THE EARTH'S CRUST

Element	*Weight %*	*Atoms %*	*Volume %*
Oxygen: **O**	46.60	62.55	93.77
Silicon: **Si**	27.72	21.22	0.86
Aluminium: **Al**	8.13	6.47	0.47
Iron: **Fe**	5.00	1.92	0.43
Calcium: **Ca**	3.63	1.94	1.03
Magnesium: **Mg**	2.09	1.84	0.29
Sodium: **Na**	2.83	2.64	1.32
Potassium: **K**	2.59	1.42	1.83

After Paton, Humphreys & Mitchell[73]

Part Three

THE LIVING SOIL:

An Organ of the Biosphere

The physical processes of soil formation

Two major groups of physical processes are involved in pedogenesis, the formation of soil.[73]

The first involves the reaction and adjustment of minerals that had been formed deep within the Earth to near-surface conditions. These processes—**epimorphism**—largely concern the silicates, all of which were formed at depth, under conditions of high temperature and pressure, with low oxygen and water concentrations. At the surface, in the presence of abundant oxygen and water, low temperature and pressure, these minerals would be expected to change to others more in equilibrium with surface conditions.

The epimorphic changes involve:

- **Weathering**—with breakdown of primary minerals;
- **Leaching**—which is controlled by the differential mobility of breakdown products; and
- **New mineral formation**—which is determined by interaction between weathering and leaching.

In addition, **inheritance** (assessment of the contribution of bedrock to soil material) has to be taken into account. A soil type can be directly influenced by the bedrock. A clearly demonstrated case exists on the Darling Downs of Queensland, where red soils rich in kaolin and hematite cover the flat, well-drained hilltops; on the gentle slopes of the adjacent pediments are the fertile black soils that make this a prime cropping region. Hard, fresh basalt underlies the hilltop soils; easily eroded, hydrothermally altered basalt flows underlie the slopes.

The second group of processes involved in soil formation comprises processes that depend on the tendency of surface topography to be reduced by gravity, involving erosion by water and wind (and sometimes ice). This causes lateral movement of material and involves **detachment**, **transport**, **sorting and deposition**, either particulate or massive. These processes are profoundly affected by plants and animals, with fauna mainly causing breakdown while the role of flora is mostly protective.

THE WEATHERING OF SILICATES[73]

Weathering of silicates releases a range of cations—potassium, sodium, calcium, magnesium, iron, aluminium and silicon—together with silicate framework fragments. Rates of weathering are determined by the different degrees of mineral stability, and by environmental conditions at the mineral surface.

Weathering of silicates is mainly due to the process of **hydrolysis**. Hydrogen ions in solution, in contact with the minerals, are responsible for the breakdown. The main source of these ions is carbonic acid produced from the reaction between CO_2 and water. The acid ionises to form bicarbonate and hydrogen ions. The small size of hydrogen ions makes their entry into the crystal lattice of the silicate rock easy and results in the replacement of the larger cations, which will then be subject to leaching. Different elements have different susceptibility to hydrolysis—potassium is easy to hydrolyse, aluminium is much more difficult under average conditions—but susceptibility changes depending on the acidity or alkalinity (pH) prevailing at the time.

Silicates are of four main types with different grades of resistance to hydrolysis:

- **Mafic minerals**, which contain large amounts of magnesium and ferrous iron which are easily hydrolysed. Their removal leads to complete structural collapse. One group, the *orthosilicates*, contain iron and magnesium. The *inosilicates*—pyroxenes and amphiboles—contain iron, magnesium, calcium and aluminium.
- **Micas** or *phyllosilicates* have a sheet-like habit, containing iron, magnesium, aluminium and potassium. The interlayer potassium of micas is easily removed, but the rest is much more stable under normal conditions.
- **Feldspars** or *tektosilicates* contain aluminium, calcium, sodium and potassium. Hydrolysis removes the potassium, sodium and calcium, leaving a much weakened framework.
- **Quartz**, a *tektosilicate*, contains no included elements. Very little happens to quartz. Its dense structure excludes hydrogen ions.

Mineral stability is influenced by conditions like the degree of acidity or alkalinity of the water in contact with the mineral, referred to as the **pH** (pH 1 is highly acidic; pH 14 is highly alkaline; pH 7 is neutral). Aluminium and silica are both far more soluble at a pH higher than 9. More leaching under such conditions leads to a greater degree of new mineral formation.

The breakdown of organic material can increase the CO_2 concentration at the interface between Biosphere and lithosphere by as much as 10 times, affecting pH in the acid direction. As this occurs where water is absorbed into the mineral surface, hydrogen ions are available in quantity to accelerate hydrolysis. Around living plant roots the concentration of CO_2 is even greater—a pH of 2 to 4 is maintained as long as metabolism continues, and aluminium becomes readily soluble at this level.

A visible example of the ability of organisms to produce acidic solutions that promote weathering occurs in lichens, which can be powerful rock weatherers. Hollowing of the rock surface to which they are attached is sometimes clearly seen. Lichens excrete oxalic acid as well as a variety of weak phenolic acids (lichen acids).

The degree to which the products of weathering are mobile depends on their stability in the watery environment into which they are released. Some recombine to form new minerals. Clay minerals, now known as **physils** because of their phyllosilicate (platy) structure, result from such recombinations.

The differential absorption of the products of weathering by plants affects the amounts of minerals available for recombination and leaching. Many plants, for instance, accumulate aluminium. Australia's prickly ash (*Orites excelsa*) secretes aluminium succinate in cavities in its tissues; when the wood is burned its ash is 75 per cent aluminium oxide. Many plants, particularly Monocotyledons, accumulate silicon as phytoliths (stony lumps) inside their cells.

TRANSPORT, SORTING AND DEPOSITION OF SEDIMENTS

Rainwash is the most common process responsible for detachment, transport, sorting and re-deposition of near-surface materials.[73] Vegetation is a deterrent to the process, while clearing, over-grazing and fire create bare surfaces open to raindrop splash and overland flow. Even shallow rainwash flow carries some material.

Materials are moved by rainwash as a **floating load** of organic debris plus charcoal fragments; as **bedload** of sand-size, mainly quartz, particles; and as **suspended load** that moves minor amounts of silt and clay-size particles faster and further.

Lines of wash debris (microterraces) after rain at Scotia Sanctuary in western New South Wales. *M.E.W*

Aerial view of mulga groving in Western Australia. *Dean Graetz*

Litter dams and microterraces result from this sorting and overland flow. They are most visible on surfaces after fire when there is a lot of debris to be spread. Like small contour ridges, they reduce sediment removal from hillslopes. Microterraces concentrate soil moisture, allowing filamentous algae and cyanobacteria to colonise, and trap seeds. The groving of mulga, often visible on aerial photographs, and other vegetation arcs like those seen in chenopod shrublands, result from rainwash-formed contour dams that form successively down-slope, even when slopes are so gentle that the landscape appears almost flat. Litter dams can persist as prominent features on slopes for many years, as in the case of some resulting from major fires around Sydney in 1994 that were still there in 2001.

On a larger landscape scale, transverse bedforms on low-angle alluvial fans like those of the Ross River in the Northern Territory result from high-magnitude sheetflow events.[84]

Mesofauna, organisms larger than micro-organisms—ranging from invertebrates to small mammals—affect the structure and the nature of topsoils by their activities, known collectively as **bioturbation**. They make the topsoil more porous, break up and distribute organic matter, bring material from the deeper parts of the profile towards the surface, add nutrients, and balance the activities of micro-organisms that are performing micro-bioturbations.

THE PINNACLE DESERT OF WESTERN AUSTRALIA[83]

Columns of limestone rise from a sandy platform in the Pinnacle Desert of Western Australia, adjacent to the coast between Lancelin and Cervantes. *Ken McNamara*

The extraordinary limestone formations known as the Pinnacles stand in a sandy desert in the Nambung National Park, about 180 kilometres north of Perth and a few kilometres inland from the coast. Thousands of huge pillars, up to five metres high, rise out of the sand. The limestone from which they are made is residual Tamala Limestone, a unit that extends from Shark Bay in the north almost continuously to the southern tip of the state, and then sporadically eastwards along the southern coastline towards Albany. The limestone is composed of wind-transported calcareous sand. It is an *aeolian calcarenite*, to give it its geological name, and its calcareous sand grains are cemented together with secondary calcite derived from the *dissolution* of sand grains. The sand grains are fragments of marine invertebrate organisms, mainly shells of molluscs, bryozoa, foraminiferids and calcareous algae, that have been broken up by wave action, mixed with quartz sand grains. The calcium-rich sand originated offshore and was brought ashore by wave action, then blown by the wind into extensive dunes, some rising 300 metres above sea level.

(Pleistocene climate change is implicated in the formation of the Tamala Limestone. The falling of sea levels during glacial stages of the ice age; the extraordinary dryness and windiness of the last glacial stage when sea level was 150 metres lower than it is today, and continental shelves and their sand deposits were exposed up to 40 kilometres beyond the present coastline; the rising of sea levels in the interglacials, and the fluctuations before it stabilised at its present level, can all be read in the dunes of the coastal plain. The creation of the dunes of marine sand far inland is testimony to the extraordinary dryness and windiness when conditions were at their most extreme at the last glacial maximum.)

Conversion of the dunes into limestone has been occurring over tens of thousands of years, with different degrees of cementation resulting from seepage of rainwater through the porous sand. Rain is slightly acidic and dissolves small amounts of calcium carbonate on its way through. The Mediterranean-type climate of the region, with winter rainfall and hot dry summers, facilitates lithification. The dissolved calcium carbonate is precipitated as a cement around sand grains at lower levels when dunes dry in summer, eventually immobilising the dune. Vegetation then grows on the dunes, further stabilising them, and the organic acids produced by roots and by decay of organic matter in the soil accelerate the leaching process. Most of the Tamala Limestone is porous and not heavily lithified, allowing water percolation, but a hard surface capping of calcrete has occurred, forming caprock. Where tree roots have penetrated deep into the dunes, seeking the watertable, they have resulted in columns of strongly cemented limestone because their acidification of their surroundings has resulted in locally increased dissolution and re-cementing. Chemical weathering, sub-surface, over time has resulted in removal of softer parts of the limestone, leaving the quartz sand component and the hardened pillars. Subsequent wind erosion of the region (deflation) has exposed the pillars, resulting in today's amazing landscape.

On a smaller scale, the localised cementing around roots is seen in the rhizoliths that are now scattered on the sand surface around the pillars. Root systems of plants that grew in fossil soils became encased in the limestone that resulted from the changed pH that their activities created; the resistant 'fossil roots' are the casings from which the root tissue has long since gone.

A yabby tunnel in a dried-up creek bed at Combo Waterhole. The yabby has retreated deep underground to find damp soil in which it can survive during the dry season. *M.E.W.*

COMBINED EFFECTS OF RAINWASH AND BIOTURBATION ON TOPSOIL FORMATION ON HILLSLOPES

Nearly all material deposited on the surface by bioturbation comes from the topsoil and is finer-grained than the topsoil. In rainwash, the fine fraction is preferentially removed in suspension and transported out of the system by rivers and streams. This leaves the coarser fraction, dominantly quartz sand and litter, to move slowly and spasmodically downhill. From time to time, the material is re-incorporated into the topsoil, and eventually, as a result of further bioturbation, it reappears on the surface to be subjected to another winnowing cycle. Repeated cycles result in the effects of winnowing penetrating the soil to the full depth of maximum bioturbation—the whole of the topsoil. The topsoil thus becomes coarser in texture and better sorted as it descends the hillslope over the subjacent material—not *en masse*, but grain by grain—so leading to a more marked differentiation between topsoil and subsoil. The process is maintained by the mesofauna mining the top of the subsoil, which provides a continuous source of fines and causes the topsoil boundary to sink into the landscape.

Clasts (fragments) larger than the maximum size that can be manipulated by fauna move down to the limit of bioturbation and form a **stone layer**. Another factor in the formation of the stone layer is the downhill movement of topsoil.

The formation of soil on hillslopes thus involves:

- Weathering of bedrock to form saprolite;
- Differential mining of the saprolite by fauna to form a topsoil; and
- Further sorting of topsoil and its movement downslope by rainwash.

This results in the formation of a **texture-contrast soil**, which consists of a mobile biomantle dominated by residual quartz, and often with a stone layer at its base, over a subsoil saprolite.

In situations where material produced by epimorphism or directly inherited from bedrock is **uniformly fine-grained**, bioturbation and near-surface processes are still operating so that a transgressive topsoil, differentiated from the saprolite not by texture but by fabric, is produced. **Fabric-contrast soils** have different arrangement of voids—mesofauna and plant roots penetrate the topsoil but do not penetrate the subsoil.

A TEXTURE-CONTRAST SOIL PROFILE

This brown sodosol from Saraji, north-east Queensland, shows a clear and abrupt change between the lighter textured, strongly bleached A horizon and the heavy clay B horizon (which in a sodosol is sodic and not strongly acid). These soils are widespread in all states except the Northern Territory and eastern and northern Western Australia. Usually of low fertility, they are used for cereal growing in winter rainfall areas of Southern Australia. © *CSIRO Land and Water. Neil McKenzie & David Jacquier*

RECOGNITION AND DESIGNATION OF PRINCIPAL SOIL HORIZONS

	Organic horizons	O1	Form of vegetable matter recognisable
		O2	Partly decomposed vegetable matter
▲ T H E	Horizons of maximum biological activity or eluviation, or both	A1	Near-surface mineral horizon, organic matter associated with mineral fraction.
		A2	Less organic matter, Fe, Al, clay; more quartz etc.
S		A3	Transitional between A and B
O L U M ▼	Horizons of accumulation or maximum development of pedological organisation	B1	Transitional: More like B
		B2	Concentration of clay etc. or different structure or colour
		B3	Transitional zone
		C	Consolidated or unconsolidated rock, little altered. Saprolite.
	LITHOSPHERE		

The diagram of a generalised soil profile, showing the different horizons that may occur and their naming (A1–A3, B1–B3, etc.), is the basis for the complex classifications that mainly apply to European soils. Australian soils have a very different history, and new classifications have placed less emphasis on profiles.

Generally, Australian soils have weak development of O horizons. Over the vast areas where tussock grasses predominate, in particular, very little litter accumulates on the surface. Low organic matter content characterises the soils of the arid and semi-arid regions that represent 75 per cent of the continent. This factor, and the low levels of biological activity, result in weak structural development. Many soils develop hard-setting crusts because of low levels of mesofaunal activity and bioturbation.

THE O HORIZON: THE LITTER LAYER(S)

The extent and nature of the litter layer covering a topsoil depends to a great extent on the biomass of aerial parts of the local ecosystem. It varies with climate, latitude, soil exposure, soil fertility, moisture, stage of development of the ecosystem and cultural treatments. Where they are well developed, litter layers are a pool for energy and nutrients. Changes in land use change litter effects.

In Australia, with its poor soils, climatic factors do not favour high productivity over much of the continent because of low effective rainfall. (In more than 75 per cent of the continent, annual evaporation is more than 2000 millimetres and rainfall less than 300 millimetres.) Therefore any nutrient pool must be protected from destruction by inappropriate management. Nutrient cycling by the litter layer and soil organisms, above and below ground, is enormously important. It is essential for sustainable land management that the processes involved in conversion of surface litter to soil organic matter are understood and the requirements of the biota are met. How fire affects the litter layer nutrient pool in different ecosystems has also to be understood. The potential for increasing greenhouse gases by burning of the litter layer is little considered.

While very little research has been carried out in Australia on the whole subject of litter layers and the processes of nutrient cycling within them, some basic information is available, mainly concerning woodlands.

Surface litter is important in conditioning the behaviour of other components of an ecosystem.[78] It creates particulate matter, modifies temperature and moisture regimes, and provides physical protection of soils. It becomes a substrate for a diverse biota. In the upper parts of the layer a range of different invertebrate organisms is involved, largely in breaking the plant material up into small bits (particulation). Specialised organisms break down the dung and carrion of vertebrate species. The presence of invertebrate carrion leads to concentration of chitinous residue in some litter, since the wings and hard exoskeletons of invertebrates are resistant to decay. (Birds that eat insects bring up pellets of this indigestible chitinous material from their crops rather than have it pass through their systems.)

This grasshopper, *Desmoptera truncatipennis*, is one of the very few grasshoppers that actually lives in rainforest. It spends the daylight hours in leaf litter where it is perfectly concealed, and moves to small shrubs at night to feed and mate. *D. Rentz © CSIRO Entomology*

Residual fine plant particulate matter increases stored soil carbon; sequestering carbon in this way is a positive contribution to curbing greenhouse warming. Decomposition, by bacteria and other micro-organisms, particularly fungi, occurs in lower litter layers in touch with the damp soil and in the surface layers of topsoil.

Interesting research carried out recently in Tasmania tests the widely held view that a change from native forest to plantation poses a threat to invertebrate litter-layer biodiversity.[236] Surprisingly, the plantations in north-west Tasmania were found to be far from 'biological deserts'; even the rare species of velvet worm (*Peripatus* sp.), a focus of the research because of its endangered status, was found to be able to tolerate replacement of native forest with plantations. Some rare snails fared less well and invasion of feral species was seen as a problem, potentially altering species composition in time. Several plantation practices appeared to have encouraged rapid re-invasion and re-population by invertebrates.

These include windrowing, where massive localised accumulations of rotting logs provide shelter; mound-ploughing, where long moist furrows accumulate leaf and twig litter in eucalypt plantations; pruning, which generates quantities of rotting wood in pine plantations; and thinning, which adds rotting logs, often with persistent bark that is a good shelter for invertebrates.

Where plantations were developed on cleared farmland, the local status of invertebrate populations improved, particularly where the plantations were close to remnant patches of native vegetation.

The ability to recolonise litter and the relative independence of the invertebrate litter fauna in its reactions to change is explained by the extraordinary antiquity of the litter-eating assemblage.[237] Ancestral springtails, mites and millipedes were active in the early land-plant associations of 400 million years ago. Litter invertebrates are influenced by the carbon pool in the litter layer, not by its specific source, and the composition of the litter fauna varies according to the sort of rotting material available to feed on. The rule for conserving litter invertebrates is 'keep it shady and moist'—that's why there are more of them on the south sides of hills. Rotting logs are important as shelter and nurseries. **Frequent fuel-reduction burning is the worst possible forest management for litter invertebrates.** While the less ancient winged invertebrates, which became abundant as land plant evolution led to forests and ultimately to flowering plants, are able to seek new abodes when fire destroys their food source, litter invertebrates are not as mobile.

Amounts of surface litter are difficult to assess. In tropical north Queensland, litter on a forest floor was estimated to be 7.4 tonnes per hectare in a wet site and 11.1 tonnes per hectare in a drier site. In a mature forest of *Araucaria cunninghamii*, the amount of litter was estimated to be between 11 and 22 tonnes per hectare. In Brigalow scrub, with its dense growth patterns and relative paucity of termites, litter stores and locks up a great deal of carbon. An estimated 75.5 tonnes of surface litter and a further 58 tonnes per hectare of still-standing dead phytomass in a brigalow scrub gives some idea of the amounts involved. Conversely, the amount of CO_2 liberated in the burning of cleared brigalow has to be significant in terms of greenhouse warming, even taking into account the fraction that becomes inert carbon and enters the soil as charcoal.

In eucalypt woodland a dense layer of tough leaves collects on the ground, and 'haloes' of annually shed bark surround the trees. Because of the fires that generally occur at intervals of less than ten years, over much of New South Wales, for instance, equilibrium is seldom reached. The decomposition of eucalypt leaves that follows particulation is largely carried out by a series of fungi that produce special enzymes. Micro-arthropods active in the litter include collembola that consume a complete range of organic matter, including the faeces of larger arthropods; and free-living nematodes feeding on micro-organisms and providing a food source for all the other saprovores (feeders on dead plant material) that belong to the Isoptera, Symphyla, Diplopoda, Diptera, etc. (Saprovores have a low assimilation efficiency—the amount of energy assimilated is low compared with the energy available in the food ingested[78]—but they have a high production efficiency. Of the energy assimilated, a high proportion, 11–17 per cent, is converted into new tissue. By comparison, herbivores convert only 2–7 per cent of their energy intake into new tissue, and carnivores only 2 per cent.)

A halo of shed bark surrounds the base of the trunk of an *Angophora costata* in urban bushland in Sydney. *M.E.W*

In more arid sites, the density of micro-arthropods increases only slightly from arid grassland to low shrubland, collembola and mites in equal numbers usually being found in these areas. In less arid areas there is greater variety of organisms. Nematode numbers are always positively related to a soil's water content. A significant increase in numbers of saprovores from semi-arid mallee and sclerophyll mallee is noted as the mean annual rainfall increases from less than 500 millimetres to greater than 500 millimetres, and the plant litter layer changes from being discontinuous to becoming more or less continuous.

Cool fires that burn litter layers may result in an ash-bed effect, where seedling growth is enhanced. Intense burning may leave the mineral soil bare and open to erosion; mineral leaching and loss of nutrients in run-off often follows. Much ash is lost in smoke, but this particulate material is probably redistributed in rain. Reduction in biological activity is known to follow fire (measurable soil respiration falls) and breakdown of residual litter is reduced. Numbers of micro- and meso-organisms, and their diversity, are reduced. Little is known about the long-term effects of regular (and prescribed) burning in native forests and woodland.

THE ROLE OF WIND IN THE TRANSPORT OF SEDIMENT

While water is the main transporter of sediment, and its downslope movement by gravity continues its distribution, wind may also play a part in its transportations, particularly in deserts and arid zones.

Wind distribution—**aeolian processes**—can only operate within limits set by terrain and sediment type.[73] Suitable particle size is the first prerequisite. In general, it is unusual for particles coarser than sand size (more than 2 millimetres in diameter) to be moved by wind across rough surfaces. But medium size and sand size particles can be transported over several kilometres by **saltation** to form sand sheets and cover sands. (Saltation is the 'dancing' or bouncing of particles across a surface, driven by the wind.) Clay dunes on the downwind side of seasonally dry lakes in Australia are made from blown pellets of clay-sized particles that have been compacted together and saltated to their places of accumulation. Gypsum dunes on the downwind margins of Lake Eyre and other large salt lakes in Australia's centre were produced by saltation of lake-bottom deposits during glacial stages of the Pleistocene ice age.

Really fine soil particles can be transported as dust. In the Lake Eyre Basin, fine-grained sediments provide the basis for the dust storms that characterise the region. The 'bedouries' of the Channel country, where the clay-rich Mesozoic sediments laid down in the Great Artesian Basin are being eroded, take their name from the remote township of Bedourie where dust storms are frequent.

Dust has made, and still contributes to, the soil mantle along the western slopes of the Great Divide where it is recognised as *parna*. Much of the dust from inland Australia moves offshore and accumulates in Tasman Sea sediments; some even reaches New Zealand, where it can be seen as a red tracer in the snowfields in years when Australia is particularly dry and dusty.

AUSTRALIAN SOILS: ORIGINS AND CLASSIFICATION

Australia is an ancient, stable landmass that has been weathering and eroding through hundreds of millions of years, becoming ever flatter. Its soils are mainly made from deeply weathered rocks and form on the accumulating bodies of sorted sediments in sand sheets, dune fields and riverine plains. There has been no renewal of the soils by the passage of ice sheets across the landscape in the last 300 million years, and little volcanism has occurred to produce fresh rock and fertile soil patches in the uniformly impoverished terrain.

To anyone who is not a soil scientist, the classification of soils as presented in textbooks and scientific papers, or even in the literature that is supposed to be for the general information of farmers and others who need to know, is nightmare stuff needing non-existent dictionaries and glossaries. So much jargon, so much assumed background knowledge, and such an example of unmanageable complexity.

How lucky it is for Australians wanting to make sense of it all that three eminent soil scientists from Sydney's Macquarie University have produced *Soils: A New Global View*[73], in which they have been able to show that what was globally accepted practice in soil classification does not work for Australian soils. It was simply another example of systems developed for the new soils that resulted from glaciations of Northern Hemisphere lands during the Pleistocene ice age being unquestioningly applied to a continent that had had no such soil renewal for 300 million years. Those generally accepted European and American based classifications were designed for new soils that had features strongly governed by their inheritance—and a different set of inheritance factors governs Australian soils.

(This sort of transfer of expectations and practices from Northern Hemisphere lands to a continent whose geological

Dunes on the Strzelecki floodplain in the Lake Eyre Basin were blown into their present form by the strong winds that accompanied the extreme dryness of the last glacial stage of the Pleistocene ice age. *M.E.W.*

The anabranching of the Diamantina River as it crosses the floodplains on its way to Lake Eyre is a consequence of the floodplains' extreme flatness and their cohesive clay-rich sediments. *Gerald Nanson*

history, particularly its comparatively recent Pleistocene history, is completely different, has been at the root of the unsustainability of Australia's agricultural and land use practices. Such practices were, and still largely are, imposed without question on a continent whose nature and resources could not be more different. In addition, the unpredictable climate, with its extremes of droughts and floods orchestrated by ENSO, the El Niño–Southern Oscillation climate warper, has compounded management problems. Settlement by people indoctrinated in Northern Hemisphere practices has led to a misreading of the capability of the continent to sustain such practices without degradation and desertification resulting. The parameters for sustainability have been set by the nature of the continent itself, as has been emphasised in my books *Listen ... Our Land is Crying* and *Running Down*. A reassessment of the Australian background for classification of its soils was long overdue.)

The whole Australian continent is positioned on a tectonic plate centre.[73] Its tensional margin lies far to the south in the Southern Ocean; its compressive margin lies north and east of New Guinea. Its bedrock comprises a great deal of granite or granite-derived sedimentary rocks (sandstones, shales and low-grade metamorphics). It has been a stable landmass through geological time, eroding to become the continent with least relief. It has suffered little volcanism, so areas of volcanic soil are very limited.

Being old and stable, there has been time for the full development of epimorphic and near-surface processes, and this has led in the main to the **texture-contrast soils** that are widespread across the continent, usually associated with middle and lower hillslopes and valley margins, but with many also found in valley-fill deposits.

Many different names were given to the different soil varieties by earlier scientists. The older names relate to texture-contrast soils with different thickness and degree of bleaching of the topsoil, different colours and fabrics in the subsoil, and a number of other variants. The categories are confusing and non-exclusive, while the acceptance of foreign terms that might or might not adequately describe a particular profile added to confusion. The hydrological properties of these soils are often complex, while low rainfall over much of the continent has led to the accumulation of massive amounts of salt in subsoils and underground water systems.

The role of living organisms in the development of topsoils, previously discussed, has also been a completely neglected area in the classification and description of Australian soils.

Great areas of Australia are covered by **siliceous and earthy**

sands. These represent the accumulating endpoint of granite pedogenesis that has been in operation for hundreds of millions of years, dating back to Pangaea and Gondwana, when Australia was part of the supercontinents. In terms of soil water, these areas have rapid infiltration, low water-holding capacity, deep drainage and very little run-off.

Apart from the shallow or deep texture-contrast soils, and the residual quartz sands that are widespread on the continent, **clay soils** form a class of epimorphic product that have both agricultural and regional significance. These are of three kinds, related to different bedrock types.

- **Clays from basaltic lavas** along the Great Divide. Two main types of clay soils have been formed from the lava flows, depending on the degree of breakdown. In one, the minerals have been affected by strong weathering and leaching and the end-products consist mainly of kaolin, and oxides and hydroxides of iron and aluminium, red, yellow and brown in colour. In the other type, that has undergone less weathering, the clay soil is smectite-rich with coarse aggregates, and is highly plastic, expanding and contracting with wetting and drying. Some of these cracking clays develop gilgais (see box).
- **Clays from fine-grained Cretaceous sediments** deposited during the marine invasion of the Great Artesian Basin. These are smectite-rich, coarse-structured grey and brown clays, frequently associated with carbonates. Their characteristics are due to direct inheritance from their bedrock. The nature of the floodplains of the Channel country, and the anabranching of its rivers, are due to these clays, as are Channel country dust storms.
- **Clays from the deposition of fines**, carried as suspended load in rainwash and floodwater and deposited on floodplains by rivers that drain regions where texture-contrast soils are developed. Such clays are particularly widespread in the Riverina, the Darling riverine plain and the floodplains of the Gulf of Carpentaria.

GILGAIS[78]

Gilgais and mukkara, commonly called 'crabholes' and 'melonholes', are features of a range of soils that are either predominantly swelling and cracking clays, or texture-contrast soils with thick subsoil clay horizons. They are most commonly found in deep bodies of clay—5 metres or more is typical—which makes them hard to study. They occur on gently sloping plains with a wet-dry climate regime, or are subject to more frequent, intense drying. They occur as mounds with closed depressions inside, and sometimes a sinkhole; or as banks with open-ended troughs between linear features on clay plains. They hold water after rain and result in lack of uniformity in conditions for cultivation and cropping. Where they are strongly developed, as in parts of the Brigalow region in southern Queensland that have been cleared, they can make cultivation and grain cropping almost impossible.

While some controversy still exists about their formation, the generally accepted view is that when the clay that characterises the soils where they occur dries, the deep cracks become partially blocked with the surface material that accumulates in them. On re-wetting, swelling results in increased pressure because of the blockages, and subsoil is displaced. As mounds or ridges develop, the depressions deepen further. (An alternative view, or perhaps an additional factor in development, involves differential wetting because of the cracks and also because of plant roots making pathways for water penetration, thus amplifying the effects of partial blocking; the exchangeable sodium percentage or ESP of the clay determining optimum gilgai formation.)

Gilgais in the Brigalow: Ridges alternating with clay-rich depressions make agricultural use of the land difficult. © *CSIRO Land & Water. Neil McKenzie and David Jacquier*

BASIC CLASSIFICATIONS OF SOILS

A more precise soil classification may be necessary for the purposes of correlation, but this is beyond the scope of this book. In addition to Paton, Humphreys and Mitchell's *Soils: A New Global View*[3], already referred to, two CSIRO texts are of value. The CSIRO Division of Soils published an enormous tome in 1993—*Soils: An Australian Viewpoint.*[78] Much simplified by comparison with the Northern Hemisphere classifications that mostly refer to new soils produced by glaciation in the Pleistocene ice age, this classification acknowledges the Australian continent's completely different geological history, particularly that of the Pleistocene.

Concepts and Rationale of the Australian Soil Classification[85], a small handbook that modifies the current American soil classification system, was published by CSIRO Land and Water in 1997. This book aims to establish the main Orders of soils, similar in status to Orders in animal and plant classification. (Unfortunately, soils are not plants or animals, and do not have evolutionary relationships that could form a solid basis for classification. Neither do they relate well to global-scale climatic or biospheric environments.)

Pictures of profiles of the most widespread soil types, supplied by CSIRO Land & Water, are reproduced in this section with permission.

A CALCAROSOL
Calcarosols are soils with calcium carbonate, which occurs as soft or hard white fragments or as a solid layer. They occur on low rainfall areas of arid and semi-arid Australia. Land use includes cereal growing in the south, sparse grazing in the north. Limitations include shallow depth, low water retention, wind erosion of sandier types, soil fertility deficiencies. Sometimes salinity, or alkalinity, or sodicity may be a problem. *Primary Industries and Resources South Australia.*

A CHROMOSOL
A common texture-contrast soil from Queensland. Chromosols show an abrupt increase in clay down the soil profile. They occur in most districts. Large areas in southern Australia have been cleared of vegetation. These soils are common in the cereal belt of southern New South Wales and Victoria. Chromosols in the tropics support cattle grazing on native pastures. Agriculture has caused degradation of large areas, with hard-setting crusting and impeded drainage. *© CSIRO Land & Water. Neil McKenzie and David Jacquier*

A VERTOSOL
Vertosols are clay soils that shrink and swell, cracking as the soil dries. Australia has the largest area and greatest variety of vertosols, including deep forms (more than 6 metres). The largest areas occur in the arid and semi-arid interior of the continent. Vertosols are used for grazing of native and improved pastures, extensive dryland agriculture where rainfall is sufficient, and irrigated agriculture. High sodium levels are present in the upper part of many profiles. *© CSIRO. Neil McKenzie and David Jacquier*

A FERROSOL
Ferrosols have high free iron and clay content. They occur along the eastern coastline, in northern Western Australia and in the Top End. In high rainfall zones they may be very deep and well drained. Land use includes dairying on improved pastures, horticultural crops, some plantation forestry and sugarcane in Queensland. The shallow and stony types in northern Australia support beef cattle grazing. Ferrosols may be degraded by erosion and compaction due to cropping practices and may also suffer from acidification. *© CSIRO. Neil McKenzie and David Jacquier*

A TENOSOL
Tenosols are weakly developed soils. In the eastern half of the continent they occur as vast red and yellow sandplains. Large areas of Western Australia have red loamy soils with red-brown hardpan at shallow depths. Poor water retention, low fertility, and their occurrence in regions with low and erratic rainfall, combine to mean that most land use is grazing of native pastures. Limited areas in better-watered parts of south-west Western Australia and in eastern Australia support forestry. *Agriculture Western Australia*

The living soil: The basis for terrestrial Life

'Soil is the Mother of all things', according to a Chinese saying.[86] The wisdom of the phrase is evident when one considers the importance of this layer of the living Biosphere's support of the photosynthetic organisms which are the basis for food chains for practically every other land-living thing on Earth. Soil is 'a living and dynamic entity that requires a unique balance between its physical, chemical and biological components in order to remain productive. Proper functioning of the soil is vital for the continued functioning of terrestrial ecosystems'.[87]

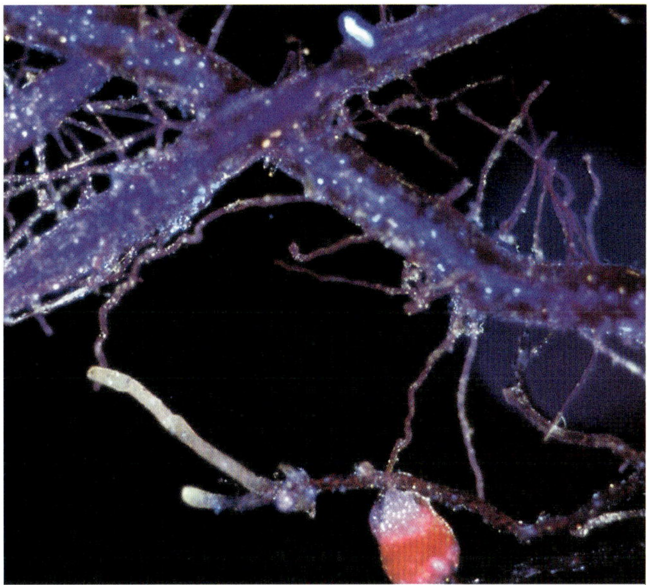

Bacteria on the surface of wheat roots appear as bright pinpricks of light due to their staining with a fluorescent dye; and a small photosynthetic alga glows red as a consequence of the autofluorescence of its chlorophyll. *Margaret McCully*

Of the six elements most important in sustaining terrestrial Life—carbon, hydrogen, oxygen, nitrogen, phosphorus and sulphur—carbon, hydrogen and oxygen are derived by organisms from air and water; phosphorus and sulphur from minerals in the soil, and nitrogen either from the atmosphere (by legumes and other nitrogen fixers) or from soil reserves (all other plants).

The living Biosphere extends below the soil, as we now know, deep into the lithosphere below the land and the living sea, and above into the atmosphere, but if soil were to die, Life on the land would die too.

Soils comprise the greatest number of biological, biochemical and biophysical phenomena of any slice of the planet. The complexity of processes and interactions that take place in soils, the predominantly sub-microscopic nature of the biota, and the mind-blowing number of unfathomable interconnections that exist as a result of 4 billion years of co-evolution of inanimate and animate components, are only now being unravelled. Technological advances in microscopy, chemical analysis, electro-chemistry and a host of other fields, are starting to explain some of the mysteries.

It is in the soil that the most intimate associations exist between organic and inorganic, between biotic and non-biotic. Myriad upon myriad of biochemical exchange processes are taking place, with electrical energy ultimately powering all the interchanges at an elemental and molecular level.

Soil organisms play an integral role in influencing soil formation, structure, fertility and plant growth. The soil biota regulates most of the soil processes that determine plant growth and ecosystem functions. Micro-organisms—bacteria, protists (protozoans and algae), fungi and viruses—occur in all environments and are interrelated with all other life forms.[2,88] They are extremely important in biocontrol mechanisms and in the transformation of food, energy and chemicals, including the mineralisation of nutrients. Here in the soil we see most clearly how it is 'little things that run the world' as E.O. Wilson described their role. Despite their importance, knowledge of their diversity and activities is limited. Only a small fraction are yet able to be cultured; new molecular techniques will enable many more to be identified in the future. It is absolutely essential that the biodiversity of this portion of the living biosphere is protected—it is basic to ecosystem function, even if we do not yet understand how the webs of life in this realm function. Although the contributions of micro-organisms to the functioning of ecological systems and the very existence of plant and animal life are now freely acknowledged, they remain largely overlooked in the debate on biological diversity and ignored in policy considerations.

Micro-organisms and microbial diversity are threatened by environmental factors and changes caused by human activities. Threats include: pollution by sewage, oil, organic compounds, pesticides and heavy metals; loss of habitat because of salinity changes, long-term or unnatural flooding or draining; over-exploitation of resources in agriculture, forestry and fisheries; introduction of competing or antagonistic species; and even local and global changes in rainfall and climate.

Micro-organisms begin the decomposition processes when organic matter enters the system. In the process of mineralisation they release nutrients and make them available for absorption by plants. They stabilise soils by helping the formation of stable aggregates—their activities produce mucins that bind grains together. They provide the basis of food chains of organisms of increasing size that continue and enhance the activities that improve soil fertility and structure.

The simplest generic definition of **soil structure** is the size, shape and arrangement of its particles.[82] This simultaneously defines the size, shape and arrangement of its pores, and the pores are the places where biological activity occurs. Continuity and connectivity of pores is important for aeration and for water relations, and therefore for root and fungal hyphae growth and for mobility of organisms such as protozoa and nematodes.

Roots use and enlarge existing soil pores and create new ones as they go down through the soil. The young wheat root in this picture lies within a biopore which contains dead roots from last year's crop. *Margaret McCully*

A root crossing pore space has long root hairs on the side where there is a large pore and short root hairs on the other where space is limited. *Michelle Watt*

Structure changes with a soil's changing water status—for instance, with shrinking on drying and swelling on wetting. Well-structured soils encourage biological activity.

As a result of the concentration of biological activity in the upper part of the soil profile, soil nutrients are also concentrated there; the loss of this layer by water and wind erosion removes or depletes nutrients. Although it is difficult to calculate such losses on a continental scale, some idea of the magnitude of the problem can be given by figures for the Indian subcontinent. Of India's 329 million hectares, 175 million are degraded, and 142 million hectares are used for agriculture. Annual soil loss by erosion is calculated to be 5334 million tonnes! How much nitrogen, phosphorus, carbon and trace elements are lost is anyone's guess.

Most soils contain a rich diversity of organisms larger than the microscopic forms that are the most numerous.[89] In fertile croplands the weight of micro-organisms may exceed 2000 kilograms per hectare; earthworms may number 500 to 1000 per square metre (even up to 2000) corresponding to a biomass of 500 to 1000 kilograms per hectare (up to 2 tonnes per hectare).

In a 1991 paper[89] it was reported that on red-brown earth in north-east Victoria, total weight of soil micro-organisms or microbial biomass from a continuous cropping site was measured as 295 kilograms per hectare, while earthworms numbered 275 per square metre, equivalent to 260 kilograms per hectare—bringing the total biomass to 555 kilograms per hectare. Stock grazing on the surface represented 400 kilograms per hectare for sheep and about 480 kilograms per hectare for young steers. Thus the weight of organisms below ground exceeded that of stock grazing above ground. In the light of new techniques for measuring microbial biomass it would seem that the figures are too low, and 2 tonnes per hectare would be a better estimate than 555 kilograms (Margaret Roper, personal communication). The fact that so few of the enormous number of different micro-organisms have been identified and studied, combined with the difficulties involved in sampling to estimate their presence and abundance, and the complexity of their largely unknown interactions, makes soil microbiology a frontier science in which just about everything has still to be learned[90,91], although great advances have been made in the last decade.

The chemistry of soil organic matter begins with plant and animal materials that are eventually converted to microbial biomass and products of microbial turnover. Microbial and other micro-organism activity is closely associated with the clay fraction and habitable pores that provide most of the surfaces available for interactions between the living and non living.

MICRO-ORGANISMS AND THE SOIL

The micro-organisms that occur in enormous numbers in soils are neither uniformly nor randomly scattered, but clumped near or within organic residues or in micropores that are accessible to soluble soil organics. Such sites must also have compatible pH levels, water potential and oxygen concentrations. Protozoans accumulate in soil regions rich in bacteria and fungi, such as beneath plant litter and in parts of rhizospheres where the root hairs and outer cells of roots are being shed.

The size of soil micropores influences micro-organism distribution. Pores of small enough size deter predators and provide safety for their residents in this ultra-microscopic world, where the natural laws of eat or be eaten apply just as they do in the macro-world above ground. The availability and nature of the fine clay fragment of soils is an important determinant of micro-organism abundance and prosperity. (When foreign bacteria are artificially introduced into soils as biofertilisers, it has been found that adding bentonite, a volcanic clay, improves their chances of survival.)

Microflora, including bacteria, actinomycetes and fungi, are the primary decomposers, and the microfauna—the protozoa (now classified as protists), nematodes and micro-arthropods—form a critical link between them and the larger fauna of the detritus food-web in the soil. The microfauna are the primary agents for the release of nutrients immobilised by the soil microflora.

The protozoa are unicellular organisms (flagellates, ciliates and amoebae) that feed on bacteria or fungi. Nematodes are tiny worms that feed on bacteria, fungi and protozoa, or are parasitic on plants. Both groups are water dependent, requiring water-filled pores or continuity of water films for their activity and movement. They are found in all soils in great numbers—more than a million per square metre—and show great diversity of taxa. They are involved in decomposition and turnover of organic matter, nutrient mineralisation, regulation of population densities of microflora (including some plant pathogens), and in decomposition of some agricultural chemicals. Their grazing on microflora keeps populations active and young, and helps to control pathogenic fungi.

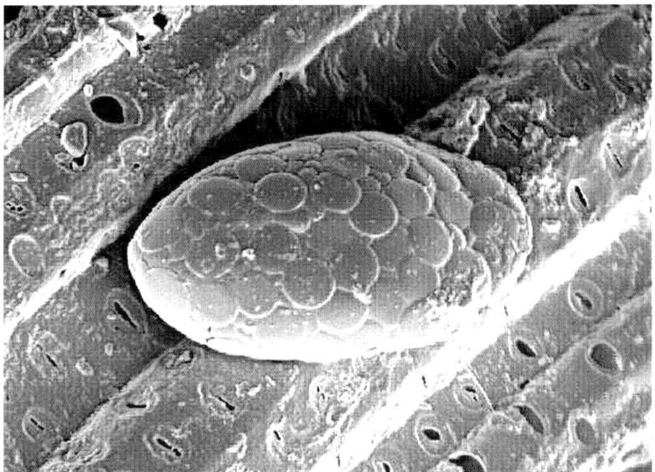

An amoeba with a patterned shell—a predatory protist. *Margaret McCully*

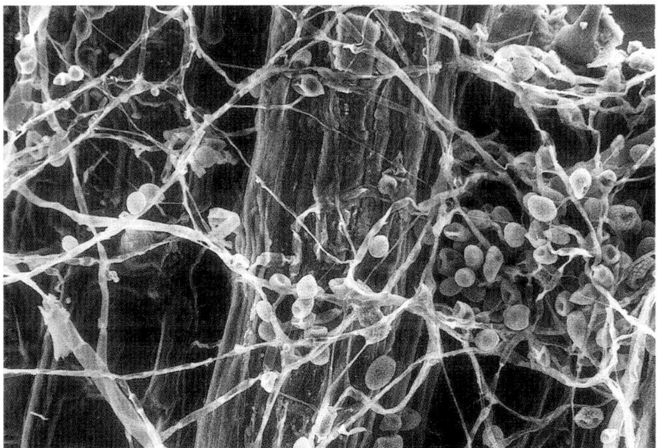

The surface of crop residues colonised by fungal and actinomycete hyphae and testate amoebae. VVSR Gupta et al. © *CSIRO Land & Water, Adelaide*

Soil micro-organisms are affected by pesticides, herbicides and fertilisers, and respond in different ways to agricultural practices. The swing to conservation tillage means that more herbicides are being used, and used more frequently.[92] Under ideal conditions, the perfect herbicide, properly used, would affect only the target organism. In the real world, herbicides may also affect non-target plants and an unknown number of micro-organisms, ranging from pathogens to the most beneficial. Increased disease severity can result, with greater pathogen damage; additionally, physiological changes can occur in the host plant when a pathogen is being targeted. These changes can include reduction in wax formation in the cuticle of leaves, changes in carbohydrate, nitrogen and glucoside metabolism, and growth retardation.

Pesticides are essential in modern agriculture. Australia spends $800 million a year on pest and weed control—65 per cent on herbicides, 35 per cent on insecticides.[93] The impact of these potential poisons on soil biota is a major concern, and much research is under way to establish such things as their long-term safety, persistence in the soil, how they affect the composition of soil biota populations, and how much leaches into waterways. Further than that, research is now aimed at using native Australian microbes to clean up contaminants in soil and groundwater. One such bioremediation system involves using microbes of a specific type to render atrazine, one of the most persistent and widely used herbicides, non-toxic in the soil—a process that is accomplished by the organisms in the laboratory in a matter of hours.

Micro-organisms and soil structure

Section of a root inside the rhizosheath showing aggregates attached to root hairs, and the pores between them. *Margaret McCully*

One of the most important functions of soil micro-organisms is their creation of micro-aggregates. By virtue of their production of mucigels or mucilages (extracellular slimes) that stick mineral and organic particles together, micro-organisms increase the stability and porosity of the soil. The polysaccharide glues

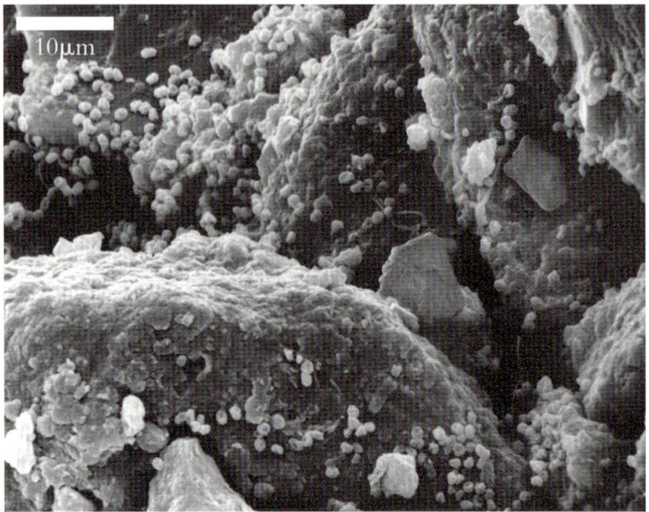

Extensive colonisation of bacteria and fungi on the surface, and in the small pores, of a soil aggregate. VVSR Gupta et al. © *CSIRO Land & Water, Adelaide*

A wheatfield after harvesting, stubble remaining and providing a valuable food source for soil micro-organisms. *Margaret Roper*

produced by bacteria attach them to mineral surfaces, while clay particles become stuck onto the sticky mucilages in processes involving calcium-ion bridges. Whole colonies of bacteria are often enclosed in sheaths of extracellular polysaccharides (ECP); such colonies often form the cores of micro-aggregates. Aggregate stability, and content of microbial ECP, are highly correlated. In micropores that are no longer accessible to micro-organisms, the ECP may persist for some time.

Plant roots and most fungal hyphae also produce these mucilagenous glues. In fungi, the gels are used to attach hyphae to mineral surfaces, resulting in aggregate formation, and colonies of gel-secreting bacteria are supported by the fungal gels, increasing the effectiveness of aggregation.

Microscopic soil fungi are abundant in pores between micro-aggregates in bulk soil. Branches of their mycelia occasionally penetrate into the micro-aggregates. This is achieved when hyphae are inserted between the various components of the micro-aggregates, and they subsequently dilate, swelling and pushing the soil minerals and organic matter aside. Macro-aggregates are often created when micro-aggregates are linked together by hyphae, especially by those of mycorrhizal fungi.

Zones of water repellance in soils always show fungal activity, the fungi in this case creating a hydrophobic organic film that coats sand grains. Water repellance is also caused by substances of plant origin (for example, glaucus from eucalypt leaves) that saturate sand-grain surfaces and render them hydrophobic.

In annual cropping systems, removal of the crop leads to a decrease in mycorrhizal fungi (although some survive in crop residues). Saprophytic fungi are not affected by crop removal. Aggregates must be recolonised by mycorrhizal fungi each year. Failure to recolonise may lead to loss of organic matter and the fine clay particles that are no longer stabilised by gels.

Extracellular enzymes produced by bacteria and fungi are important in the breakdown of plant residues in soils. Cellulose is easier to degrade than lignin, but a great deal of particulate matter that persists for very long periods in soils does so because the cellulose is protected by a coating of lignin. White rot fungi degrade cellulose through the action of the oxidative enzyme cellulase.

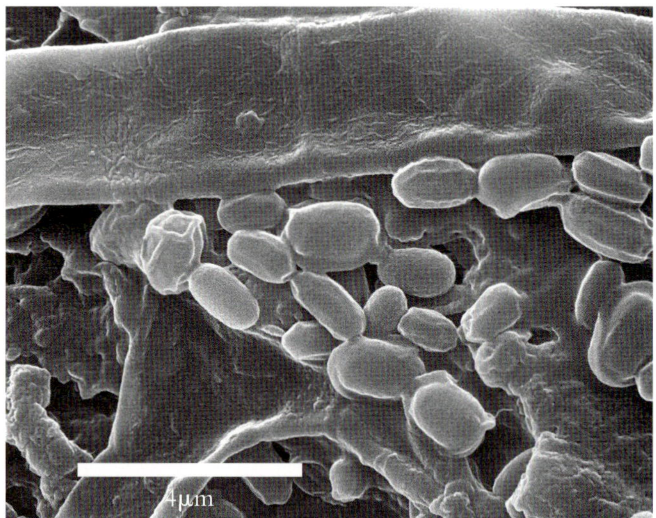

A large colony of soil bacteria growing on the surface of crop residue. *VVSR Gupta et al. © CSIRO Land & Water, Adelaide*

Agricultural practices and micro-organisms in southern Australian dryland farming systems[2,94,94a–i]

Cropping management practices have a significant impact on all soil organisms, both microscopic and macroscopic. When **stubble retention** is involved, the crop residues provide an energy source for growth and activity, and research has shown that the numbers of organisms of all sorts increase, including protozoans and invertebrates, apart from nitrogen-fixing bacteria. The higher the nitrogen levels in the stubble, the greater the increase in numbers, and levels of organic matter in the soil increase. (A huge amount of organic carbon as crop residues is produced annually in Australia—estimated at more than 15 million tonnes.)

DETRITUS FOOD WEBS UNDER TWO TILLAGE SYSTEMS

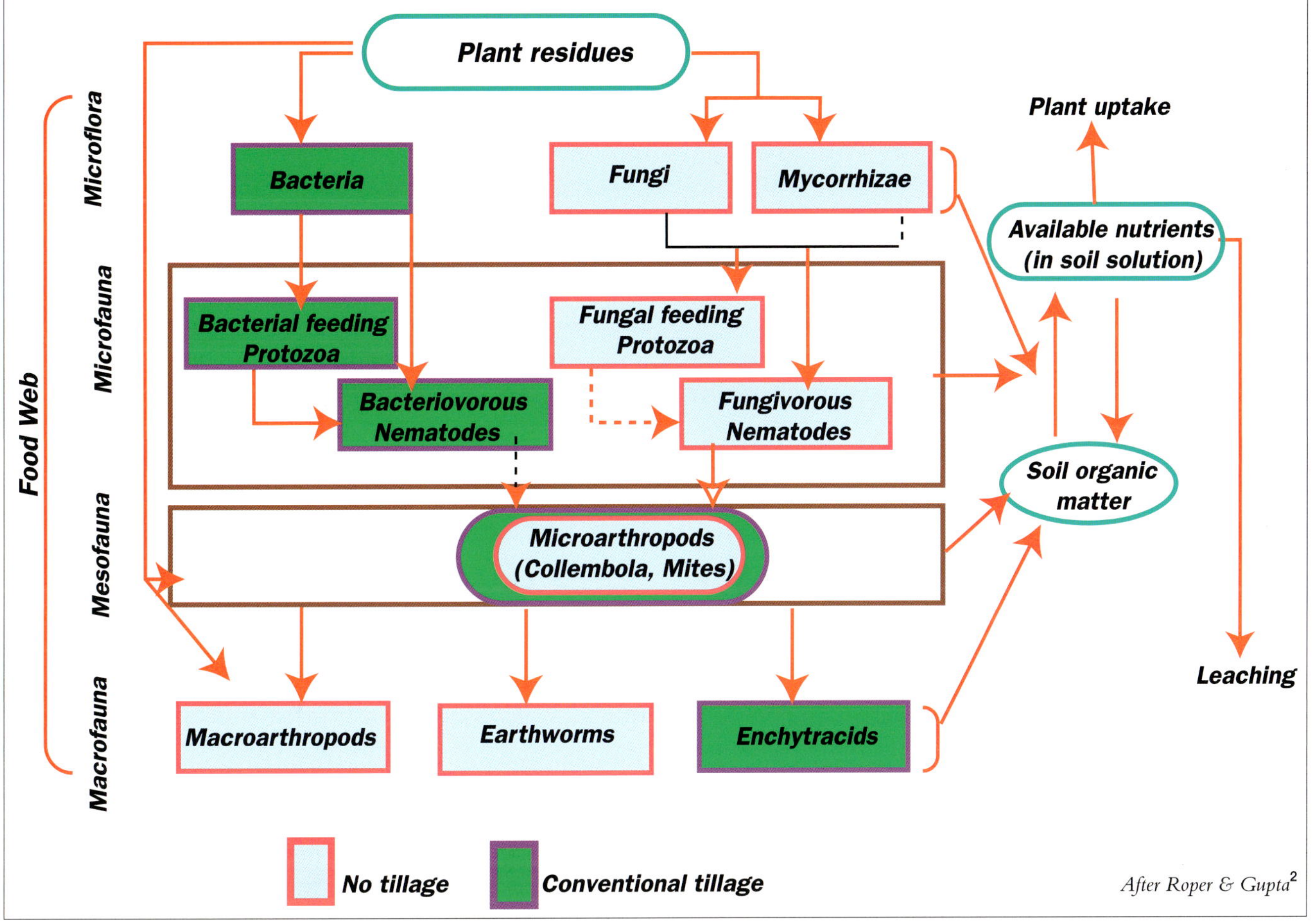

After Roper & Gupta[2]

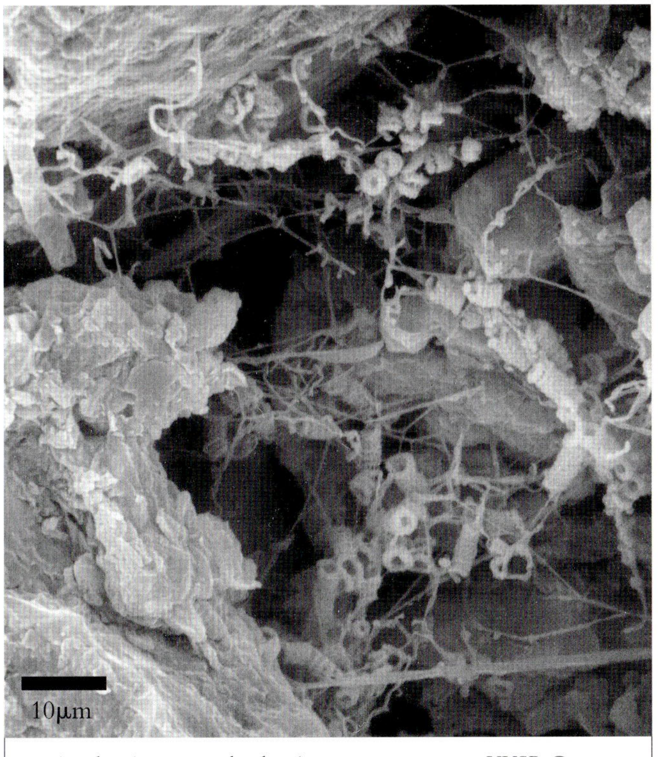

Fungi and actinomycete hyphae in an aggregate pore. *VVSR Gupta et al.* © *CSIRO Land & Water, Adelaide*

How the stubbles are managed may impact further on the soil biota and may alter the population balance. In general, no-tillage results in increased concentrations of organisms closer to the soil surface and less alteration to soil structure. Increases in fungi and bacteria that break down cellulose are, predictably, observed. Some fungal plant diseases tend to increase but generally the increased numbers of competing, predatory and parasitic organisms tend to maintain balance, and crop rotations can be used to break disease cycles. The use of herbicides in stubble retention systems is widespread, and continued use has been shown to depress some groups of micro-organisms. Their effect on soil bacteria is largely unknown.

The micro-, meso- and macro-fauna of nematodes, arthropods and earthworms that feed on micro-organisms and soil organic matter increase with stubble retention. Nematodes that feed on fungi became more numerous than those that feed on bacteria, in response to the increase in fungi.

Crop rotations that include legumes and/or pastures are reported to increase populations of micro-organisms, particularly in the rhizosphere, and of nitrifying and denitrifying bacteria. Rotations including pastures have been found to have greater number and diversity of collembolans compared with annual crop rotations.

Root diseases of crops

Three main root diseases are prevalent in cereal crops in Australia. Cereal cyst nematode (CCN), Take-All and *Rhizoctonia* are costing hundreds of millions of dollars in lost production and much research has been undertaken to control them. Crop rotations in which non-host species are used to break the cycle of infection; changed cultivation methods; and suppression of the pathogens by introducing micro-organisms, are all used.

Increasing awareness of the harmful effects of chemical fungicides and pesticides and desire for 'clean green' products is leading to greater emphasis on biocontrols of root diseases in which naturally occurring soil micro-organisms are cultivated for use as soil inoculants. Chemicals affect the beneficial soil organisms as well as the pathogens. In a number of instances it has been shown that herbicides have increased the damage from root diseases[101], and the increase in fungicide-resistant pathogens is making the search for biological alternatives more urgent. The persistence of herbicides in some soils presents a potential problem for increased herbicide–root disease interactions in sensitive crops. The more that is learned about the microbiology of soils, the more obvious it is that interactions of all sorts are occurring at that level, and that further interactions are occurring with plant roots. The whole picture is highly complex. The food chains in the soil are such that not only the micro-organisms but the small invertebrates are directly or indirectly affected by use of herbicides.

The filamentous fungi are a plentiful resource of control agents for plant pathogens.[101] Some are fungal parasites that directly attack the pathogens; others produce antibiotics; some secrete enzymes that damage the cell walls and tissues of harmful organisms without the fungus having to make physical contact; all act by competing for nutrients. The fungal genus *Trichoderma* has been a focus for research. It occurs widely in soils, including the root zone, and also on the plants' above-ground tissues. *Trichoderma* species have been shown to produce more than 70 inhibitory substances, including antibiotics. Progress in biotechnology now allows modification of different strains and at least 11 formulations are on the market for treatment of soil-borne pathogens. Trials with *T. koningii 7a*, a naturally occurring beneficial fungus isolated from West Australian soils by CSIRO and University of Western Australia scientists, showed that applications of the organism boosted wheat yields by 8 per cent. Joint trials sponsored by the Australian Centre for Agricultural Research in China have boosted yields in irrigated wheat there by 13 per cent.

There are obvious risks in the use of biocontrol agents, however, more so in those that are not natural but have been produced by mutating or genetically altering species. Going feral is as much a risk in the microscopic world as in the macroscopic; upsetting natural balances brings its own problems; and there is always the risk that non-targeted organisms may be adversely affected.[102]

Because there is always the risk of gene transfer between microbes, it is possible that introduced genes for superior competitive ability, for instance, could be transferred into pathogenic species, making them even harder to control. Meddling in areas that are potential minefields because so little is really known about the interactions and complexities within the soil ecosystems is hazardous and caution has to be the watchword.

Take-All

Take-All is a serious root-rot pathogen of wheat and barley all over the world.[103,104] In Australia it is present in 80 per cent of cereal paddocks. It can reduce yield by as much as 75 per cent, the magnitude of the loss depending on the amount of spring rainfall.[105] It is caused by the fungal organism *Gaeumannomyces graminis* var. *tritici*. Take-All blocks up the plumbing in plants, preventing the absorption of water by roots. Blackening of roots occurs in severe infestations. The fungus survives parasitically in the living roots during the growing season and saprophytically on dead plant debris after crop harvesting.[106] The level of viable hyphae of virulant *G. graminis* strains present in soil organic matter will determine how severely subsequent crops will be affected. It is now possible to quantify the amount of the pathogen in a soil and to predict how serious the infection of the next crop is likely to be. A rapid, routine DNA-based assay has been developed in which the DNA specific to the pathogen can be recognised and measured in soil samples, indicating potential yield loss, which allows farmers to decide on the most appropriate management. Software has been developed to facilitate decision making. Because the disease may be patchily distributed in a field, techniques for accurate sampling and reliable predictions have also been established.

When wheat is grown as a monoculture year after year, Take-All decreases in severity; this decline syndrome is due to natural micro-biological control. An assortment of different soil micro-organisms is believed to be implicated, varying in different situations, and Take-All suppressive soils are a potential source of organisms that can be cultured and used as biocontrol agents. Pseudomonads, in particular, have proved to be efficient in controlling the pathogen in laboratory trials by CSIRO scientists at Glen Osmond in South Australia, but their effectiveness in field trials is limited by their slow rate of unaided dispersal in cereal soils in southern Australia.[107] It has been shown that *Aporrectodea* earthworms, common in these soils, were efficient dispersers of the organisms in the laboratory, and they may be used to distribute the bacteria. (Earthworms are already used to disperse *Rhizobium* nitrogen-fixing bacteria in soils.)

Yield-increasing bacteria that both control pathogens and actively promote plant growth are potentially valuable aids to agriculture. Strains of *Bacillus* isolated in China[108] have been used to suppress Take-All and rhizoctonia root rot and promote seedling growth. Produced commercially in China, where they are now used on over 7 million hectares annually, they are giving improvements in yield ranging from 11 per cent for wheat to 25 per cent for vegetables. Some trials with strains of yield-increasing bacteria in Australia have shown promising results in glasshouse experiments, and research continues in this important field. A number of bacterial and fungal biocontrol agents for the Take-All pathogen have been identified in Australia.

Another root-rot pathogen, *Pythium irregulare*, is highly pathogenic to a wide range of cereal and legume hosts, causing germination failure and post-germination stunting of crops and pastures.[109] This organism is particularly abundant in soil subjected to minimum-till practices.

Rhizoctonia bare-patch disease of cereals

This disease, caused by *Rhizoctonia solani*, was prevalent in Australia in the 1920s and 1930s when the use of horsepower and old methods of cultivation did not provide sufficient disturbance of the soil for its control.[110] With the mechanisation of farming that followed and the increased soil disturbance that was involved in multiple cultivations prior to sowing, it became much less of a problem. Today, with the widespread adoption of minimum-tillage and soil conservation cropping systems, *Rhizoctonia* again has the potential to be a serious problem in the southern cereal growing regions, particularly in those with a Mediterranean climate. The pathogen results in thinner roots, broken spear tips and damage to germinating seedlings.

Rhizoctonia's most visible symptom is the production of bare patches in cereal fields where seedlings have died. These patches may later be invaded by weeds. Less severely affected crops may show patches of very stunted plants rather than bare patches. The pathogen has the potential to reduce yields by up to 25 per cent. It is not confined to the areas of visible patches, where it is concentrated, but occurs as a root parasite throughout the crop, with the potential to induce yield losses overall that may be greater than those resulting from the visible patches (because the diseased roots result in less productive plants). Several strains of the organism have been identified and no single control method is effective. All available control methods have to be used simultaneously for best effect. These comprise:

CEREAL-GROWING REGIONS

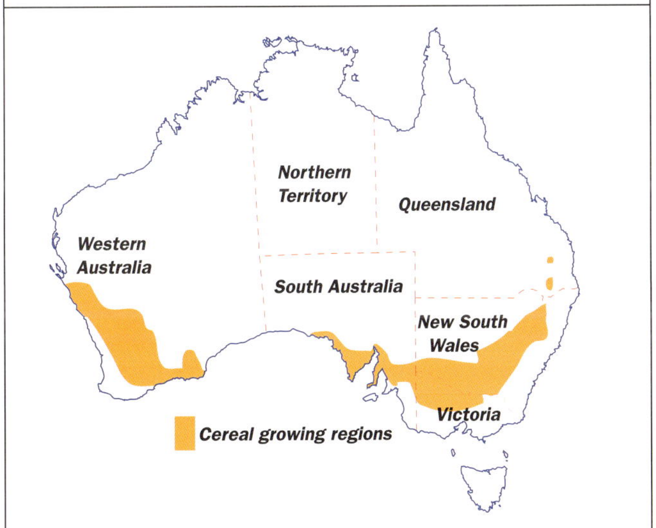

A *Rhizoctonia* bare patch in the field. *CRC Soil & Land Management, Adelaide* © *CSIRO Land & Water, Adelaide*

- Soil disturbance and cultivation. Modified seed sowing and cultivating machinery has been developed to provide the disturbance necessary while maintaining surface stability. The use of narrow-sowing point-tillage to prevent soil degradation and to control the disease proved successful in trials by the CSIRO at Glen Osmond in South Australia[111,112], and is now used in farming systems across southern Australia.
- Good nutrition of the germinating seeds enables them to resist severe infestation by the disease, so correct fertiliser usage is beneficial.
- Rotations of cereals that are susceptible with crops like oilseeds and legumes that are less susceptible.[113] This only decreases severity.
- Grazing crop residues heavily to reduce decomposing material prior to sowing; or use of herbicides to give a bare fallow. With the latter, removal of host plants like annual grasses and wild oats that carry the disease through to the next crop can be helpful. These methods have variable success, probably mainly determined by soil type.
- Volunteer pasture, comprising barley grass and ryegrass that grows over several weeks between autumn rains and seeding, promotes *Rhizoctonia* damage to roots of direct-drilled wheat. A chemical fallow, in which the volunteer pasture was sprayed with herbicide 3 to 6 weeks before seeding, was used in experiments at Glen Osmond. It was found that it greatly reduced root damage and did not affect the yields of wheat sown following cultivation.[114]
- Chemical control with fungicides has not been shown to be effective.
- Biocontrol agents. Applications of bacterial agents in the form of *Streptomyces* and *Bacillus* inoculants have not proved effective, but some species of *Pseudomonas* that occur naturally in South Australian wheat fields have potential as biological disease control agents.[115] Introducing earthworms was found to be helpful in experimental trials, but few are found naturally in the soils where the disease is prevalent.
- Natural suppression of the disease sometimes occurs after four

or five years of heavy infestation following implementation of conservation farming practices, with decline over following years to negligible levels, as in the Take-All decline syndrome. The suppression has been demonstrated in experiments to be biological. Its mechanisms as yet are not fully understood, although the level of disease suppression has been shown to be directly related to changes in microbial and microfaunal composition resulting from continuous stubble retention and increased carbon turnover.

Cereal cyst nematode (CCN)

Minute nematode worms are part of the microfauna in soil. The nematode causing CCN disease, like the Take-All and *Rhizoctonia* pathogens, is present in 80 per cent of soils in the cereal growing regions of southern Australia. The nematode *Heterodera avenae* enters the roots just behind the root tip (the growing area), causing them to thicken, knot and branch.[113] White cysts as big as pinheads appear behind the knots.

It has been shown that the severity and carry-over of CCN is less in direct-drilled (minimum-till) wheat than when the soil is cultivated.[92] Cultivation appears to make the spread of worms, cysts and eggs through the soil easier.

A natural decline syndrome in CCN, in which natural antagonists build up in the soil to a point where, after many successive wheat or barley crops, they cause populations of worms to decline sharply, has been attributed to the fungus *Nematophthora gynophila*, which is an obligate parasite on cyst nematodes. It colonises the developing female worm inside the root tissues.

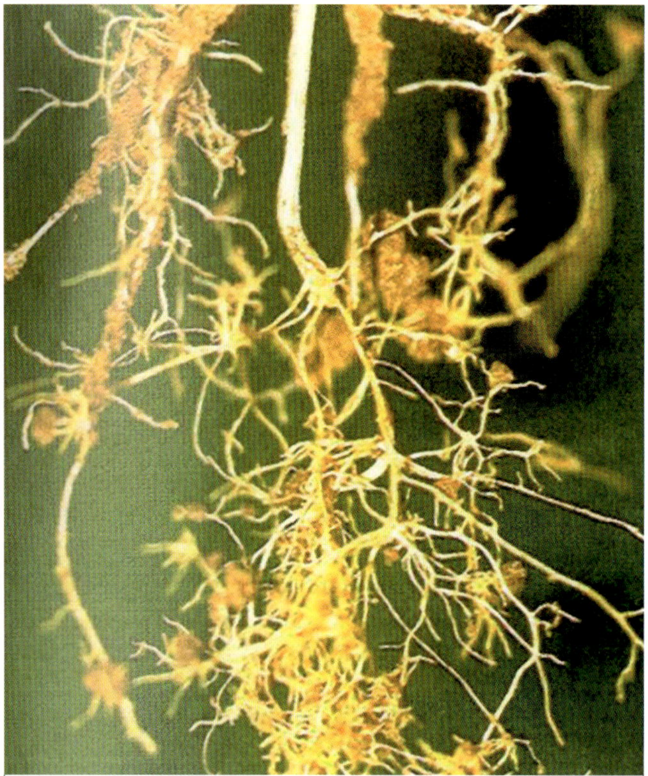

CCN-infected roots of wheat, showing root knotting and thickened roots. © *CSIRO Land & Water, Adelaide*

Damping-off diseases in nursery seedlings

Microscopic fungi of the genera *Rhizoctonia*, *Pythium*, *Phytophthora* and *Fusarium* are the usual causes of the damping-off that presents such problems in plant nurseries. Damping-off can be prevented by using pasteurised soil and pots, fungicides and good hygiene. Adequate soil drainage and environmental conditions that favour fast seedling growth also reduce the risk. Increasingly, biocontrol by introduction of micro-organisms into potting mixes is being seen as a satisfactory solution to the disease, and commercial firms are supporting research. Incitec Ltd has been involved in a joint venture with CSIRO Division of Soils at Glen Osmond to develop a commercial biocontrol product for use in nurseries and horticulture, and will market the final products.

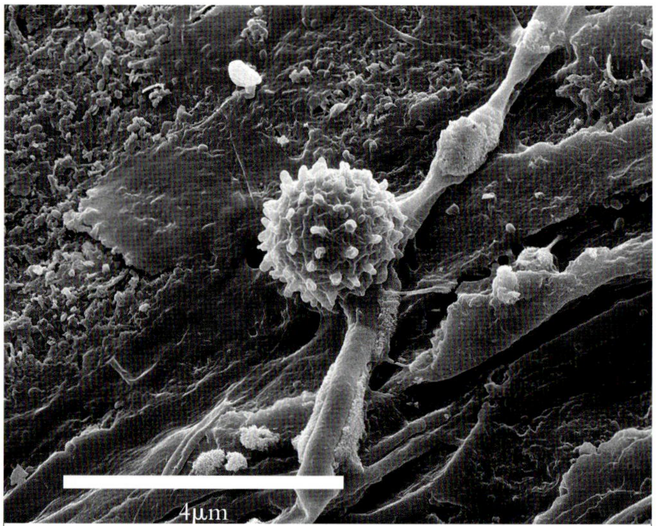

Hyphae and spore of plant pathogen fungus *Pythium* sp. on the surface of canola stubble. *VVSR Gupta et al.* © *CSIRO Land & Water, Adelaide*

THE KINGDOM OF THE FUNGI (MYCOTA)

The Kingdom of the Fungi includes moulds, mushrooms, truffles and similar organisms, yeasts and a multitude of microscopic organisms, from minute and unicellular to amazingly elongated and thread-like. The threads are **hyphae**, elongated tubes with or without septa that may divide them into cell-like sections, each of which usually contains more than one nucleus and several mitochondria. (Fungi are eukaryotes that evolved, like all other living things, from prokaryotes.)

Soil has been described as a largely fungal phenomenon because of the importance of fungi in soil processes and because microscopic soil fungi are present in such astronomical numbers in fertile soil. (For their function in soil see page 88.) Australia is thought to be home to possibly 250 000 species of fungi, of which about 13 000 have been identified and named.

Fungi are the most dominant component of the microbial network that picks up and recycles nutrients. They are also a food source for many marsupials. Symbiotic truffle relatives that

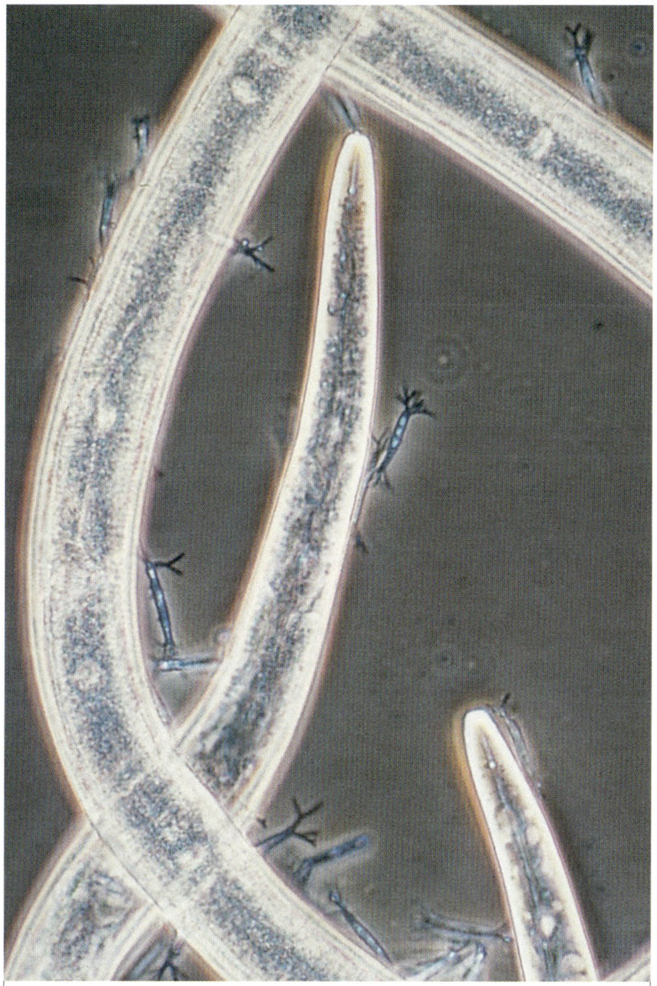

Plant-parasitic nematodes. Image Publication CRC Soil & Land Management, Adelaide © *CSIRO Land & Water, Adelaide*

live on the roots of trees and shrubs and facilitate their nutrient uptake may also protect their hosts from harmful soil organisms – as well as producing the underground fruiting bodies that are scratched and dug for by rat-kangaroos and bandicoots. In return, the small marsupials distribute their spores.

All fungi reproduce by spores that germinate into hyphae. The hyphae in a handful of rich forest soil, unravelled, could stretch two kilometres—which gives some indication of the significance of invisible fungi in the scheme of things.

The growing form of most fungi is a mass of hyphae called a **mycelium**. Though most fungi are microscopic, the fruiting bodies of some, also made of hyphae, can attain considerable size, as in mushrooms and the woody bracket fungi that grow on trees. Although individual hyphae are far from robust, the extent of fungal proliferation in soil from a single source can be enormous. A case has been documented in a new conifer forest in Michigan, USA, where a single *Armillaria bulbosa*, a tree-root coloniser, has a coverage of 37 acres; the weight of fruiting bodies and soil hyphae combined has been estimated to be 11 tonnes. The infestation was genetically sampled, and shown to represent one 'organism' that has spread through the soil of the whole forest.[7]

All fungi are heterotrophs, that is, they are unable to synthesise their own carbohydrates. They may be either saprophytes or parasites, secreting powerful enzymes that break down dead or living material outside the fungal body into a solution that they can absorb; or enter into symbiotic relationships with plants that photosynthesise the food they need; or provide the thallus in lichens, where they are in symbiotic relationship with (photosynthetic) algae or cyanobacteria. The sugars and other molecules that they require for metabolism but are unable to manufacture themselves are absorbed through their cell membranes.

Most fungi are terrestrial, but a few are known to live in the sea or under water. In our daily lives, yeast fungi make the alcohol in wine and beer, raise bread, ripen cheese, flavour soy sauce. Moulds are responsible for the souring of milk, producing isobutyric acid; they grow on our walls and other surfaces in humid weather, coming, it seems, from nowhere.

Symbiotic fungi were among the first organisms to colonise the land, probably about 450 million years ago. The emergence of the Plant Kingdom, when plants evolved from aquatic green algae and made the move into the terrestrial realm, would not have been possible without mycorrhizal fungi, and fungi are important members of that suite of invisible 'little things that run the world' enabling the recycling of materials that keeps the Biosphere alive.

Fungi are able to survive in an extraordinary range of conditions, and are part of almost every ecosystem. Some are extremophiles that can tolerate extreme heat, cold, desiccation and even strong acids. There are some that inhabit toxic mine dumps and appear to have the added advantage of protecting the plants that grow there by inhibiting their toxin uptake.

As partners in lichens, and no doubt as saprophytes in alga-infested swamps in the 'green scum' phase of early land life before the first vascular plants emerged, the ability of fungi to survive under conditions of high ultraviolet radiation and other hazards, and their need for symbiotic relationships, were fundamental to the progress of evolution—not only of the Plant Kingdom, but of animals beyond the aquatic stage. Without fungi, plants could not have made the transition and set up habitat and a food source that allowed animals to follow.

Fungi are as important to plants today as they were to the ancestral vascular plants. Their nutrient cycling in the soil, their activities in the rhizosphere facilitating nutrient uptake, and their further enhancement of nutrient availability within the tissues of plant roots, are all essential. Ninety per cent of plants form mycorrhizal associations with fungi that infect their roots. The competetive advantage this brings in terms of efficient nutrient extraction from low-fertility soils has ensured that mycorrhizal symbioses have spread widely (and are just about ubiquitous in Australia, with a few notable exclusions like *Banksia*).

Research by scientists of the CSIRO Division of Soils was among the first in the world to identify fungal species that were efficient in enhancing nutrient uptake and offering protection against pathogens.

Mycorrhizal associations

(Professor Sally Smith, a specialist in mycorrhizal symbiosis at the University of Adelaide, contributed to the writing of this section.)

Four main types of mycorrhizal association, differing in the nature of the interface that forms between the host plant and the fungus and in the way that the symbiosis works, are recognised: arbuscular mycorrhizas, ectomycorrhizas, ericoid mycorrhizas and orchid mycorrhizas. All play vital roles in the nutrition of their plant partners and are increasingly recognised as regulators of biodiversity, as well as having beneficial side-effects in stabilising soil. In all types except the orchid mycorrhizas (see below), these fungi increase accessibility of nutrients from soil and receive sugars from the plant, avoiding competition for scarce carbohydrates in soil. The different associations are generally formed by a characteristic group of plant species but there are occasional examples of groups that have more than one variety. Eucalypts, for example, have both arbuscular and ectomycorrhizal types. The types differ in the structures that they form in the host roots, and they differ in the ways they enhance host-plant growth. Each type appears to have evolved to suit a particular set of environmental conditions, but it is worth noting that the range of potential partners available for each fungus is huge, compared with the very narrow choice available to many plant parasites.

Enlargement of a portion of the image below shows arbuscules and hyphae inside the cortex of the root. *Margaret McCully*

Arbuscular mycorrhizas

Arbuscular mycorrhizas are the most common and widespread, involving 80 per cent of plant species, including trees as diverse as monkey puzzles, bananas and maples; many shrubs; and the vast majority of herbaceous plants, including ferns. They are common in a huge range of habitats, from tropical rainforests to deserts and alpine meadows. Many of the soils are relatively low in phosphorus, and the fungal symbionts play key roles in helping the plants to get this essential nutrient. The very large

Vesicular arbuscular mycorrhizal (VAM) fungi in association with wheat roots are stained red in this SEM image. *Margaret McCully*

range of plant partners probably reflects the ancient origin of the group as symbionts of the earliest vascular land plants.

Arbuscular mycorrhizal associations are formed by zygomycetes. During the infection process, fungal hyphae traverse the epidermal layer of roots and form highly branched structures known as arbuscules within individual cells of the cortex, without penetrating deeper into the vascular conducting tissues. The arbuscules, together with hyphae, provide a major interface across which nutrient and sugar exchange occurs. Although the host's cell wall is penetrated, its inner plasma membrane remains functional and proliferates to surround the arbuscular branches as they develop. The interface between plant and fungus thus involves membranes of both partners through which the nutrient traffic can be controlled, separated by a region of apoplasm (where solutions move between, not within, the cells). The highly branched nature of the arbuscules increases the surface to volume ratio in the host plant about fivefold, providing an extensive interface for the exchange of nutrients. A mycelium of hyphae is developed in the soil surrounding the infected root, extending for at least 12 centimetres and probably more, and greatly extending the volume of soil that the symbiotic root system can explore for scarce nutrients.

Ectomycorrhizas

Ectomycorrhizas are most common in tree species of the families Betulaceae, Pinaceae, Fagaceae, Dipterocarpaceae, Leguminaceae and Myrtaceae. This is the most important mycorrhizal type in vast areas of boreal conifer forest, and in some of the huge canopy trees in tropical rainforests. This type of association is often common where nitrogen availability is the major nutritional limitation to plant growth. The fungi help to overcome this limitation, absorbing ammonium, to a lesser extent nitrate, and sometimes also accessing organic forms of nitrogen in the soil. Additionally, in some habitats they also increase phosphate uptake.

Ectomycorrhizal symbioses are largely formed by basidiomycetes and a few ascomycetes, many of which are important as food (such as cepes, truffles, chanterelles). The fungi form mycorrhizas with the short lateral roots. The fungi do not normally penetrate root cell walls, instead forming an entirely intercellular interface, with extensively branched hyphae penetrating between epidermal and cortical cells, forming a network known as the Hartig net. This net can provide a vast area of symbiotic interface for transfer of nutrients. The fungus also forms a dense mantle round each lateral root, essentially isolating it from the soil around it, so that the root becomes entirely dependent on the fungus for its nutrient supply. A mycelium of hyphae extends into the surrounding soil, penetrating metres from the root and acting as an efficient nutrient-collecting network. There is evidence that some ectomycorrhizal fungi cause weathering of soil minerals.

Just like the tree-root-infesting fungus in Michigan, a single ectomycorrhizal fungus can infect several trees and its mycelium can spread over many hectares.

Ericoid mycorrhizas

Ericoid mycorrhizas are confined to the Ericales, which include the Ericaceae of the Northern Hemisphere and the Epacridaceae of the Southern Hemisphere. They predominate in the heathlands of polar and alpine regions where temperatures and rates of evapotranspiration are low and much of the soil nitrogen is locked up in inorganic form. They are also typical of some Mediterranean-type heaths, such as those found in parts of Australia and South Africa. In the Australian examples, like the Kwongan sandplain flora of Western Australia, nitrogen is again the limiting factor because of the poor soils and the restrictions on availability of the water that is required for decomposition of plant debris.

Ericoid mycorrhizal fungi are largely ascomycetes. Fungi penetrate the epidermal cells and form dense hyphal coils that are surrounded by the host's plasma membrane proliferations, again providing an extensive interface for absorption. Development of hyphae in the soil is relatively limited, but a combination of dense proliferation of the very fine roots allows the associated mycorrhizal fungi access to large volumes of soil where they break down the organic nitrogen and pass some of it on to the plants.

Orchid mycorrhizas

Orchid mycorrhizas are rather different, because the orchids themselves have tiny 'dust seeds' that are non-photosynthetic, either for a short seedling stage or, for some, throughout their lives from seed germination to flowering. This means that the plants cannot provide their mycorrhizal symbionts with carbohydrates, but instead depend on the fungi for both soil nutrients and sugars. As with other mycorrhizal types, the fungi penetrate the root cells to provide an extensive interface in which coiled hyphae are separated from the plant cell plasma membrane by a narrow zone of apoplasm. Some orchid mycorrhizal fungi are saprophytes and obtain sugars by breaking down litter in soil; others form mycorrhizal associations with other plants, such as trees, and obtain sugars in that way. In both cases some are passed on to the orchids to support their growth. The underground orchid of Western Australia is one of those linked to a photosynthetic plant, in this case a *Melaleuca*, by its mycorrhizal fungus. The plant spends its whole life underground and its tiny flowers are pollinated by slugs. It remains a mystery why any of the orchid mycorrhizal fungi bother to associate with the orchids; we still do not know if they receive anything in return.

Lichens

The ability, and the need, for fungi to enter into mutually beneficial relationships extends beyond symbioses with plants to those with algae, bacteria and even invertebrates.

Lichens are striking examples of symbiosis between fungi and green algae, or cyanobacteria, which photosynthesise for the fungi in return for safe housing and a supply of nutrients. The

Lichen on a rock at Wilpena Pound is subject to high temperatures when the sun heats the rock in summer; to cold in winter; and to a great deal of aridity. Many lichens are extremophiles. *M.E.W.*

thallus of a lichen is composed of fungal hyphae in a closely packed mycelium, surrounding the endosymbionts, providing once again a contact layer across which nutrients move. The endosymbionts feed themselves and their host with sugars and sugar alcohols. Many lichens grow in very inhospitable places with huge fluctuations of temperature and moisture and high exposure to ultraviolet light. When rains do come, nutrient supplies can be infrequent and very dilute. The symbiotic thallus is perfectly adapted to survive these extremes and the fungal partners are highly efficient at accumulating mineral nutrients from the dilute solutions—and some can also obtain them by weathering the rocks on which the lichens grow.

(In Antarctica, endolithic fungi grow with green algae or cyanobacteria inside rocks. The algae photosynthesise using light that passes through rock crystals; water comes to the fungi from melting frost and carries nutrients from the algae. Though lacking the well-structured mycelial thallus, and not yet a lichen, a symbiosis like this may be similar to that which preceded the formation of lichens long ago.)

Like fungi in general, the contribution of lichens to the Biosphere is mostly overlooked and underestimated. They play key roles in rock weathering and soil formation. In terms of biomass, their global weight is greater than biomass in the oceans.

Symbiotic relationships between fungi and ants

Ants of the genus *Atta* and some termites cultivate fungal gardens. Leaf-cutter ants of the New World tropics, whose colonies may have a million members, cultivate fungal gardens to feed the colony. They have developed depressions on their backs to carry spores into their nests. They feed the spores with leaf cuttings and remove the weeds when the spores germinate. The fungi have responded by changing their morphology to suit their masters. As basidiomycetes they used to produce mushroom-like fruiting bodies, but long association with the ants has resulted in change. They now produce staphyla, which are edible bulbous tips on hyphae, instead of caps—and the gardens are like miniature fields of grain, ready to be harvested by the ants. The gardening activities involve production of antibiotics to keep down bacteria that might damage the fungi; the harvesting keeps the fungal colonies 'young' and helps to prevent decline due to disease.

CASE STUDY: AUSTRALIAN RESEARCH INTO MYCORRHIZAS

(The mycorrhiza research group at the University of Adelaide, South Australia, is led by Professor Sally Smith, who is also the Director of the Centre for Plant Root Symbioses. She has kindly supplied a brief account of current research so that it may be seen in the Australian context as well as part of the global programme that is trying to understand the vital processes involved.)

In common with many groups around the world, we investigate the way in which plant roots are colonised by beneficial ('arbuscular mycorrhizal') soil fungi and how this symbiosis affects nutrient uptake and plant growth and survival. Our results are applied to natural ecosystems and vegetation re-establishment. Ecological studies have shown that if mycorrhizas are suppressed in a semi-arid herbland the diversity of the plant community is reduced, a finding that differs from work in other plant ecosystems. The contrasting results emphasise how little we really know about function of mycorrhizas in the field and how dependent the effects are likely to be on the different plants, their environments and probably on differences in effectiveness between different mycorrhizal fungi. In order to understand these variations we need to understand the way in which the fungal and plant symbionts interact at the cellular level, so we use specialised fluorescence microscopy to follow details of the ways the fungi grow through roots and how the root environment changes to promote the essential nutrient transfer between the symbionts. At the same time we apply molecular techniques to find out how the fungi manage to invade the plant tissues without provoking the defences that keep out damaging fungal parasites. One of the 'tools' we have is a tomato mutant that—surprisingly—does prevent invasion by some mycorrhizal fungi but not others. By comparing what happens when different fungi are paired with the mutant and with a normal tomato, we hope to find out how the mutant blocks invasion and hence what processes are 'switched off' in normal plants. With a group in the University of Western Australia, we are also trying to find out which gene has mutated, so that we can learn more about the way it operates and whether similar genes are present in all mycorrhizal plants. We anticipate that this will be the case because the symbiosis is so widespread and because it has been evolving for so long.

CASE STUDY: THE RED, YELLOW AND GREY EARTHS OF NORTH-EASTERN QUEENSLAND

Termite mounds in over-grazed land in central Queensland. *Photo M.E.W.*

Landscapes of low relief and lacking integrated stream networks are common over large areas of the low-rainfall zone of the Australian continent. Characteristic red and yellow earths are commonly developed on them, derived from quartzitic parent material and with weakly differentiated soil horizons. They cover about 20 per cent of the continent. In tropical Queensland, where the land is used primarily for beef cattle grazing at low stocking rates (1 beast to 15–80 hectares), they amount to 37 per cent of land areas. These soils are of low fertility. Deep drainage through the Queensland section has potential for recharge of the Great Artesian Basin. The permeable soils of the Denna Plain in north Queensland are believed to be a major contributor to this vital recharge.[117]

Originally regarded as ancient, relict soils on ancient landscapes, research has shown that soil forming and changing processes remain active in them today.

The red and yellow earths are characterised by non-calcareous profiles; massive structure; earthy fabric; and mineralogy dominated by quartz, kaolin, and iron oxides and hydroxides. No simple relationship exists between soils and their topographic positions in the landscape.[117] The shallow yellow earths, often with ironstone nodules, have formed in the erosional parts of the landscape on broad low ridges where weathered bedrock is close to the surface. The much deeper loamy red earths have formed in the lower depositional parts of the landscape in wide drainage lines where thick alluvial sequences have accumulated since the Tertiary. Some red earths occupy areas where deep weathering in palaeochannels has occurred to depths of as much as 118 metres.

The present drainage lines operate only in response to rare rainfall events of large magnitude. Fluctuating water levels develop in the shallow yellow earths, 0.5 to 2.5 metres deep, while the adjacent deeper red earths, 5 to more than 20 metres deep, behave as freely draining profiles in which water infiltrates on occasions beyond the zone of water extraction (3 to 6 or more metres depth). Free water levels in the yellow earths vary in response to evapotranspiration and rainfall. Differences in colour of the soils relate to water relations, in particular to the degree of saturation at watertables: on the slopes with shallow soils the water can't drain away in the wet season and the soil remains waterlogged. The iron undergoes transformation giving the yellow colour. The related and often adjacent grey earths are the end product of more prolonged waterlogging. Red earths are very deep and capable of absorbing just about all the wet season rainfall. In wet years when plants do not use all the water, deep drainage can let water enter the sandstone bedrock.

Microbial activity is involved in changing the structure of the hydrolysed iron in the soils. Strong linear relationships are found between depths of the ephemeral watertables and a number of features of the soils—red or grey mottles, ironstone gravels, maximum content of clay, depth of the soil.[118]

Although the soils have formed on surfaces of great age, biological and hydrological processes are currently active in them, and they cannot be regarded entirely as relict.[117a–e,119] It is known that termites play an important role in the genesis of these soils. Development of one generation of termite mounds (200 to 600 mounds per hectare) may result in the reworking of 20 tonnes per hectare of soil. The 20-centimetre thick sandy loam A horizon of the soils may have accumulated from erosional degradation of the termite mounds over the past 8000 years. (See Termites, pages 108–112.)

The extensive plains without integrated stream channels in the Torrens Creek area[116] are underlain by gently dipping coarse-grained sandstones, and some mudstones, in a Permian to Jurassic sedimentary sequence. Exposed since the Late Mesozoic to

DISTRIBUTION OF THE RED, YELLOW AND GREY EARTHS IN QUEENSLAND

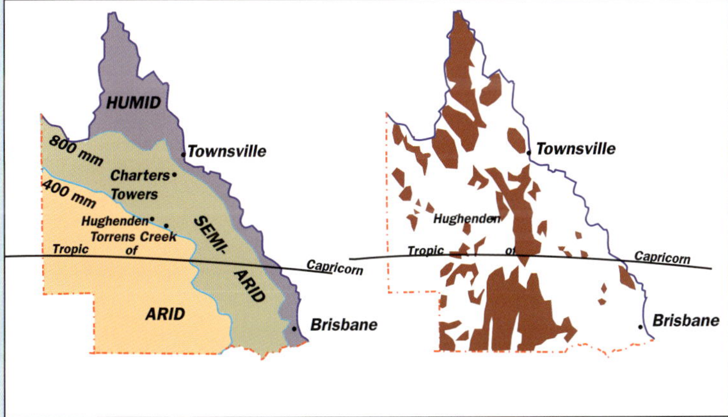

weathering and erosion, these strata have yielded a widespread mantle of Late Cainozoic sediments up to 120 metres thick. In general, yellow and grey earths (0.4 to 2.5 metres deep), many with strongly developed mottles and ironstone gravels, have formed in the erosional parts of the landscape on broad low ridges where weathered bedrock lies close to the surface. Red earths with much thicker sola (3 to more than 20 metres) have formed in the lower, depositional parts of the landscape in wide drainage lines where thick alluvium has accumulated. Research in this region has shown that in spite of semi-arid situation, watertables develop in the soils as often as one year in three. The distribution of mottles and of ironstone gravels was found to coincide with zones of temporary saturation in the soils, indicating again that although the soils are Late Tertiary to Quaternary in age and lying in a semi-arid tropical landscape, they display hydromorphic features that appear to be in harmony with their current hydrology.

Brocks Creek: Termites and landscape equilibrium[120]

Research was undertaken at Brocks Creek to challenge the notion that soils developed on the younger of three planation surfaces recognised in the north of the Northern Territory are ancient 'lateritic podsols' associated with a stable Late Cainozoic weathered surface. Results of the work emphasise that any realistic model of landscape evolution in this region must take into account present-day erosional and biogenic processes as well as possible landform inheritance from late Cainozoic times.

Comparison of two localities (A and B) on deeply weathered granite showed that modification of the weathered surface by erosion, termite activity and soil development had been active for at least 20 000 years—so that what has been interpreted as a stable Late Cainozoic surface is still in fact being weathered and lowered. Each Brocks Creek granite soil profile has one or more of the following horizons: M (mineral sand), S (stony layer), W (weathered rock).

- Mineral sand surface horizon that may overlie an S or W horizon, which since the work of Nye in 1954[157] and Watson in 1961[121] has been explained as having been produced largely by termites;
- Stony or pebble layer that may be overlain by M, but never by W;
- Weathered rock layer that has 3 zones showing increasing alteration further away from the unaltered bedrock base.

The localities A and B are 5 kilometres apart, both on granite that has suffered only minor dissection since the last phase of deep weathering. Valley aggregation dominant in area B; stream incision in A. Four species of mound-building termites active in A; termites rare in B. Soils at A are 3-layered—sandy topsoil (M), intermediate quartz stone layer (S), weathered granite subsoil (W).

B locality soils have no stone layer, and deep sand (M) rests directly on weathered granite (W).

It is postulated that the stone layer in A is largely developed by termites. The absence of termites in B is thus the reason for lack of stones. The W horizon in both is formed by weathering *in situ*. In A, the colour and textural transition from the stony layer to the uppermost weathered W1 layer is so gradual that without the concentration of quartz stones, the zones would not be separable. In B there is no such gradual transition—M overlies W, resulting in texture-contrast soils. When all factors were considered, the conclusion was reached that termites were in large part responsible for the development of the stone layer.

Mastotermes darwiniensis, a primitive termite species which is known from Oligocene fossils in Europe (20 million years old) and which is common in coastal tropical Australia, may well have been widespread in the continent for many millions of years. (Fossilised frass of Kalotermitidae has been found in petrified Oligocene rainforest wood from Queensland[122], and we know termites have been active in Australia for as long as they have been present elsewhere—in fact from Cretaceous times.)

The termites at Brocks Creek replenish topsoil as fast as it is eroded, suggesting that in the wetter parts of tropical Australia, termites may long have been in equilibrium with the local ecosystems. For a relatively stable plant community to develop and persist, a delicate natural balance must exist between plant growth, plant consumption by animals and insects, and soil development through rock weathering and recycling of nutrients from broken-down plant residues.

The equilibrium between termites and their environments is geomorphological as well as ecological. By promoting seasonal moisture penetration along their galleries, the Brocks Creek termites enable the granite bedrock to weather at about the same rate as they are removing weathered material from the subsoil. The topsoil is constantly lowered by slopewash to be continuously replaced by mound disintegration. The weathered rock is repeatedly mined by worker termites for fresh building material, but the progressive advance of the weathering front keeps pace with removal of weathered products. A state of equilibrium exists. This implies that the weathered surface of Brocks Creek granite is not an unmodified deeply weathered surface of Late Tertiary or early Pleistocene age, as assumed in purely geological assessments of the region.

Tropical northern Australia was arid during the very late Pleistocene[123,123a], so the present rate of lowering of the area was probably not maintained.

MOUSE PLAGUES, STRYCHNINE IN SOILS AND MICROBIAL DEGRADATION OF TOXINS[95–100]

A serious mouse plague in southern Australia in 1993 resulted in the loss of $64 million in agricultural production. Another in northern New South Wales and Queensland in 1995 was also devastating. Rodenticides (rat and mouse poisons) provided the only quick and effective method of bringing the plagues under control, Strychnidin-10-one being used. In 1993 in South Australia alone, wheat grains laced with strychnine were spread at a density of approximately 3 treated grains per square metre over 350 000 hectares of cropping land. A great deal of the strychnine was absorbed into the soil, the rate of absorption depending on soil type, and experiments were conducted to determine its ultimate fate, and whether it was later absorbed by crops grown on the treated soils. It was already known that strychnine does not degrade abiotically in soils; it was necessary to find out if it was degraded biologically in Australian soils, and if so by which organisms.

Strychnine is a plant alkaloid containing carbon, hydrogen, nitrogen and oxygen ($C_{22}H_{22}N_2O_2$) obtained from seeds of *Strychnos nux-vomica*. Fortunately, certain widely occurring soil bacteria are capable of using it as their sole source of carbon and nitrogen, and laboratory experiments have shown that the toxin is removed, at different rates depending on soil type and how closely it is bonded to clays, and does not present an ongoing threat. While little is known about the different micro-organisms involved and their interrelations in the field, laboratory experiments in other countries with cultured strains of a number of bacteria including *Arthrobacter* and two closely related genera, *Corynebacterium* and *Brevibacterium*, leave no doubt that biological breakdown of the poison occurs. Cultures of bacteria isolated from South Australian agricultural soils, in which three species of *Arthrobacter* were identified, confirmed the ability of the soils involved in the mouse plague treatments to rid themselves of the poison. At the present rates of application recommended—3 grains per square metre—it was also concluded that potential crop uptake does not pose a health risk.

It is the fact that strychnine is an organic substance that enables it to be broken down and detoxified by soil organisms. Where toxins in agricultural chemicals and fertilisers are metals, no such easy cleaning up is possible. The accumulation of heavy metals and metalloids in soil, especially arsenic, cadmium, chromium, copper, lead, mercury, nickel, selenium and zinc, is of concern in agricultural production systems.[97] Theses ubstances have the potential to affect the safety and marketability of food; crop growth, through phytotoxicity; and environmental health. Once in the soil they may persist indefinitely, and metal pollution may only be reduced by transporting the pollutant downslope or downwind by erosion; to surface or groundwaters by leaching; to atmosphere by volatilisation; to animals and humans in agricultural crops; or to disposal areas by physical removal of soil or plants. (There are some bacteria that are capable of oxidising, by enzyme production, reduced forms of many metallic elements, and also some that can reduce enzymatically a similar range of oxidised forms.[28])

Pollutants in agricultural soils can be sourced from the atmosphere, from fertilisers, from agricultural chemicals, and increasingly from the wastewater and biosolids from sewage treatment. Even more worrying than the pollutants in biosolids and recycled sewage waste waters is the array of toxic organic compounds that are present in the effluent from pulp and paper mills. Where once these effluents were released into waterways, with environmentally damaging consequences, they are now increasingly discharged on land where their high organic matter residues are useful fertilisers. However, they contain dioxins and a whole range of toxic organic compounds, and just about nothing is known about their degradability, or whether and how soon they accumulate to dangerous levels. In industrial land and in urban situations it is mainly the potential of the pollutants to end up in waters or atmosphere, or to affect people directly, that is the problem. Contamination of foods in agricultural lands is just as much a concern.

Examples of the build-up of contaminants in soils and their sources include: copper from fungal sprays; lead arsenate in orchards where it was used to spray insect pests (seen in soils in Tasmania and South Australia); arsenic in soils in banana-growing districts of Queensland and northern New South Wales; cadmium in market garden soils that had received superphosphate from Nauru; cadmium, copper, lead and zinc in high concentrations in soils downwind of the Port Pirie lead-zinc smelter in South Australia (and presumably downwind of Mt Isa in Queensland before pollution controls eliminated the smoke plume that had characterised the area for decades); lead pollution of soils from motor vehicle exhausts and high levels of lead in the blood of children living in congested cities.

The role of plants and animals in soil formation and function

Living organisms from all five Kingdoms play a vital role in soil processes. Micro-organisms (Bacteria and Protists) affect structure and produce chemical substances, act as micro-bioturbators and as primary cyclers and recyclers of material, as well as being agents of weathering that produce sediments from rocks in the first place. Fungi influence structure and break down organic matter, and they have an absolutely vital role as mycorrhizae, facilitating uptake of nutrients—and, as we have seen, forming connections between soil and plants. In the Animal Kingdom, Invertebrates are major bioturbators, changing soil structure and in some instances, like termites in the tropical savannas, are responsible for creating distinct horizons by their activities. Vertebrates influence soil processes by their bioturbation and nutrient cycling, and by activities on the surface that alter plant biomass, and promote erosion or cause compaction. Members of the Plant Kingdom are soil stabilisers, bioturbators, providers of materials for recycling, and through photosynthesis are suppliers of the basic essential nutrients that almost all the other organisms need to be alive at all. Plant roots have an important role in pore formation and soil structure and in their massive contribution to soil carbon, and so much more. The root-soil interface is a zone where the most complex chemical, physical and biological interactions of any experienced by plants occur.

Soil as a living organ of the Biosphere depends on the webs of life created by members of all the Kingdoms.

CASE STUDY: THE EFFECT OF GREAT ARTESIAN BASIN BORE WATER ON SOILS[179a-b]

Wierbolla bore (now piped underground) was sunk in 1921. It flowed more than 1 million gallons a day to water the old property Thrushton, which is now National Park. Jim Gasteen's father planted pumpkin seeds in the drain as the water moved on. *Jim Gasteen*

(An edited account by Jim Gasteen of the sinking of a bore on his father's property, the making of bore drains and the changes that the bore water brought to soils on the property is included here because it gives a valuable perspective and helps us to understand some of the present-day problems of the Great Artesian Basin, particularly its soils.)

W. Gasteen took up 'a waterless, unfenced and trackless mulga subdivision of the 500 square mile Cypress Downs Holding in south-west Queensland' on returning from the 1914–18 world war. Wierbolla bore was sunk in 1921 to a depth of 3338 feet, with a daily flow of 1.169 million gallons, and ran 70 miles of drain after soakage had been taken up in the drains. Water travelled in the drain at walking pace close to the bore, but its pace decreased with distance to a trickle at the end of the drains. The soakage in the 'soft' mulga loamy red earths and sandy loams was enormous.

Mr Gasteen planted melon and pumpkin seeds in the soak of the new drains as the water moved on. Their prolific vines draped nearby vegetation, and monstrously large melons and pumpkins were produced in the first years but never again subsequently because the bore water caused hardening of the soil and nutrients were insufficient in the strongly acid, phosphorus- and nitrogen-deficient soils. Fish and yabbies appeared in the drainwaters, brought in as eggs on the feet of waterbirds, and the yabbies caused severe problems within a few years by boring holes in the banks of drains and riddling the ground around culverts and divisor plates. Wastage from seepage was enormous and surrounding areas were bogged and drains dried back for miles. Much effort had to be put into mending banks and trying to control the yabbies, and feral pigs became an additional problem because although they ate yabbies their activities were damaging and caused further loss of water. Over a few years the banks and saturated areas became so hard that problems from burrows and animals digging settled down. Regularly watered garden beds became hard and unproductive and only addition of great amounts of manure enabled anything to grow.

We now know that more than 95 per cent of the artesian water delivered via open bore drains since the resource was first tapped in the 1880s has been lost to seepage and evaporation. All bore flows have been reduced, and many have ceased flowing altogether, because underground pressure is insufficient to force the water to the surface. Water today has to be pumped to the surface instead of gushing out. (An account of programmes to pipe bores and reduce wastage is given in *Running Down*, pages 97–104. How making permanent surface water available to feral animals, stock and native grazers in fragile naturally arid regions results in heavy grazing pressures that lead to land degradation is also discussed.)

An evaluation of the Bore Drain Replacement Project in south-west Queensland has been undertaken by scientists in the Department of Natural Resources in Charleville.[179c] They concluded that: 'Extraction of artesian water has caused decline in artesian pressure, while use of water at the surface has altered surface hydrology, as well as degraded land resources and influenced social and economic development.' They noted that land managers had observed changes in soil and vegetation in the last 10 to 30 years in areas surrounding bore drains. Scalded soil, soil erosion, reduced cover, increased woody weed establishment, increased salinity and tree mortality were all visible problems. Follow-up preliminary work, not yet published, has started to investigate the effects of artesian water on soil chemistry in the sort of mulga land system that occurs on the Thrushton property described above.

For over a century it has been known that artesian water is unsuitable for irrigation because of the effect it has on soil structure. The seepage and wall breaches along bore drains described by Mr Gasteen resulted in the same sort of alteration as irrigation would cause on a massive scale.

Preliminary scientific analysis of changes to soil chemistry following treatment with artesian water provide answers to some of the questions as to why such changes occur:

- Artesian water changes the chemical characteristics of the soils that occur most commonly in south-west Queensland.
- Artesian water is alkaline (pH greater than 8.0) and is dominated by sodium bicarbonate with some chloride and minor sulphate. This accounts for the increase in pH, electrical conductivity, chloride and exchangeable sodium.
- In soil treated with artesian water the exchangeable sodium percentage (ESP) increases dramatically, affecting the soil clay particles which swell and disperse, significantly changing soil structure.
- Physical deterioration of the condition of soils due to sodicity involves hardsetting and increases erosion. Reduced plant growth follows.

CASE STUDY: TOWARDS SUSTAINABILITY IN RESOURCE MANAGEMENT

(Margaret Bryant, Director of External Affairs, CSIRO Land & Water, has kindly supplied the following account of the programmes and philosophy of the organisation.)

CSIRO Land & Water is seeking solutions to complex land and water management problems.

The key focus [of research] is on delivery of innovative solutions to Australia's natural resource management problems. We are about developing options for land and water management that are economic, socially acceptable and benign to the environment.

CSIRO Land & Water has 500 research staff and a budget around $50 million a year, a third derived from external and private sector sources. We have laboratories located in Canberra, Adelaide, Perth, Townsville, Griffith, Albury, Atherton and Brisbane. CSIRO Land & Water's research and consultancy activities span key sectors of the economy: water, environment management, agriculture, energy, mining, petroleum and built environment.

Key issues for CSIRO Land & Water are: water use, water re-use, water use efficiency, water management, and water quality. If we continue with the range of water uses and water practices we have now in Australia, where will we be in 25 years' time?

What are the impacts from land use and land management on our rivers, estuaries (particularly in urban locations), lakes, wetlands, reservoirs and groundwater? How do we manage rivers and landscapes to restore water quality and maintain critical environmental flows?

How can we better match the way land is used and managed to its biophysical capability in ways that are economic, socially and environmentally acceptable? Can we significantly increase agricultural yield and profitability in the right areas of the landscape? What are the water balance and groundwater implications of re-assigning land use for productivity, profit and longevity? What are the catchments and the accompanying institutional incentives for areas to be retired and re-vegetated to advantage for perennial cover, for agroforestry, for biodiversity, conservation and the aesthetic characteristics that we want in Australia?

What are the risks of contaminants (toxins, metals, organics) in our urban and rural areas and the impacts on ecological systems and human health?

Reshaping agriculture

Australian agriculture has been very successful for over 200 years, and has produced substantial wealth to support development of the nation. However, this wealth has been generated at a significant environmental cost, and evidence of depletion and degradation of our natural resources is now widespread.

'Our farming systems are leaky. We lose too much water, and it leads to salinity. We lose nutrients that acidify soils and pollute rivers. The answer to all these things is to stop the leak—to design an agriculture that captures water and nutrients and turns it into wealth' (Dr John Williams, CSIRO Land & Water).

Clean, green produce

Agricultural production will be increasingly expected to comply with accredited environmental management systems such as ISO14000. By developing and putting into practice environmentally sustainable farming systems, Australia has a unique opportunity to capitalise on its clean, green image and supply world markets with quality assured products.

CSIRO Land & Water's Dr Mike McLaughlin is investigating the use of plants to 'mine' toxic heavy metals such as cadmium from soils where they are a problem—one of an array of tactics aimed at a cleaner, healthier food supply.

Cleansing contaminated sites

There are more than 100 000 contaminated sites in Australia affected by petrochemical, manufacturing, military, urban, agricultural and mining activities, with overall clean-up costed at between A$5–8 billion. CSIRO Land & Water's Dr Greg Davis is confident that lower cost and more environmentally sound options exist for the clean-up of these sites—making greater use of naturally occurring organisms and bio-leaching compounds to break up the pollutants.

Microbiologist Dr Peter Franzmann is using soil bacteria to purify toxic waste-streams, while researcher Amanda Tilbury has discovered a native Australian microbe that eats the chemical pesticide atrazine in mere hours. This knowledge is also being used to clean up our precious underground water, where it has become polluted by oil, chemicals, nutrients or toxic industrial spills—by using the right bacteria to break down the toxins. In time, researchers expect the same army of soil microbes will also be used to leach valuable minerals from urban wastewater, and even to mine ore bodies for metals and energy.

Improving irrigation and river health

Australia has the potential to be a leader in productive, well-managed irrigated agriculture. Accounting for more than 70 per cent of Australia's developed water resource, irrigators are in effect the custodians of most of Australia's water resources. Competing demand for water from other users and for environmental purposes, coupled with increasing pressure to reduce the adverse effects of irrigated agriculture, is prompting the development of improved water delivery mechanisms, cropping systems, drainage and disposal practices and effluent reuse options.

'Astute water management, development of leafier, more productive crops and planting systems, and more efficient conversion of energy to food are critical if Australia is to create an irrigated farming system that is sustainable in the long term' (Dr Wayne Meyer, CSIRO Land & Water).

A vision for renewal

CSIRO Land & Water is dedicated to creating the knowledge, the strategies and the tools to manage land and water in Australia and internationally. We are committed not only to environmental sustainability, but also to building new, sustainable industries and jobs for Australia in the twenty-first century.

PLANTS AND THE SOIL

Plants form the basis of food chains for terrestrial ecosystems through their photosynthetic activities. Plant structure has evolved to facilitate that life-giving process with above-ground parts organised to optimise exposure of photoreceptor tissues to sunlight, and mechanisms for the transport and conservation of water and nutrients. Below ground, root structures anchor the plant and extract the inorganic soil resources that the plant needs, requiring a large area of contact between roots and soil. This vast interface is much more than a neutral zone. Concentrated in a narrow sleeve surrounding the roots is a zone of intense chemical and biological activity—the rhizosphere.

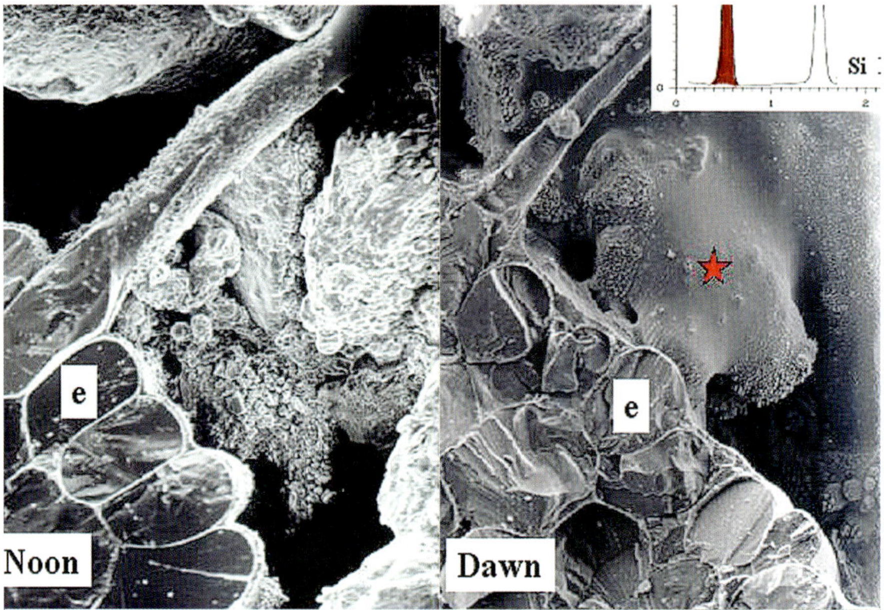

WATER EXUDATION FROM ROOTS
Left: Epidermal cells (e) and a root hair traversing pore space, dry at noon. *Right:* Pore space full of water at dawn after overnight refill.[225] (The aluminium peak in the graph is a product of the preparation of the specimen.) *Margaret McCully*

The rhizosphere

The interface between roots and the soil known as the rhizosphere is the narrow envelope of soil that surrounds every plant root where soil life is concentrated. As a general rule, the surface area of a root system exceeds that of the leaf canopy it supports, and when the surface area of root hairs is included in calculations the area is surprisingly large. The important role of root hairs, their abundance and enhanced development under different conditions of nutrient and water availability are now better understood as a result of cryo-SEM technology, as we shall see (below). Rhizosphere chemistry and physics differ from the adjacent soil matrix and the root tissues.[43] Gradients in solutes, water and gases combine with the biological activity in the zone to produce a unique compartment that extends not more than 3 millimetres from the root axis. Products of photosynthesis diffuse from roots into the rhizosphere, to feed micro-organisms or to be deposited as organic carbon. The exudates from roots contain antibiotics, organic acids, hormones and vitamins, as well as carbohydrates and enzymes[92], and it is now known that water is also exuded, to refill pores that have been emptied during times of high water use.[222,225]

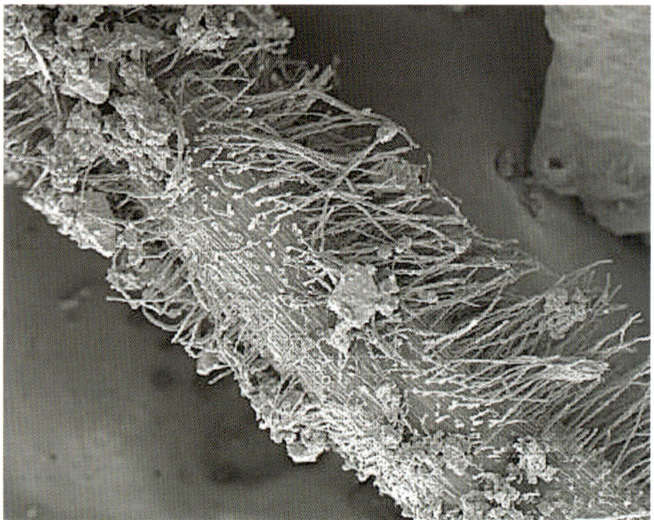

ROOT WITH ATTACHED RHIZOSHEATH
Root system of a young wheat plant. Soil clings intimately to these young roots which have been shaken vigorously before being photographed. The region of clinging soil has been called the rhizosheath and is a good visualisation of the rhizosphere. *Margaret McCully*

Signalling occurs in this zone between roots and soil microbes; roots and soil fauna; roots and other roots, as well as detection of incoming information on the local chemical and physical properties of the surrounding soil. Cells sloughed off by the root surface add to the food supply for organisms that are making available the inorganic nutrients that the plant needs. Compounds exuded from roots interact with soil components to facilitate absorption, and polysaccharides of root and microbial origin form mucilages that associate with soil particles and form a hydrated **rhizosheath** enabling hydraulic continuity between soil and roots.

How the rhizosheath functions is better understood as a result of observations that became possible when cryo-SEM microscopy became available.[225] In this technology, roots and the undisturbed soil in which they are growing are frozen in liquid nitrogen, enabling examination of the conditions that exist in the growing plant in the soil (rather than under the altered conditions obtaining in laboratories). The frozen soil and root mass can be sectioned, and details of internal root structure, root hairs, soil pores, their state of hydration, and the micro-organisms that inhabit them, all become visible under the powerful microscopes designed to operate with the frozen material.

Before cryo-SEM, the question remained unanswered as to

ROOTS AND THE SOIL[221,222]

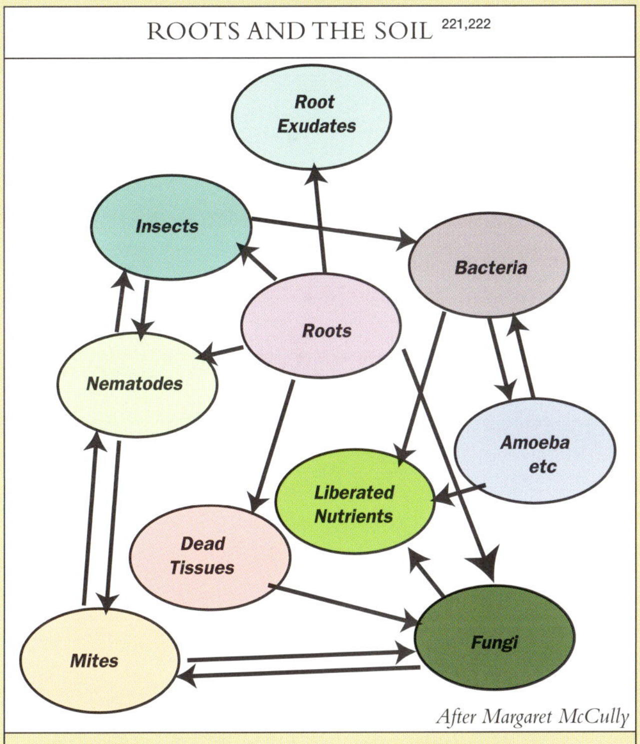

The central role of roots in the web of life in the soil is emphasised in this diagram.

Two roots, of different ages, using the same pore space to penetrate down into the soil. Root hairs can be seen curling around and holding onto their micro-aggregates; the finer hairs in the pores are actinomycetes. *Margaret McCully*

An actively growing root tip, penetrating old decaying root material, its root hairs feeding on it. *Margaret McCully*

Section of a root inside the rhizosheath, showing dense aggregations (*left*), root hairs with attached aggregates, and pore spaces. *Michelle Watt*

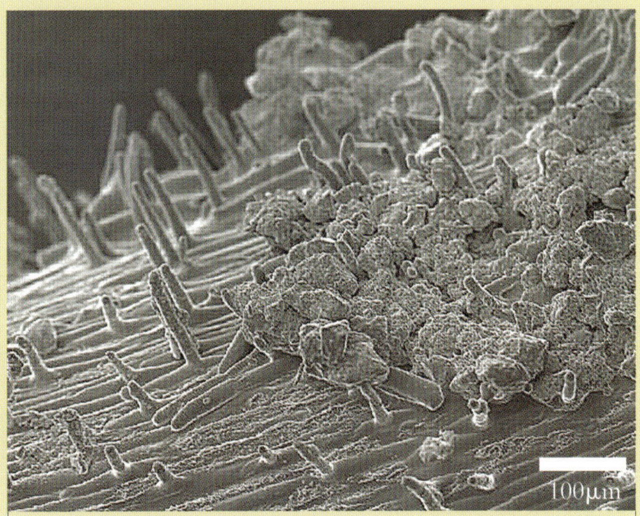

Hair roots, aggregates and the elongated cells that produce the mucilage that gums the soil particles together into the rhizosheath. *Michelle Watt*

The fine threads of actinomycetes with hair roots in the rhizosphere. *Margaret McCully*

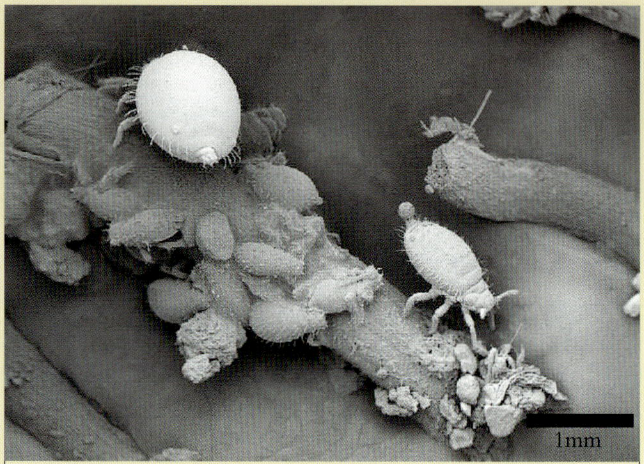

Underground aphids feeding on rhizosheath material. *Margaret McCully*

how the largest and most coherent sheaths, always found in grasses growing in dry soil, were formed. Where did the water needed for their formation come from? Sheath formation involves mucilage that is originally wet expanding into the surrounding soil. Its subsequent drying results in the strong adhesion of the soil particles to form the sheath. It is now known that the wetting of the mucilage is accomplished by water exuded by the root, facilitated by root pressure. During the night, when transpiration loss is minimal, the sheath is hydrated. It dries during the day, maintaining the essential contact between root surface and soil; when it is rehydrated the plant can access nutrients in solution.

Thus the rhizosheath is actively maintained by the plant!

Some of the materials produced and lost by roots have been found to have significant impact on soil pathogens. For instance, germinating cotton seeds produce exudates that promote *Rhizoctonia* proliferation; species of bean that are highly susceptible to the root-rot fungus *Pythium* exude greater quantities of certain substances from their roots than less susceptible species do. It is now known that some of the commonly used herbicides intended for control of fungal parasites actually promote some of them. (See also Root diseases, pages 90–92.)

Complex reactions between root-derived compounds and soil particles result in enhanced availability of phosphorus; solubility of other elements is affected by other chemical products of the root; local pH is influenced by the concentrated biological activity of the root itself and its attendant micro-organisms. Mycorrhizal associations between soil fungi and roots, as a result of infection that takes place in the rhizosphere, confer benefits on the host and symbiont. (See Mycorrhizas, pages 94–95.)

Biological concentration in the rhizosphere involves extraordinary numbers of organisms, particularly in areas adjacent to actively growing root tips (sub-apical zones). Figures like 1 billion bacteria, including millions of actinomycetes; millions of fungi; and thousands of protozoa and algae (protists) in just one gram of habitat are recorded[124]—densities 50 to 100 times more than in the bulk soil away from the interface zone. In the rhizosphere bacteria can be seen invading root cells to feed on contents; long filaments of fungi tap nourishment from the plant; organisms preying on each other and some damaging root cells and causing leakage that in turn feeds more organisms.

A significant proportion of the food manufactured by the plant in photosynthesis is used to support the rhizosphere, reflecting the huge cost of microbial activity and polymer exudation. Possibly as much as 20 per cent of plant carbon is consistently lost by roots.

Soils become more stable after plants grow on them, and soil aggregates are more stable as a result of the gels produced in the rhizosphere.[125] Soil stability is directly related to the length of roots. Grasses produce more roots than legumes and release greater amounts of water-soluble material and sticky mucilage and gums into the soil. Brome grass (a pasture grass) releases twice as much organic material as does maize, so soils under this pasture are more stable. As tillage breaks up root systems and oxidises organic matter, direct-drilled soils are more stable than tilled soils.

Roots, root hairs and fungal hyphae, especially the hyphae of vesicular arbuscular mycorrhizal (VAM) fungi, bind micro-aggregates (less than 0.25 millimetres in diameter) into stable macro-aggregates. VAM fungi belong to different groups of fungi and are widespread in most soils. The hyphae grow inside the root and in the soil, taking nutrients from the root and supplying phosphorus and other nutrients to the plant. Some VAM fungi suppress some pathological fungi.

Macro-aggregates stabilised by the growth of ryegrass contain many fine roots and fungal hyphae, in a network like a sticky string bag. Each fungal hypha is about 1 micron in diameter, with clay sticking to it through a mucilage of gums, keeping the aggregate intact and stable. Fungi that stabilise aggregates can produce up to 50 metres of microscopically thin hyphae per gram of soil.

A collembolan in a soil pore. © *CSIRO Entomology*

INVERTEBRATES AND THE SOIL

In Australia, because of the almost universally poor soils, nutrient cycling by soil biota is of greater importance than it is in lands with better soils. The low effective rainfall over at least 75 per cent of the continent, where evaporation exceeds precipitation by an order of magnitude, does not favour high productivity. (In the dry 75 per cent, rainfall does not exceed evaporation loss from a free water-surface in any month of the year, and arid and desert conditions have prevailed in much of the inland for the last 6 million years.)

A gradient of declining rainfall exists from the coast towards the arid centre of the Australian continent and is accompanied by a simplification of the soil fauna. Some higher taxa are lost and there are fewer species in the main groups like springtails and mites. Two groups are not as generally affected—termites and ants. Density of ant species is high over the continent, but particularly in semi-arid regions; diversity of both is reduced in the cool wet end of the climate range, particularly in the highlands of south-eastern Australia and Tasmania.

The richer the soil, the greater the diversity of micro-organisms and larger fauna, including earthworms and ants. In arid lands, termites largely assume the role played by earthworms in well-watered lands.

Bioturbation and nutrient cycling by invertebrates

Ninety-five per cent of bioturbation activity is confined to the topsoil, usually to a narrow zone about 5 centimetres deep at the topsoil–subsoil boundary.

Because invertebrates play many important roles in the ecological succession relating to soil development, nutrient cycling, plant growth and reproduction, and the establishment of food webs, they can often provide a more sensitive indication than plants can of the overall state of the ecosystem in which they occur.[126] Because they have shorter life histories than plants and vertebrates they can respond to environmental changes more rapidly. They also have great potential for use as indicators of restoration success (or the level of degradation of an ecosystem).

Soil fauna comprises three groups:

- Microfauna: the most abundant, but at less than 0.1 millimetres size; this group has been little studied.
- Mesofauna: organisms of intermediate size, between 100 microns and 1 centimetre long, including small insects, spiders and millipedes. Mites and springtails (micro-arthropods) are the most abundant in this group. The root-feeding springtails (collembola) in this group are pests. They feed on roots and can seriously damage crops, even preventing total seedling emergence.
- Macrofauna: earthworms, woodlice, millipedes, centipedes, snails, land shrimps, dung beetles, spiders, scorpions, etc.

('**Soil microflora**' is the label often given, somewhat inappropriately, to the microscopic mixture of bacteria, fungi, actinomycetes, and algae.)

Springtails and mites

These **micro-arthropods** are extremely abundant in soils carrying native vegetation in Australia.[126] Average densities of 40 000 to 130 000 per square metre have been recorded for eucalypt forest in various parts of southern Australia. Drier soils contain lower numbers, but even in arid shrubland, densities of 2000 to 3000 per square metre have been recorded. A very large number of different species is involved—in a single square metre of forest soil, tens of species of springtails and hundreds of species of mites are found. The total Australian fauna for these groups runs to thousands and tens of thousands of species respectively.

A wide range of different life strategies and feeding types are represented. Many springtails feed on soil fungi, algae and bacteria, but some like the lucerne flea eat plants, and some are predatory on other springtails, their eggs or other small creatures like nematodes. Mites also have many specialised habits, many feeding directly on dead leaves. The micro-arthropod communities facilitate colonisation of fresh organic matter by fungi and bacteria, and stimulate the growth of bacterial colonies by grazing on them. They break up organic matter, providing larger surface areas for attack by micro-organisms. Nutrients are released and in this way a significant contribution is made to soil fertility. Their activities in the soil contribute to soil structure, promoting aeration and water retention, and preventing run-off, and their waste products help form micro-aggregates.

The effect on micro-arthropod numbers of clearing land and stocking with sheep was measured in the Armidale district of New South Wales. In winter, numbers dropped from 500 000 per square metre at a low stocking rate to 150 000 at a high stocking rate; in summer from 160 000 to 87 000. Sheep trample and compact soil and reduce plant cover by their

A predatory mite, length about 1 millimetre. Species like this are abundant in soils and feed on springtails (collembola) and other small creatures. Some springtails secrete a distasteful substance to repel mites and other predators. *R. Southcott*

Another aggressively predatory mite. Length about 2 millimetres. Related species have been trialled as biological control agents for an introduced springtail that is a pest of pastures, particularly in Tasmania. *R. Southcott*

The collembolan (springtail), belonging to the family Hypogastruridae, is common in soils and is found on fungal fruiting bodies. (Adult from 0.3 to 1 millimetre long.) It feeds on fungal spores and hyphae and is considered an important part of the decomposer cycle in soils. It can disperse propagules of micro-organisms, its grazing prevents senescence in microbial colonies, and it contributes to soil structure through its faecal deposits. *J. Forrest*

A springtail, member of the family Onychiuridae, length about 1 millimetre. This species is common in gardens and horticultural situations where the soil is moist with high organic content. They can reach plague proportions and damage crops by feeding on roots and seeds. This can result in complete non-emergence of the crop. Poppy crops in Tasmania have been particularly badly affected. *J. Forrest*

feeding, making the soil environment less suitable for the activities of the micro-arthropods. Changes also occurred under agriculture, showing similar reductions in abundance, and under alteration from native to introduced grasslands as well. Changes in species were found to follow the introduction of non-native grasses, revealed by the presence of introduced micro-arthropod species that had arrived from Europe with early settlers. Tillage is known to reduce the numbers of mites and springtails by a third. Herbicides reduce soil mite populations drastically, while little is yet known in Australia about the effect of pesticides on them. It is known that degraded and eroded areas have few micro-arthropods and once the top few centimetres of soil have gone, no recovery of numbers has been observed even after 20 years.

Arthropods and micro-arthropods of the arid centre

Soils of the arid centre of Australia are characteristically poor with low organic content. Survival of the biota in the deep, coarse-textured red earth, sandplain and dune landscapes has required adaptation in habits, physiology and morphology.[78]

A rich fauna of burrowing arthropods is found. By tunnelling down the soil profile they are able to avoid the surface heat and dryness. The establishment of permanent subterranean nests and burrows even allows some to have access to mobile dune crests.

Those that can't burrow have adapted in other ways. Their most extraordinary adaptation is **anhydrobiosis**—the ability to become desiccated and to rehydrate and suffer no ill-effects when water is available. This adaptation parallels that of 'resurrection plants'. While in a state of anhydrobiosis or cryptobiosis the arthropod is inactive and does not feed or reproduce.

Pellets of sand pushed out of underground tunnels by cicadas on their way to the surface. Red Ochre Lake, western New South Wales. The cicada lives underground as a nymph for most of its life, occupying an extensive network of tunnels, emerging eventually as a winged adult. *M.E.W.*

The mallee moth caterpillar pupates underground. When it is ready to emerge it burrows its way out of its sand-filled tunnel before inflating its wings and flying off to mate. *M.E.W.*

Ground lichens that look like splashes of dung are abundant in the mallee at Red Ochre Lake. *M.E.W*

Earthworms

Worldwide, more than 3100 species of earthworms have been identified.[127] They occur from permafrost to the tropics, concentrated in humid or seasonally humid areas. Australia has approximately 400 species, mainly of Gondwanan origin but with a small oriental component. A number of exotic species have been introduced, mainly into the southern and eastern agricultural environments.

(Exotic species in New Zealand are known to have arrived in soil used as ballast in ships, and in soil on roots of imported trees in the early days of colonisation. Many probably came to Australia in the same way.)

Charles Darwin in 1881 produced a whole book, *The Formation of Vegetable Mould Through the Action of Worms*, documenting research, in which he was aided by his sons, on the formation of soil by earthworm castings.[128] He showed that in the deep, damp soils in Britain, worms caused disturbance to a depth of 2.5 metres and that over time sufficient material was brought to the surface in castings to 'bury the remains of Roman villas'! He recorded that when first produced, the casts are water stable—a condition now known to be due to the presence of short-lived fungal hyphae, whose death after a few weeks renders the casts susceptible to dispersion by wind and water.

Subsequent work by others emphasised that as well as the surface castings, a great deal of casting occurs below ground. While it was difficult to estimate the full extent of earthworms' contribution to the soil in England, it was estimated that in suitable conditions, over a 70-year period, the whole soil surface to a depth of 10 centimetres could be ingested and produced as casts where maximum concentrations of worms existed.

Earthworms are bizarre eaters. They throw their throat through their mouths to grab a lump of dirt, then slide the lump back to swallow it. Their castings are concentrated pellets of soil and organic matter, full of iron, calcium, magnesium, potassium and phosphorus, laced with nitrogen from urine and mucus, and laden with gut bacteria. Work in New Zealand has shown a threefold increase in plant-available phosphorus in soil that has passed through earthworm guts. Anyone with a miniature worm farm for recycling kitchen waste will be aware of earthworms' ability to break down organic matter and produce fertile soil. Their activities promote microbial activity. They are valuable bioturbators as their burrowing aids water infiltration, aeration and root penetration. Their slime causes aggregation of soil particles.

The native species of earthworms in Australia have largely been replaced in agricultural soils by a few introduced species of lumbricids from northern Europe. Dryland cropping soils in South Australia have been colonised by four immigrant species—*Aporrectodea rosea* and *A. trapezoides*, *Microscolex dubius* and *M. phosphoreus*.[129] So we have ferals and introduced species below as well as above ground, changing the natural balance.

Trials in South Australia showed that under field and

glasshouse conditions the *Aporrectodea* earthworms were able to decrease the severity of Take-All infections in wheat crops.[104] While the mechanisms involved are unclear, a number of factors are probably involved. The worms increase nutrition to the plants by breaking down more organic material; they reduce the microbial biomass and suppress development of various fungi in the soil; and their activities cause soil disturbance which deters spread of fungal hyphae.

Little research has been done on the subject of earthworm castings and the function of worms in bioturbation in Australia, and has been confined to eastern Australia. In the Snowy Mountains, above 1100 metres in sub-alpine woodland, worm castings have contributed substantially to soils with a very high water retention rate so that there is minimal surface flow, even in high rainfall events. Mineral soil material and well-humified organic matter are mixed in the casts to form 'mould' like that described by Darwin in his book. (Very limited and restricted environments exist in Australia for this sort of soil formation, compared with humid Northern Hemisphere lands with deep, fertile soils that were renewed by glaciations in the Pleistocene ice age.) At elevations below 1100 metres in the Snowy Mountains, in the wet sclerophyll forest zone, earthworms are not the major bioturbators—presumably because organic matter is much less abundant.

At Cordeaux, in the Sydney Basin near the coast in sandstone country[130], earthworm activity was regarded as significant in soil formation because the fine material of the castings contributed greatly to the suspended load in the overland flow that occurred on the slopes.

On the widespread red-brown earths of eastern Australia, earthworm castings have been found to reduce crusting and increase water infiltration, while concentrations of nitrogen, phosphorus and potassium are greater in and around casts than in surrounding soil. More than half of the large pores in soil may be created by earthworms, although they may create only 5 per cent of total pore space.[89]

The Rutherglen Research Institute has studied the responses of earthworms to different cropping and tilling regimes. Earthworm concentrations have been found to increase greatly under **conservation cropping**. Worm biomass was greater under direct-drilling than under conventional cultivation and seeding, while there was a considerable increase in earthworm numbers when stubble was retained rather than burnt. Tillage *per se* did not affect the total number of worms. It was the increased food and decreased evaporation when stubble was retained that encouraged the increase in numbers, and the more earthworms, the more decomposition of stubble occurred, which in turn reduced the mechanical problems encountered in direct drilling. (Stubble retention is not without a downside, however: chemical substances, some unfavourable to seed germination and seedling establishment, are leached from the stubble and can be a disadvantage for the following crop.)

Using earthworms to introduce micro-organisms to degraded soils

Degraded soils have low levels of soil micro-organisms, especially fungi—a catch-22 situation, because interaction between roots and soil organisms is a requirement for plant growth, and healthy vegetation protects and enriches soil.[131] Tillage promotes aerobic micro-organisms (typically bacteria) with high metabolic rates and leads to increase in organisms that feed on bacteria, like nematodes, amoebae, enchytraeids. This promotes rapid breakdown of plant residues. In minimum till systems, the retention of plant residues near the surface results in fungal growth and nutrients tend to be immobilised. The condition can be remedied by inoculating with rhizobia and mycorrhizal fungi and adding earthworms.

Earthworms can, in fact, be used as a dispersal mechanism for beneficial micro-organisms through the soil.[107,132] Special earthworm food is distributed, laced with the required organisms, but this practice is only possible where rainfall exceeds 400 millimetres a year. Successful trials have been carried out in South Australia[131], using earthworms to distribute a bacterium that is a biological control for the Take-All pathogen that affects wheat crops.

(Earthworms could inadvertently be carriers of foot-and-mouth disease, and the fact that introduced species dominate in agricultural lands in Australia emphasises the need for stringent quarantine regulations and vigilance, because their introduction to this continent in the first instance was not deliberate.)

Termites

Termites form the broad base on which the food chains of arid Australia rest. Ants are almost as important. Ants and termites have different and complementary roles in the soil system and depend on different resources. The only shared resource is seeds.

Termitaria in savanna at Mulgrave Café, Cape York. *M.E.W.*

Fossil evidence indicates that termites have been around from at least Cretaceous times, probably before the arrival of flowering plants 130 million years ago.[122] In the Australian fossil record, silicified Kalotermitidae frass 30 million years old has been found in conifer wood from an Oligocene rainforest in central Queensland. Termites' nearest relatives are cockroaches, highly successful creatures that were among the first to colonise the land after land plants had made it habitable about 425 million years ago.

Termites are social creatures found in colonies in enormous numbers. Nearly 2500 species of termite have been described globally. Distribution is widespread in tropical and subtropical regions, but some are found to latitude 45°, both North and South, in cool temperate regions.[50] Vaclav Smil gives an estimate of the number of termites in the world at 250 quadrillion—if one is able to comprehend such a figure! Termites process between a third and a fifth of all phytomass in tropical and subtropical habitats, which is **a quarter of all terrestrial primary production**. Their respiration returns at least 12 gigatonnes of carbon to the atmosphere each year. High CO_2 levels prevail in their nests, from 2 to 4 per cent, or up to 15 per cent in some cases, higher than in the atmosphere.

As a result of the symbiotic nitrogen-fixing activities of their gut residents, termites produce methane (CH_4), which is 20 times more effective as a greenhouse gas than CO_2. The global production of methane per annum by termites is estimated at between 20 and 80 million tonnes, the higher figure being as much as that produced by all the world's sheep and cattle.

The role of termites in drier parts of the world is like that of earthworms in wetter regions. They consume and break down vegetation to recycle nutrients, particularly nitrogen. Termites process surface litter, feed on it, break it up, metabolise it or make it partially soluble. Their digestion is similar to that of ruminants (cud-chewing mammals like cattle), effected with the help of micro-organisms. The more primitive genera have flagellated protozoa as endosymbionts (protists that are highly motile, using beating filaments); higher order termites have symbiotic bacteria in their guts. Some even more highly evolved species have gone a step further, cutting up leaves and taking them into special chambers in their nests where they become the base for fungal gardens. The fungi have responded by evolving special bud-like tips to their hyphae that are nutritious food for the termite gardeners. (Some attine ants have also perfected this arrangement.)

Debris-feeding termites literally rework tropical soils. Where litter is scarce, some species have to swallow tens of thousands of kilograms of soil to get enough cellulose. Termites can partially degrade lignin, but only white-rot fungi can decompose it completely. Many termite species eat wood only after it has been degraded by fungi.

Magnetic termites at Mulgrave Café on Cape York make castle-like structures whose flat, thin segments are aligned north-south to avoid exposure to the sun during the heat of the day. *M.E.W.*

There are about 200 species of termites in Australia, with 80 to 90 occurring in the Northern Territory. Mound densities vary from less than 2 per hectare in southern New South Wales to 210 per hectare in Northern Territory savannas. Perhaps the most interesting are the magnetic termites which are capable of sensing the Earth's magnetic field. They align their tombstone-shaped termitaria precisely north-south, thus avoiding overheating during the Sun's daily transit by exposing minimal surface areas to direct sunlight during the hottest time of the day. Australia has two species of magnetic termites, one occurring around Darwin, the other in Arnhem Land and northern Cape York.

Many termites are fossorial (living within the soil), bringing material to the surface but not building mounds, and their galleries are extensive. Tree-nesting species can have galleries radiating out for up to 30 metres.

In the Pilliga scrub of central New South Wales, surface-foraging, non-mounding termites were shown to be capable of turning over the surface 10 centimetres of soil within a period of 260 to 1000 years.

Research at Landsdown Research Station near Townsville showed that local species of termites consume about 20 per cent of the above-ground vegetation. Most of the remaining surface vegetation is broken down by soil-dwelling bacteria and fungi. It was also shown that termite galleries improve soil structure and the rate of water entry, thus reducing surface run-off.

In open woodland near Charters Towers, mounds are made from clay-rich subsoil brought up from as much as 4 metres depth. Nest populations can be as high as 1 million, and termites probably exceed 24 million per hectare. Their live weight was calculated to be about 105 kilograms per hectare, four times the weight of cattle feeding on the pastures.

Termite mound material is usually B-horizon material, excavated at depth and brought to the surface. It has a higher proportion of fine particles (clay and silt) than the horizon from which it is derived. The composition of mounds gives an indication of the saprolite deep below the surface, a feature which has been used in geochemical prospecting in Zimbabwe, where Kalahari Sand overlies mineral-bearing rock in large areas. Analysing material from the abundant large composite termitaria that form islands (often densely vegetated) in a sea of sand can lead geologists there to likely commercial prospects.

Termite mounds in semi-arid woodland, on red and yellow earths[119a], were shown to take up to 30 years to be dissolved by rain when abandoned. One generation of mounds involving 200 to 600 per hectare was capable of reworking 20 tonnes of soil per hectare—up to 12.5 cubic metres of soil. The 20 centimetre deep sandy loam A horizon is now considered to have accumulated from erosional destruction of termitaria over the past 8000 years.[119,133] The topsoil has been enriched with coarse material and depleted in clay as a result of the subsequent sorting of eroded termite mound materials by wind and water. Thus the uppermost parts of the red and yellow earth profiles owe their textural differentiation, at least in part, to the long-term effects of termites in the landscape. The A horizons overlie a landscape where the soils have been considered previously to be of the order of 2 million years old.[117e]

The mounds of three species of mound-building termites were involved in these red and yellow earths[134], and they were estimated to involve 1 per cent of the total mass of soil in the A horizon but to contain 5 to 7 per cent of plant nutrients in the system. An estimated turnover of 300 to 400 kilograms per hectare of soil was occurring annually, with nutrient levels 2 to 7 times that of the A1 horizon. The forager termite *Amitermes vitiosus* prefers the red earths, the harvester termites *Tumulitermes pastinator* and *Drepanotermes rubriceps* prefer the yellow earths. While soil texture seems to determine the distribution of species elsewhere, this does not apply here, as texture is the same in red and yellow earths. Red and yellow have different hydrological regimes, however, and this is what determines distribution. Red earths are freely draining; yellow earths have shallow groundwater tables in wet seasons. The mounds of the foragers and the harvesters both contain more clay than either soil does naturally, indicating that material is derived from the whole soil profile. While the termite colony is in the mound, constant repair counteracts erosion; afterwards the mound is eroded and materials are redistributed with an estimated 20 to 40 years life for the mound and 5 to 10 years for redistribution. (Soils in West Africa have also been shown to have been formed from termitaria over about the same time span.)

Nest structures of the funnel ant *Aphaenogaster barbigula* and of subterranean harvester termites (*Drepanotermes* spp.) were studied in respect of their effect on infiltration in a semi-arid woodland in Yathong Nature Reserve, 140 kilometres south-west of Cobar in New South Wales.[135] Red earths characterise the region, and it was found that the nests and activities of the ants and termites resulted in the development of moisture-rich patches. The sub-circular pavements created by the termites had the effect of shedding water and creating a well-watered annular zone; the tunnels increased water penetration. (See Red and yellow earths, page 97.)

In the Northern Territory[136] termites are most dominant in the interzone between wet-dry savannas and the arid zone, such as in the Larrimah area where there are hardly any mammals. They harvest nutritious grass in the early wet season and store it to feed on throughout the year, along with recently harvested dead grass and litter. Their ability to fix atmospheric nitrogen—because of the symbiotic micro-organisms in their hindgut—enables the use of the nutrient-poor forage. Savanna vegetation is generally found on nutrient-deficient soils[137] and

Nasutitermes triodiae mound, Pine Creek aerodrome, Northern Territory. *N. triodiae* is a grass-eating termite species found in northern Queensland, the Northern Territory and north-western Australia. This species constructs spectacular mounds up to 7 metres high. These huge nests are unique to the genus. © *CSIRO Entomology*

Nasutitermes triodiae soldiers and workers. © *CSIRO Entomology*

Termite castings (dark matter at the base of trunks) provide nutrients for this mallee tree at Yookamurra, compensation for hollowing out its branches. *M.E.W.*

A mallee trunk completely piped by termites at Yookamurra. *M.E.W.*

depends on rapid recycling of nutrients in the biomass. Soil micro-organisms play a major role in the decomposition of organic matter, but their activities are confined to the wet season. Termites are responsible for up to 20 per cent of organic matter decomposition. (Fire is also an active nutrient-cycler in some systems.)

Termites are a source of food for insectivorous birds, mammals, reptiles and frogs. The nuptial flight of winged reproductives early in the wet season provides a feast for other creatures that kickstarts their reproductive cycles. In the arid zone, an extraordinary diversity of lizards feed on termites throughout the year. Few Top End vertebrate species are specialist termite eaters, however, most eating the spiders and ants that eat the termites. The only termite-specialist vertebrates are the short-beaked echidna (*Tachyglossus aculeatus*) and tiny *Uperoleia* frogs.

Some termites, like *Coptotermes acinaciformis*, eat wood, dead or alive. They enter through a fire scar at the base of a nursery tree, building their colony both inside and outside the trunk. When the tree collapses, their structure continues as a free-standing termitarium. By that time the colony will have feeding tunnels running to another 6 or 10 trees scattered over a quarter of a hectare. The initially colonised tree always houses the nursery, and can be recognised by protruding bumps on the bole. These have small holes that can be opened and shut to release alates (winged reproductives) and may also be used as temperature regulators.

Some wood-eating termites provide a service to the trees that they hollow out. They free up nutrients locked in the heartwood and provide homes for many animals which inadvertently provide fertiliser to the tree. In the Top End, about one-fifth of bird and amphibian species, half the mammals and a quarter of the reptiles use termite-hollowed trees for some purpose. How much nutrient they contribute to the tree depends on their size, toilet habits, mortality of young, levels of metabolic activity, and whether they use the hollow for nesting or roosting. (Aboriginals are not alone in the use of hollowed trunks for musical purposes, making didgeridoos—the northern tree frog's croaks are amplified by their hollow log residences in the wet season, and large tree rats and possums use the acoustics to assert their territorial rights.)

Termitaria are homes for numerous and diverse lodgers. Huge numbers of invertebrate species—comprising aphids, silverfish, slater-like crustaceans, beetles and flies and their larvae—share the accommodation. A number of termites never build their own termitaria, two genera being completely dependent on cohabitation with *Coptotermes acinaciformis*. *Ahamitermes* species eat only the waste 'carton' material that *Coptotermes* deposits in the nest as internal structure; the other, *Incolitermes pumilis*, is restricted to the nursery region where they eat only the internal partitions. The host and the dependent species live in separate parts of the termitarium. *Ahamitermes* species have separate exits, but *Incolitermes* has lost the power of releasing its own alates—this is done by the normally hostile workers of the host species when they open the holes to release their own winged generation.

Ants inside termitaria have a variety of relationships with termites, ranging from aggression to indifference. Generally the termites wall out foreigners but some ants, dominant *Iridomyrmex* species in particular, take over parts of the termitaria.

As parts of older termitaria are taken over or abandoned, internal structure starts to decay and larger spaces become available for lodgers. Geckoes, frogs, legless lizards and small snakes like the red-naped snake (*Furina ornate*) and the anthill python (*Liasis stimsoni*) are found in reasonably intact termitaria.

The mass of termites in tree hollows and termitaria generates heat. A constant temperature is maintained in the nursery; cold-blooded animals can maintain themselves in an active state using the warmth in a termitarium; nesting birds have help with incubating their eggs.

Termites are abundant in the **mulga lands**[138] that are the dominant vegetation type over nearly a quarter of the Australian continent. Different species exploit their environment in different ways, many showing an ability to adapt their feeding and nesting habits to the prevailing conditions. The biomass of harvester termites may be comparable to that of introduced grazers, but only under conditions of food shortage, as in drought, is there direct competition. Under drought conditions, the combination of grazing pressures and the nesting habits of termites can lead to lasting degradation, irreversible where the mulga cover has been lost.

In undisturbed mulga there is a dynamic balance between insects and the mulga. Grasshoppers, caterpillars and beetles defoliate, sometimes severely; galls develop through infestation by thrips; seeds may be destroyed by weevils or wasps; sap-sucking bugs attack; roots and trunks are damaged by termites and other wood-borers. Disruption of natural patterns by human activities, either directly, as in thinning and clearing, or indirectly, as in land management that leads to over-grazing and erosion, results in permanent damage.

In south-western Queensland near Eulo it was found that under pastoral management, mulga loss through fire combined with sheep eating the regenerating plants led to increased grass pasture when climatic conditions were unusually favourable. Such conditions have only appeared a few times in the last hundred years, when they resulted in increased numbers of termites and increased carrying capacity for sheep. Then, under the normal, drier conditions, sheep and cattle denuded the grass; termites damaged the remaining tussocks, making them unable to regenerate, and erosion followed. If the mulga had been undisturbed, the termites would have fed on mulga debris and would not have killed the remaining grass.

Under conditions resulting in denudation, the hard caps of nests just below soil level, which can be 2 to 3 metres in diameter and occupy 20 per cent of the area in dense infestations, can impede water penetration and seed lodgement. This inhibitory effect can last for 70 years or more.

SYMBIONTS ON SYMBIONTS: *TRICHONYMPHA*

Trichonympha, which lives in the hindgut of termites, is a wondrous example of symbiosis.[7] It is an Archaean bacterium that has provided 'holdfasts' on which free-swimming spirochaetes can dock, propelling the bacterial cell through the water. The bacterium enables the termite to digest lignin and in return gets its food from its host—a symbiont with its own private symbionts!

Spirochaetes are proton-powered bacteria that ferment carbohydrates and whip about with a corkscrew motion. When they attach themselves to bacteria like *Trichonympha*, those which operate in the sac at the back act like rotary motors pushing it along, those attached to the other end act like aeroplane propellers, pulling it along. The spirochaetes feed on the metabolic leftovers of the cells to which they attach. So here we have a symbiosis between bacteria that enables mobility and feeds the organisms that are providing the service, and a symbiosis between the complex of bacteria and the termite that provides the basic nitrogen resource that they all need.

The use of termite mound material in medicine

Aboriginals eat termite mound material for its medicinal value.[139] The clay-rich material is analogous to the kaolin used in Western medicine to cure diarrhoea. Communities from Central Australia to Bathurst Island prescribe termite clay for diarrhoea; women use it as relief from abdominal pains, menstrual pain, upset stomach, or during pregnancy. In other regions it is used to alleviate symptoms of mineral deficiency, or to retain flavour in meat and fish. It can be used simply crushed, or baked in a fire before being soaked and crushed, and the mixture taken as a drink. Mounds of *Nasutitermes triodiae* at Daly River were found by scientists at the Northern Territory University to have more calcium, potassium, sodium and magnesium than adjacent soils. The saliva and faecal material incorporated in the mound material results in the enrichment.

Material from mounds of *Ahamitermes vitiosus* is used by women of the Elliott community in the Northern Territory to stimulate milk production in lactating women. Analyses showed that the water extract used made 38 per cent of the calcium available—and that amount might have some benefit.

Ants

Worldwide there are 15 000 known species of ants, distributed from the polar regions to the tropics, and particularly abundant and important in ecosystems in dry regions. They can form large colonies; many burrow extensively within the top 2 metres of the soil; some have surface mound nests, but most species excavate under stones or logs and do not have superstructure.[73]

A cement-filled nest of *Camponotus* from which the surrounding soil has been washed, revealing the patterns of galleries. *Peter Mitchell*

TYPE II MOUND OF *CAMPONOTUS INTREPIDUS*

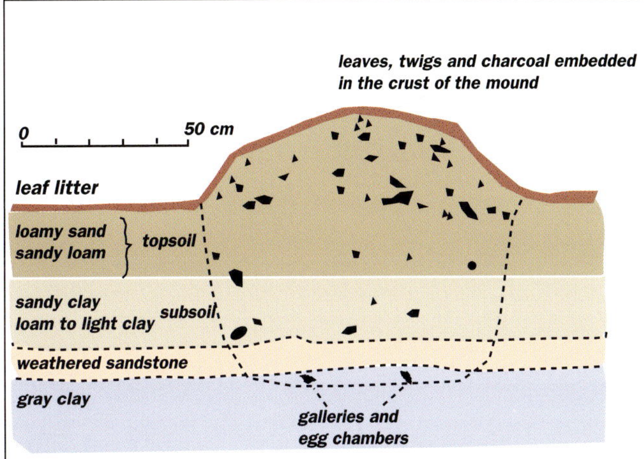

After Paton, Humphreys & Mitchell [73]

Section through a mound of *Camponotus intrepidus* from which the diagram was made. *Peter Mitchell*

Ants are as abundant and important in bioturbation in arid regions as are termites. Australia has 1275 described species of ants and unknown numbers yet to be described. With so much of the continent under arid regimes, Australia has aptly been described as the 'continent of the ants'.

Ants perform other functions as well as bioturbation, interacting with with soil processes in many ways.[140] They:

- increase soil porosity and infiltration;
- alter size and abundance of certain soil particles;
- reduce bulk density;
- regulate erosion by depositing subsoil on the surface;
- regulate plant and animal litter; and
- concentrate organic matter round their nests.

Some desert ants are scavengers of seeds, fruits or animal remains; some are carnivores, some prey on termites. All provide a source of food for other invertebrates, as well as for the thorny devil (*Moloch horridus*) and other reptiles, for birds, and even for some mammals. Ants have differing times for above-ground activities. Some are diurnal, out at the beginning and end of the day; others are nocturnal; and some are only active in the middle of the day.

Many of the mound-building ants construct mounds only when their underground nest is well established, carrying the soil pellets away and scattering them irregularly on the surface until they are ready to build a superstructure. (Some of the bulldog ants, *Myrmecia* spp., appear to operate in this way.) There are two types of mound: Type I, in which material is simply deposited on the surface and very susceptible to erosion; and Type II, which is larger, compacted, often cemented, resistant to erosion and repaired when damaged.

Camponotus intrepidus mounds, Type II, near Sydney average 20 centimetres high, are elliptical and cover about a square metre. They are formed around clumps of grass or sedge and covered by a thatch of leaves, twigs and charcoal fragments worked into the surface. The nest underneath extends down almost a metre. The mound and topsoil beneath it are enriched with silt and clay. Chambers and galleries are concentrated in the mound with only two or three spiral galleries extending below into the largely undisturbed clayey subsoil. Nests persist for decades; although mounds may involve about 50 kilograms per hectare, their slow breakdown means an almost negligible contribution to soil formation. (The sugar ant *Camponotus consobrinus* has no signs of its nest on the surface, but creates large underground disturbance.)

The meat ant *Iridomyrmex purpureus* makes larger mounds, covered by coarse particles (fine gravel size), and near roadsides uses construction aggregate and sometimes windscreen glass; it will even use superphosphate pellets near farmland. The capping prevents erosion and aids absorption of water. Meat ant mounds have been known to persist for up to 70 years, probably longer. The hard mounds create surface diversity and a better-watered perimeter.

Seed collecting (granivory) is largely a function of ants and birds in Australia. Other continents have many rodents in this niche, but Australia has an impoverished fauna of seed-eating mammals with few rodents, and those that occur are variable in numbers and mostly rare. Harvester ants are present in greater richness of species and in numbers similar to those found in the North American arid zone, where rodents and harvester ants are of equal importance in ecosystem function.

Seeds are one of the most stable resources available to consumers in arid ecosystems. They are often abundant, persisting through dry times, and are a nutritious food that can be collected and stored.

In the tropical savannas of northern Australia, ants are probably as important in granivory as birds are. Most of the species involved are in fact omnivores with seeds as a part of their diet when available. However, there has been one specialised endemic radiation of exclusively seed-eating ants of the genus *Meranoplus*, comprising many species, mostly undescribed.[141] The different species are mostly very specific in their seed diet. For instance, one species only collects the seeds of *Sorghum intrans*, the dominant annual grass in the Kakadu region. Colony activity is timed to seed availability, the first foragers appearing at the end of the wet season when seed fall occurs, and activity declining during the second half of the dry season as supplies are depleted. It has been estimated that about a third of the seed produced by this grass species is taken by these ants, but more than enough remains for regeneration, except perhaps in unusually dry years or when introduced animals increase grazing pressures and decrease seed-set.

Ants and bioturbation

The funnel ant *Aphaenogaster barbigula* is common on sandy soils dominated by cypress pine (*Callitris*) and mallee communities in semi-arid eastern Australia, where it builds nests with conspicuous collared entrances. It has been studied in the Yathong Nature Reserve[140], about 140 kilometres south-west of Cobar in western New South Wales, to assess its role in bioturbation and other soil processes.

The soils are massive red earths of aeolian origin and the Reserve has not been grazed by domestic livestock since 1977, so by 1993, the time of the study, it was in fairly natural condition. It was found that nest entrances remained active for about nine months and changed location twice a year, resulting in an annual bioturbation of 336 grams per square metre, which was equivalent to the annual development of a layer 0.28 millimetre thick. It was estimated that 92 per cent of the total soil profile would be removed in construction of entrances over 100 years; 100 per cent within 200 years. This is regarded as proof that these soils are largely biogenic. The incorporation of organic matter around entrances and in tunnels adds to soil fertility.

Studies of *Aphaenogaster barbigula* to estimate the effect of its funnels and nests on water penetration showed that by controlling infiltration through the construction of surface soil features, the development of moisture-rich patches was accentuated in an environment where moisture is limited.[135,142] Another species of funnel ant, *A. longiceps*, a nocturnal insectivore, makes crater rims of pure sand grains without the gravel, clay and coarser particles occurring in the surrounding soil. Each funnel leads to a chamber with horizontal galleries extended to connect adjacent mounds and shafts.

Ants are important bioturbators in the mulgalands.[143] Two species of mulga ants in the region north of Alice Springs build fences of mulga phyllodes 10–15 centimetres high around nest openings. These ants are nocturnal and omnivorous, scavenging dead insects, seeds and fruits. The protective rings prevent nest flooding when overland flow results from rain.

Ants as environmental health indicators

Because ants are abundant and diverse and interact in many ways with other parts of ecosystems, they are functionally important; and because they are easily sampled and highly sensitive to environmental variables, they are useful in evaluating the status and health of ecosystems.[144,145,146]

The composition and structure of ant communities were used to assess the success of a preliminary restoration at the Ranger Uranium Mine in the Northern Territory. A clear succession of ants over re-vegetated sites enabled comparisons to be made between sites where restoration was of different ages, taking into account variables like the effects of fire and the distance of the restoration site from undisturbed control sites that were a source of re-colonisation.

Feral ants causing havoc

The accidental introduction of South American **fire ants** into Australia has a serious potential for ecosystem damage; it has been compared with the cane toad in the extent to which its spread and activities could threaten our biota and affect our lifestyles.

The **African yellow crazy ants** that were accidentally introduced into Christmas Island and are currently decimating the land crab populations are an example of a feral species causing extinctions before our very eyes. The ants are seriously upsetting the island's unique rainforest ecology, in which crabs are the major bioturbators and controllers of understorey composition, and an estimated 15 million of the 130 million crab population have already disappeared. A combination of the further spread of the ants and a few years in which the hazardous stages of the breeding cycle, when larval crabs are

LOCATION OF THE YATHONG NATURE RESERVE

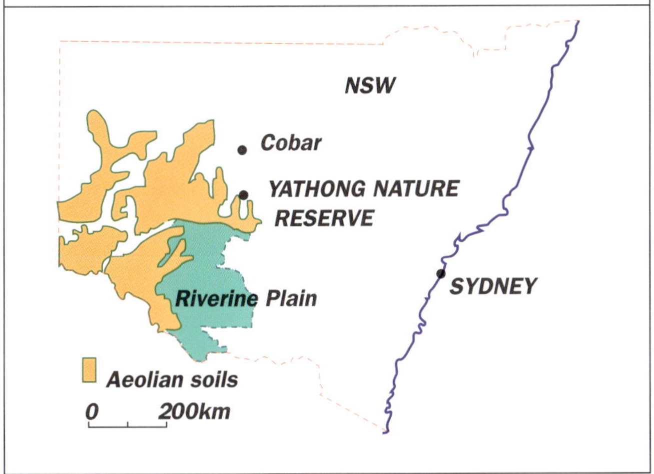

Dense *Aphaenogaster* mounds in sandy A horizon of a texture-contrast soil on the New South Wales south coast. *Peter Mitchell*

Mulga ants build a dome-shaped structure with a funnel entrance and attach mulga phyllodes to the whole structure. *M.E.W.*

maturing in the ocean, result in no replacement of populations, could see a devastating extinction event in the near future. The ants were a hidden menace until their numbers increased to a point where they formed super-colonies cooperating over vast areas. Now they are blinding the crabs with their formic acid sprays, interrupting the genetically programmed annual migrations in which the crabs have to travel to the coast to breed, and also killing off the enormous robber crabs that are part of the food chain.

Dung beetles[147,148,149]

Dung beetles are members of the scarab family. A dung beetle rolling a ball of dung away from a cowpat to bury it is an interesting sight, but hardly as inspiring as it was to the ancient Egyptians, for whom it symbolised resurrection. Illustrations of the pharaohs' after-life depicted the sun's rays radiating from the scarab's head while the dung ball it was rolling was the world, caught in a daily cycle of renewal, with the emergence of the next generation of scarabs from the dung ball perpetuating the cycle. The ancients were right about the Biosphere being solar-powered and everything being constantly recycled, at least.

Australia has 400 native species of dung beetle, scrub and woodland dwellers, specialising in coarse kangaroo and other marsupial droppings and adapted to the prevailing conditions. Inland species need to be strong fliers and have good sense of smell to find scattered kangaroo droppings. Some species even follow kangaroos, attaching to the fur and ready to hop onto the pellets as they drop. A completely wingless species in Queensland lives in vine thickets where the rock wallaby comes to roost—it has no need to fly. However, the native species simply could not cope with the dung that arrived on the scene with the introduction of cattle, sheep and horses in huge numbers. The massive proliferation of flies that breed in dung which followed the introduction of hoofed animals was reason enough to seek a solution. The sheer volume of dung produced is daunting. An adult bovine produces about 12 cowpats a day; a herd of 50 cattle can cover 50 square metres with dung in a day. Each pat can produce 300 bushflies in two weeks. While there were only native dung beetles around, the pats dried out and remained on the surface for months or even years.

Continents like Africa with its herds of ungulates have seen co-evolution of a suite of dung beetles about 2000 species strong that recycles the dung efficiently, greatly increasing soil fertility.

Between 1968 and 1982, CSIRO introduced 50 species of dung beetles into Australia from Asia, Europe and Africa,

LIFE CYCLE OF DUNG BEETLES

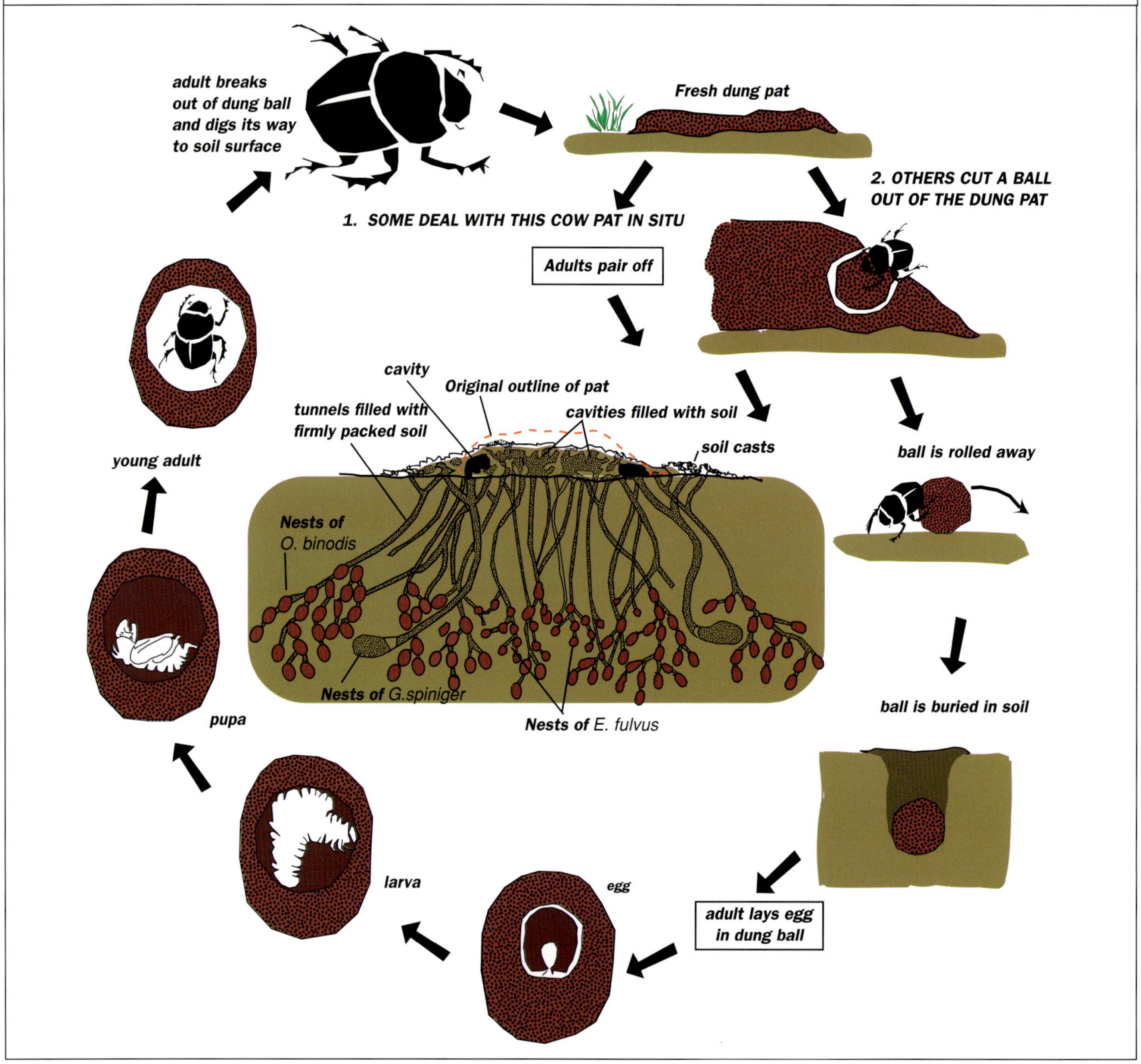

aiming to match them to different climatic conditions. About 26 species have become naturalised and are spreading from their introduction sites. Some deal with the pat *in situ*, simply tunnelling below it and carrying the dung down to make chains of small balls that become egg-chambers and house the young while they metamorphose. Several different species can be involved in processing of a single pat in this way. Others cut out portions of the pat as balls, rolling them away and burying them in soft soil at some distance from the pat, laying a single egg in each ball.

In the pastures where their introduction has been most successful, dung beetles perform a number of important soil functions. Their tunnelling increases aeration and infiltration; burying the dung and breaking it up fertilises the soil, distributing valuable organic matter; and they prevent fly breeding and the development of localised patches in pastures with rank vegetation, such as tends to occur round unburied old pats. In a continent with soils that are inherently poor with low organic content, their activities in pasture improvement are most significant.

Use of dung beetles for disposal of dog droppings

Dog droppings are an urban problem and contribute to high phosphate levels in waterways. Coliform bacteria from the droppings abound in stormwater run-off from urban streets. Some innovative councils, like Warringah on Sydney's northern peninsula, where there are more than 15 000 registered dogs, have introduced dung beetles into the lawns in their parks and reserves, with considerable success. Beetles will fly up to two kilometres to reach other places where their favourite food is available.

Vertebrates and the soil

Australia has a known mammal fauna of 245 species, of which 17 species (8 per cent of terrestrial mammals) are believed to have become extinct over the last 200 years.[150] A further 43 species (20 per cent) have disappeared from at least 50 per cent of their former ranges. (Twenty-six of these occur only as remnant populations, occupying less that 20 per cent of their former ranges.) The geographical pattern of this decline has been markedly uneven, with a high rate of extinctions in the south-eastern and south-western woodlands and grasslands, areas that have been heavily modified by agriculture, but with the highest rates of decline and regional extinctions reported from arid and semi-arid central Australia, where there has been little deliberate modification of the environment since European settlement. The biota of the monsoonal tropics had been considered, until recently, to be immune from the declines, but recent surveys and a report to the Kakadu Research Advisory Committee[151] has shattered this comfortable assumption. (See Part Four, page 142.)

The mesofauna, animals ranging from 0.035 to 5.500 kilograms, and mainly marsupials and rodents, has suffered most in the decline.[152] Regions where the animals have been worst affected are the southern arid zone (spinifex country and marginal grazing lands) and the wheat belt of Western Australia; least affected are the tropical north (where this may no longer apply), and south-west Queensland, the subtropical north-east, the coastal fringe of New South Wales, and Tasmania.

Ecological disturbances over 200 years of European settlement have included hunting; habitat clearing and fragmentation for agriculture; grazing (and over-grazing) by sheep, cattle, rabbits and other feral and domestic stock; changes in fire practices; and increased predation by foxes, cats and dogs.

Intensive hunting of native 'vermin' occurred from the 1850s to 1930s—in some cases to control 'pests', in others for skins. In New South Wales alone, according to Government Gazette information[153], 21 million mammals were killed under bounty in the period 1883–1920, while in Queensland over the much the same time span (1877–1917), 26 million kangaroos, wallabies, rat-kangaroos and bandicoots were killed by bounty-hunters. Invasion by rabbits and foxes, and changing land-use practices were, at the same time, proving the final straw in fragile ecosystems. In the light of such figures, it is remarkable that numbers of extinctions were not even higher.

The very early decline of numbers of bettongs and stick-nest rats was recorded by the zoologist G. Krefft[154], who wrote in 1866:

> This burrowing *Bettongia* has long retreated before the herds of cattle with which the plains bordering the Murray are now stocked; and it is no longer to be found south of that river … [and the stick-nest rat] is one of the many species which will soon be extinct, as I found that it had already retreated before the herds of sheep and cattle across the Murray …

PAST AND PRESENT DISTRIBUTION OF THE WESTERN QUOLL

WESTERN QUOLL
Dasyurus geoffroii

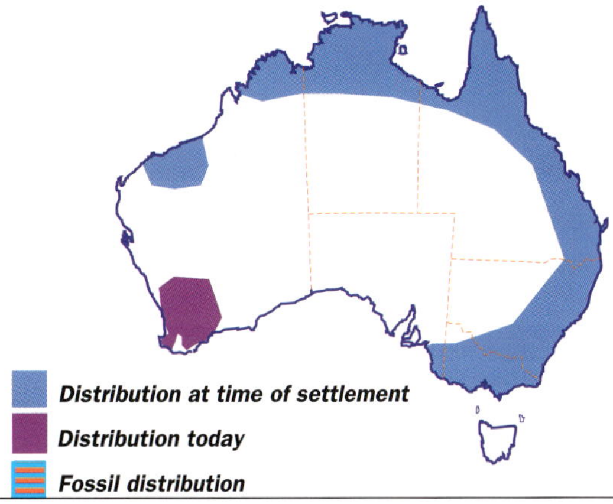

PAST AND PRESENT DISTRIBUTION OF THE EASTERN QUOLL

EASTERN QUOLL
Dasyurus viverrinus

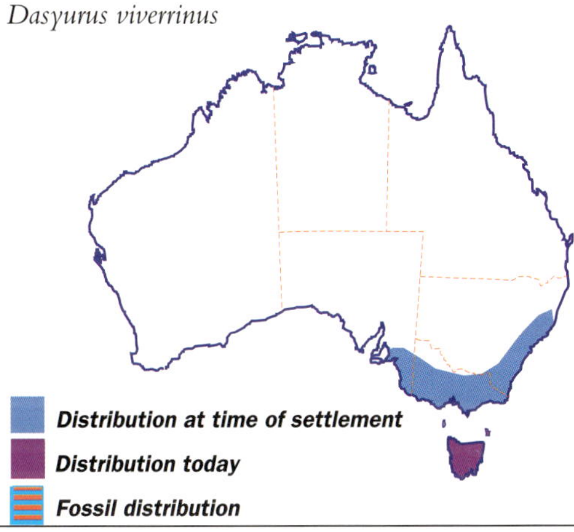

The rare, endangered or increasingly threatened status of two of Australia's four quoll species illustrates the plight of one of the kinds of small marsupials. (Two other species are native to New Guinea.)[156]
All the distribution maps in this section are reproduced from Strahan[166]

Now virtually confined to Tasmania, the eastern quoll, the 'dreaded enemy of poultry', will not thrive there either if foxes, recently illegally introduced, become widespread predators. *Earth Sanctuaries Ltd*

What has been little recognised is the effect of the loss of the burrowing, scratching and soil disturbing mesofauna that played a significant role in Australian ecosystems.

Archival records of the museum collector Charles Hoy, who collected mammals for the US Museum of Natural History between 1919 and 1922, have recently been studied[155] and provide a valuable snapshot of the fauna of the time, continent-wide. He recorded a range of threats to mammals in temperate Australia that had operated over the previous decades—introduced foxes and cats, poisoning and trapping to control rabbits, land clearing and hunting, and regular burning of forests. The arid land faunas were noted as being especially vulnerable. In contrast, he saw few threats to mammals in tropical regions. The information he provided on the sequence of decline in the mammals is enlightening. The larger native rodents were the first to decline, and mammals the size of small wallabies were next. Medium-sized mammal decline coincided in several cases with times of rapid fox invasion, while disease and predation from feral cats explained the decline of specific mammals.

The eastern quoll was one of the commonest bush animals, according to Horace Wheelwright in 1861. In 1863, John Gould said quolls were the 'dreaded enemy of poultry' and one of the settlers' greatest pests. They were poisoned with strychnine, shot, or killed by dogs; their habitats were lost progressively, but they were still abundant until the end of the nineteenth century. Then disease and foxes almost removed them from the mainland. Some recovery was possible in Tasmania after the disease died out, because there were no foxes. The western quoll, or chuditch, once widespread, is now threatened. It used to occupy 70 per cent of the continent, from forests to deserts. Now its last stronghold is in the south-western corner of Western Australia, where its survival has been aided by an active fox-baiting programme. Male quolls only live for a year, females for two, but they probably breed only once, producing more babies than can be accommodated by their four teats, and many newborns die. The northern quoll has fared somewhat better. It is not afraid of people and has been known to invade shops and houses in the Kakadu National Park.

MAMMALS AND BIOTURBATION

Worldwide, there are 777 genera of terrestrial mammals.[73] Of these, 58 per cent (447) could be regarded as potentially significant in soil disturbance. Of the 447:

- 31 per cent disturb soil by surface feeding (humans, cattle, sheep, goats, etc. are in this group);
- 58 per cent disturb by feeding and burrowing; and
- 11 per cent disturb because of their subterranean lifestyle.

Very little research has been done on the role of mammals in bioturbation in Australia (and very little, in fact, on bioturbation by other biota). Australia has no mole or similar mammal to occupy the subterranean niche. In America and Eurasia, small mammals like squirrels, chipmunks, marmots and prairie dogs are major contributors to upper-level soil disturbance. In Australia this scratching, burrowing and soil-disturbing niche was occupied at the time of European settlement by widespread populations of rat-kangaroos, bilbies and other small marsupials, and by a few native rodents. As we will see, this vital sector of ecosystem biodiversity had virtually disappeared from most of mainland Australia by the 1920s. The consequences of the loss are only now being fully realised. It has been the activities of Earth Sanctuaries, in particular, in returning these essential animals to degraded ecosystems and noting the changes that have occurred following their return that have been a revelation. A full account of this follows (pages 124–126) in a case study on Yookamurra Sanctuary.

The distribution maps for several small marsupials, showing their extent at the time of European settlement and their present distribution, illustrates just how fundamental and serious their loss has been in terms of natural ecosystem function. Brief accounts of some of the animals involved are included here because biodiversity loss in Australia (and worldwide) is the most serious consequence of human expansion and we should know what it is we are destroying (or, hopefully, working to regenerate).

Wombats

Wombats disturb the soil by surface feeding as well as by burrowing. In Victoria, burrows of the common wombat have been shown to extend for 10 metres, involving extraction of 1.2 cubic metres of soil. Measured over a 32-hectare plot, the total volume excavated by all the burrows was 18 cubic metres, but this amount was spread over the many years in which they were in use. Thus in terms of soil formation the volume from the diggings is not very significant. The same sorts of figures were found to apply in the Snowy Mountains. The huge wombat burrows that are found in Yookamurra Sanctuary, where the soil

A wombat warren at Yookamurra Sanctuary. The shallow calcrete layer on the soil surface forms an excellent roof and allows the development of large cave-like chambers beneath it. *John Gitsham*

calcrete layer forms a convenient roof for the diggings, cause great localised mounding and disturbance. It is during surface feeding that wombats cause the most overall soil disturbance, however. By pulling up roots or actively digging them out they cause 50 times more disturbance than by burrowing.

Programmes to protect the rare and endangered hairy-nosed wombat in South Australia by the South Australian Natural History Society have been successful and great increases in numbers, and improvement of vegetation in the reserve areas fenced for their protection, have rewarded their efforts.

Numbat

PAST AND PRESENT DISTRIBUTION OF THE NUMBAT

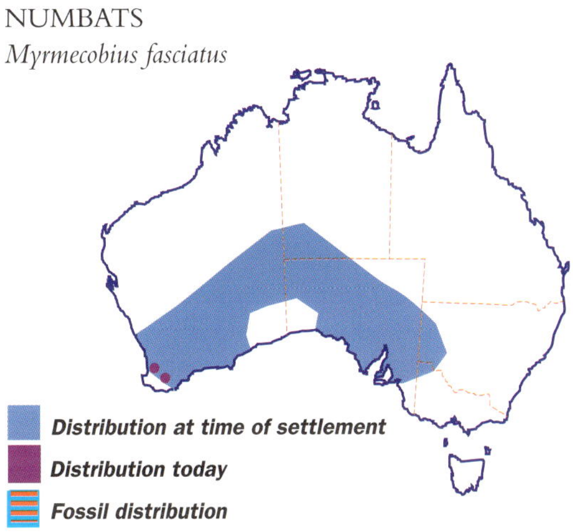

The numbat is a termite-eating marsupial.[157] Numbats were once widespread over much of southern Australia, but disappeared at an alarming rate after settlement. They have not been seen in the wild in South Australia for 140 years. Two colonies have survived on the mainland, in the south-western corner of Western Australia. There, the ubiquitous presence of *Gastrolobium* bush made poisoning of ferals easier than elsewhere, because native animals are immune to its poison (which has the same toxins as synthetic 1080) while introduced species are readily killed. However, in spite of fox-poisoning programmes, numbats were on the verge of extinction in the 1970s. Dryandra Woodland, a conservation reserve 170 kilometres south-east of Perth, had a remnant population, and there the habits and diet of numbats was researched. It was found that they were too small to open termitaria, so they feed from subterranean galleries, doing a great deal of soil disturbance in the process.

The numbat has a long tongue which it flicks into galleries opened by digging. Termites stick to the tongue, and are pulled back into the part-open mouth, and held there by the ridged palate while the tongue is flicked again. Numbats are active by day when most termites are active in their galleries. The presence of termites in near-surface soil is dependent on

Numbats nest in hollow logs, hunt for termites by day and are active diggers and scratchers and cultivators of the soil in their search for food. *Earth Sanctuaries Ltd*

temperature. Because the termites retract into deeper layers in the hot middle of the day, numbat feeding occurs early and late in the day in hot weather. In winter, they can feed all day; in spring, the greatest numbers of termites are available. A lactating numbat has been found to eat up to 20 000 termites a day. This gives some indication of the amount of scratching and digging that is involved in feeding by populations of numbats.

Numbat breeding patterns have evolved to match the season of termite abundance to the weaning of offspring, their most vulnerable time. Termites, apart from the alates (winged reproductive phase), are low in nutrition, which explains why numbats have poor temperature control and have to make warm nests in logs. Their babies grow very slowly, also as a consequence of poor nutrition. While bandicoots are weaned less than three months after birth, young numbats are not weaned till after nine months. Females give birth in January so that weaning is not until spring when the food supply is best. The young have a difficult time, it would seem, from the time they are born. The mother has no comfortable pouch. Small pink 'haricot beans', just a centimetre long, the newborn young attach themselves to teats in a depression on the abdomen, if they are lucky enough to find one. They are carried till July, when, now furred and 4.5 centimetres long, they are put in a nest and left while the mother feeds during the day, and suckled at night. A female may move to another log home, carrying her young on her back, to be closer to the food supply. The young disperse only when fully weaned and independent.

Bilby[158]

A hundred years ago, the bilby was to be found over 70 per cent of mainland Australia. Its range is now a fifth of the original range and it is classed as a vulnerable species. None have been known in the wild in South Australia for 50 years. Western Australia has a breeding programme at the Kanyana Wildlife Rehabilitation Centre, Gooseberry Hill, and another on the Peron Peninsula. The products of these programmes are

PAST AND PRESENT DISTRIBUTION OF THE BILBY

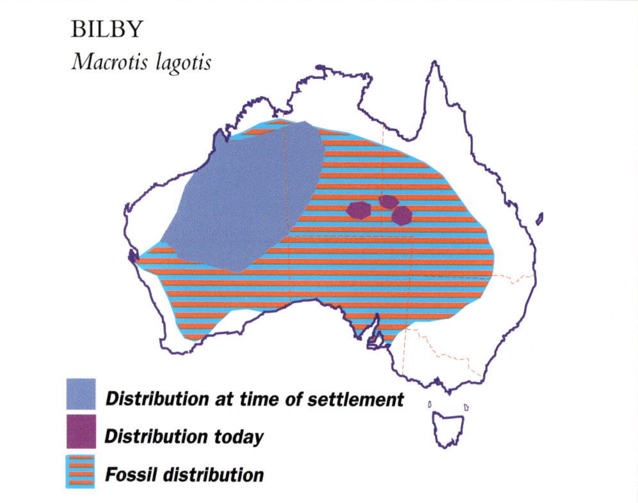

The bilby, the Australian Easter Bunny

intended for eventual release in a protected enclosure at Dryandra Woodland Reserve. Some of the progeny have been exchanged with progeny from the Yookamurra breeding programme to stop inbreeding.

The Save the Bilby Fund, run by Coles Supermarkets, the Australian Conservation Foundation and the Wildlife Preservation Society of Queensland, has raised nearly $200 000 in five years. The Easter Bunny's replacement with a charismatic Australian marsupial has been a winner for this engaging creature and has hopefully contributed to its survival.

Isolated populations of bilbies still exist on the mainland between Birdsville and Boulia. It is estimated that 600 to 700 bilbies are scattered over 90 000 square kilometres of Mitchell grass plains and stony downs stretching from the Diamantina floodplains to the edges of the Simpson Desert. Astrebla Downs National Park houses most of Queenland's bilbies. It lies 400 kilometres south-east of Mt Isa on cracking clay plains and is a reserve for scientific research only. The decrease in plant and insect biodiversity as a result of grazing had affected the bilby. Introduction of artificial watering points had increased overall grazing pressures, and dingoes and cats had increased predation. Removal of stock and programmes to control predators are giving the bilby a chance of survival. Bilby breeding can be rapid, with three or four litters a year possible, each of one to three babies. Young stay for two or three months in the pouch. Females breed at six or seven months, and are productive for four or five years.

Bilby burrows spiral down, executing a complete turn, to 1–2 metres depth. Each bilby builds and maintains several burrows—one Tanami Desert family of mother, father and one juvenile was found to have excavated 28 burrows and always kept within their home range. Where numbers are sufficient, bilbies are obviously to be reckoned with as bioturbators.

Eastern pebble-mound mice[159]

Pebble-mound mice were thought to be extinct in eastern Australia, but one species has recently been found in Queensland in dry, exposed and eroding woodlands and grassy uplands along the Great Divide from Townsville to Kilkivan; a second on open spinifex grassland and gibber plains from near Cloncurry to Camooweal. Both species are absent from the Great Artesian Basin. They build shallow burrows under their piles of 'marbles', with some pebbles also packed throughout the burrow system. The pebbles provide protection and insulation and also serve to mark the territory, identifying the occupants of the burrow. Pebbles are carried in the mouth, and rubbed on the body, acquiring trademark scents. Pebbles may be dew collectors, particularly in the very dry areas near Camooweal. Water-collecting of another sort occurs near Clermont where burrows are built among roots of eucalypts. The bark is stripped from the roots below ground and water drips into the burrows, cooling and humidifying them.

Rat-kangaroos and bettongs

Rat-kangaroos and bettongs are small marsupials within the family Potoroidae, a subgroup of the kangaroo superfamily. They are smaller than kangaroos, most species being less than 2 kilograms in weight, have shorter hind feet, construct grass nests or burrows for shelter, and have a somewhat prehensile tail. They are nocturnal and typically solitary when feeding. Four of the nine species that were widespread over the continent at the time of European settlement are either extinct or surviving only on offshore islands.[160] In New South Wales alone, about 3 million rat-kangaroos were killed by bounty hunters between 1883 and 1920. They had survived the introduction of the rabbit, the build-up of sheep numbers to a record 55 million in the state by 1892 and the major vegetation changes that resulted, and they had survived the major drought of 1888. Their decline is closely associated with the south to north advance of the fox.

It has been suggested[161] that bettongs selectively grazed shrub seedlings that proliferated after fire and that the disappearance of rat-kangaroos in the arid environments of the Western Division of New South Wales was a contributing

PAST AND PRESENT DISTRIBUTION OF THE EASTERN (TASMANIAN) BETTONG

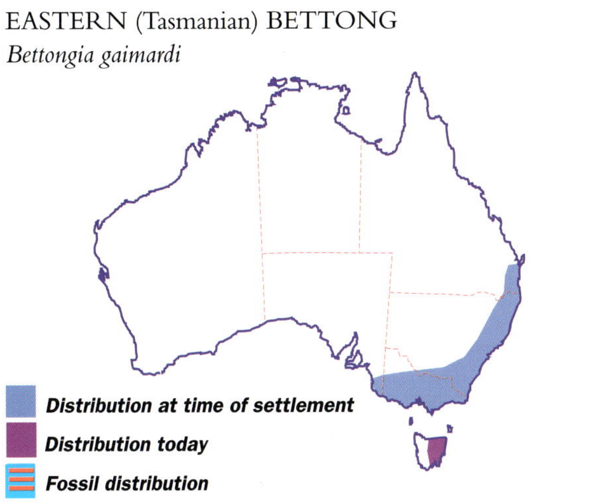

factor to the proliferation of 'woody weeds'. The traditional explanation for the increase in density of native shrubs like *Eremophila*, *Senna*, *Dodonaea* and *Callitris* in the mulga lands has been a decline in fire frequency following excessive grazing pressures, but the removal of a vital link in the ecosystems, further upsetting the balance, may have contributed. In fact, the regional extinction of both the burrowing and brush-tailed bettongs throughout the mulga lands probably affected a number of fundamental landscape processes, including patch dynamics, nutrient cycling, dispersal of seeds and mycorrhizal fungi, and plant population regulation. Biodiversity of the rangelands was ultimately impaired.

The eastern or Tasmanian bettong, *Bettongia gaimardi*, originally occupied the narrow coastal strip and adjacent tablelands of New South Wales and Victoria, just reaching into southernmost Queensland and South Australia. It was recognised as a pest of agriculture in early days of settlement.

PAST AND PRESENT DISTRIBUTION OF THE BRUSH-TAILED BETTONG

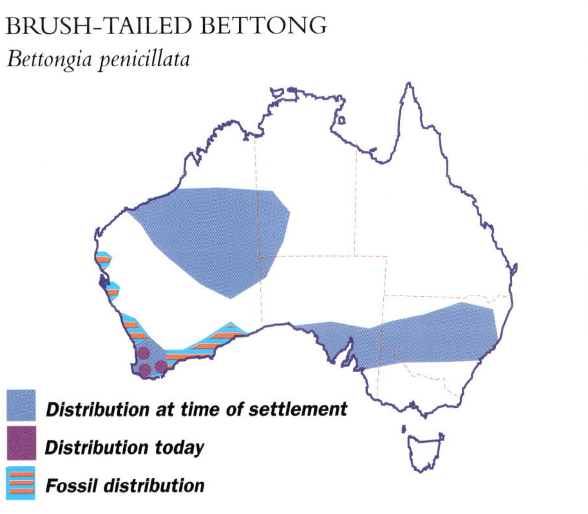

A brush-tailed bettong at Warrawong Sanctuary. *Earth Sanctuaries Ltd*

Its last record in New South Wales is prior to 1906, and there is no record of its survival in Victoria after the 1870s. It persists in Tasmania today, where it occupies open dry forest with a grassy or heath understorey in all parts of the island except for the forested western section. It feeds on underground fungi, seeds, roots and bulbs, and is strictly nocturnal, resting during the day in a woven nest of dry grass and bark, usually located under a fallen limb or among tussock grasses or low bushes. It does a lot of digging and scratching in its search for food. Its survival in Tasmania is probably due to the absence of the fox there. (The recent discovery of foxes in Tasmania may pose a threat to its ultimate survival.)

The brush-tailed bettong (or woylie), *Bettongia penicillata*, was an animal of open forests and woodlands in the better-watered parts of its range, and of sandplain and sand dune country in drier regions. Now it occurs in a few isolated remnants in south-western Western Australia, having been absent from New South Wales since 1857, last seen in South Australia in 1924, and believed to persist in the Great Sandy Desert until the 1960s. Its decline and extinction in South Australia has once again been shown to have coincided with the introduction and spread of foxes. Its food consists largely of fruiting bodies of fungi, particularly in summer and autumn, supplemented with tubers, bulbs, seeds, insects and resin. Fungi are not a satisfactory source for amino acids for mammals, but the brush-tailed bettong has a sacculated foregut that harbours special bacterial symbionts that digest the fungal material. The bacteria and their by-products are then digested in the hind-stomach and small intestine and the mixture is far more nutritious than the original fungi.

Studies of the brush-tailed bettong in the wild in Western Australia have demonstrated it plays a significant role in the dispersal of ectomycorrhizal fungal spores in the sclerophyll forests.[161] (Potoroos, close relatives, also eat fungi, particularly the truffles that have underground fruiting bodies, and are also known to be instrumental in the dissemination of the fungi, whose spores germinate more readily after passing through the animals.)

PAST AND PRESENT DISTRIBUTION OF THE BURROWING BETTONG

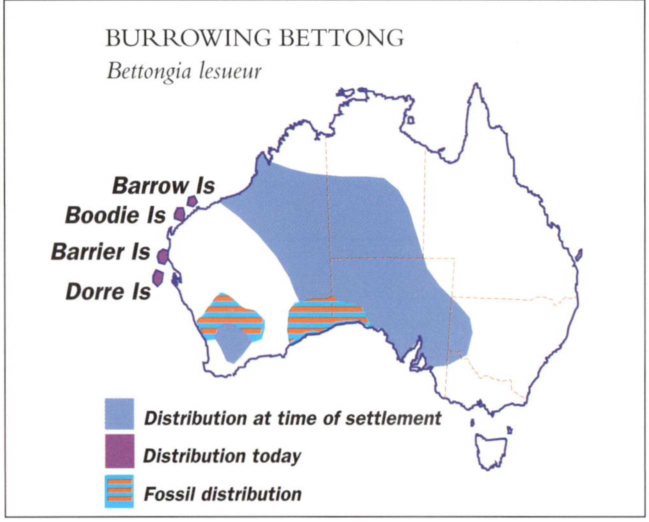

A burrowing bettong at Yookamurra Sanctuary. *Earth Sanctuaries Ltd*

Burrowing bettong warrens still form landscape features in regions where the animal is no longer found. This huge warren in the mulga lands near Glenora in the Western Division of New South Wales shows how much soil disturbance the species was capable of making when it was widespread over much of mainland Australia. *Jim Noble*

The burrowing bettong (boodie; silver boodie), *Bettongia lesueur*, exists in the wild today only on four small offshore islands (Bernier and Dorre off Shark Bay and Barrow and Boodie off the Cape Range Peninsula). It used to be one of Australia's most widely distributed mammals, found over two-thirds of the mainland. It is unique among kangaroos in that it lives underground in large communal warrens.

Warren systems with many entrances and high mounds of soil around them are still recognisable in many parts of the mainland even now, although their occupants are long gone, and they are often occupied by rabbits. The persistence of the warrens has enabled an assessment of the original density of populations.[162,163] The burrowing bettong feeds on tubers, seeds, nuts and green herbage, even climbing into low bushes to eat. It is recorded as having disappeared from Victoria by 1863 and New South Wales by 1882, persisting in south-west Western Australia until the 1930s, and perhaps in parts of Central Australia until the 1960s. Rabbits occupied its warrens but the two species coexisted for several decades, often being found in the same warren.

The burrowing bettong was successfully reintroduced to the Australian mainland in Western Australia in 1992 after an absence of 50 years.[164] Brought from Dorre Island, 90 kilometres north of Heirisson Prong in Shark Bay, to that narrow peninsula that has the same vegetation and climate, the 42 individuals had multiplied to 260 by the year 2000. The choice of the site was crucial—a peninsula that could be fenced off from the mainland and from which ferals could be eliminated, yet accessible for maintenance and monitoring; an area suitable for a captive breeding facility with the right conditions within it; and a buffer zone established outside the fence where feral control could decrease risks of re-invasion. The project is a cooperative venture between the local community of the Useless Loop mining company, the CSIRO and the Agricultural Protection Board of Western Australia.

PAST AND PRESENT DISTRIBUTION OF THE RUFOUS BETTONG

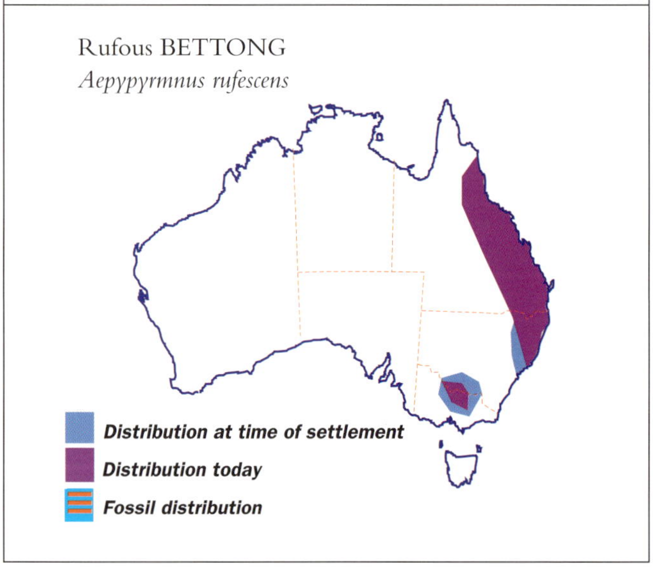

The greater stick-nest rat and the Western barred bandicoot have also been successfully returned to the mainland at Heirisson Prong from island refuges. The Shark Bay mainland used to have 31 species of native mammals and now has only twelve.[165] A future plan hopes to clear predators from the neighbouring Peron Peninsula and introduce endangered species there.

(In South Australia the Venus Bay Peninsula, 400 kilometres north-west of Adelaide, is planned as a sanctuary for release of endangered species. The Department of Natural Resources has already released brush-tailed bettongs and stick-nest rats.)

The rufous bettong, *Aepyprymnus rufescens*, remains relatively common in eastern Queensland and survives in localised areas in New South Wales, where its range has contracted from about 60 per cent of the state at the time of European settlement to about 2 per cent today. It is larger than other bettongs and occupies forested and wooded habitats with sparse or grassy understoreys. It rests in a grass nest during the day and feeds on herbs, grasses, roots, tubers and underground fungi at night.

A rufous bettong at Warrawong Earth Sanctuary. *Earth Sanctuaries Ltd*

BIRDS AND BIOTURBATION

Of the 700 species of birds in Australia, only a few are serious bioturbators: ten species extensively probe the soil during feeding; 11 rake the forest litter; 26 nest in soil burrows; 44 nest in shallow scrapes in the ground, often in colonies; 3 build incubating mounds; and 10 construct mud nests in trees and caves. Even where large colonies are present, nearly all research has been on guano, not bioturbation[73], with a few exceptions.

A huge malleefowl nest near Ongerup in Western Australia. *M.E.W.*

A study of the lyrebird, *Menura novaehollandiae*, was undertaken at Blackheath in the Blue Mountains, on a site at 900 to 1000 metres elevation, in open eucalypt forest with moderate understorey. Lyrebird display mounds had been established within 15–20 metres of the plateau edge, with feeding areas confined to the slopes and benches below. In all, 22 mounds, each about 20 centimetres high and covering a square metre, comprised this display ground. The mounds were each surrounded by a band of disturbed soil 10–20 centimetres wide and 10 centimetres deep. About 25 kilograms of soil was involved in each mound and no more than two mounds were replaced in any one year—so mound building contributes little to soil detachment. However, feeding causes great disturbance. On the steeper slopes, a vertical face was created 5–10 centimetres high, advancing to form a zone of excavation with a debris plume, up to 1.5 metres long, thrown out behind—resulting in actual terracing of the slope. On gentler slopes, feeding patches were more circular and covered several square metres. A disturbance of 6.3 kilograms per square metre per year was calculated, of which 4.5 kilograms was mineral soil.

Brush turkeys of the vine thickets and wet forests build great mounds of forest-floor litter and soil and their scratching during feeding also contributes to soil disturbance. Malleefowl, close relatives left behind in areas that became arid as the forests retreated during the drying of the Australian continent, make huge sand and debris mounds. (See pages 70–76 in *Listen … Our Land is Crying.*)

CASE STUDY: EARTH SANCTUARIES IN THE MALLEE—YOOKAMURRA AND SCOTIA

Earth Sanctuaries' Yookamurra Sanctuary in South Australia lies between the Barossa Valley and the Murray River, 100 kilometres north-east of Adelaide, in mallee country. Programmes of captive breeding of a number of small mammals that were endangered and nearing extinction, and had been absent from this region for most of the twentieth century, have been carried out here since 1988. Now that sufficient numbers have been released into the sanctuary, it is possible for an assessment to be made of how their loss from ecosystems had changed the mallee environment, and to witness the impact of their restoration.

Kangaroos in the mallee at Yookamurra. *M.E.W.*

Scotia Sanctuary in New South Wales, situated about halfway between Broken Hill and Wentworth (on the western side of the Silver City Highway) was started in 1997. It comprises a range of vegetation types including saltbush plains, belah (*Casuarina*) woodland, arid grassland and some mallee. It is involved in the same sorts of programmes as Yookamurra.

LOCATION OF YOOKAMURRA SANCTUARY

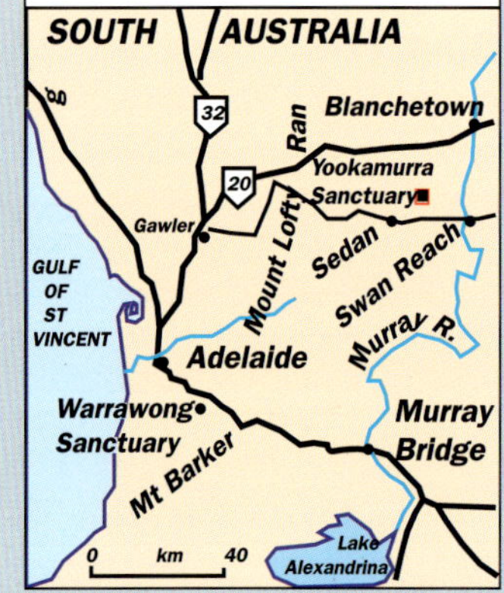

Yookamurra and Scotia, like all Earth Sanctuaries' properties, are surrounded by vermin-proof fences, and feral animals have been eliminated from the enclosures.

Much research has gone into the structure of the 1.8 metre fence in order to ensure that ferals are kept out, the captive-bred animals are safely contained when they are released, and kangaroos and emus can come and go. The lower metre of the fence has mesh fine enough to stop rabbits, foxes and cats getting through. Skirts of mesh on the ground on either side prevent burrowing under the fence. Alternate electrified and non-active wires above the mesh rise to 1.8 metres, and it has been found that emus can part the wires and roll through on their chests, while kangaroos perform a sideways dive to get through. The current in the live wires does not deter these bigger animals but eliminates entry by foxes and cats. There are now two such fences around the perimeters, with a 40-metre buffer zone between, because it was found that malicious people threw cats over the original single fences. Constant surveillance and repair of fences is a priority and the buffer provides an extra line of defence. Double fencing is to be the model for all new sanctuaries.

Spinifex and mallee at Scotia Sanctuary. *M.E.W.*

THE VERMIN-PROOF EARTH SANCTUARY FENCE[167]

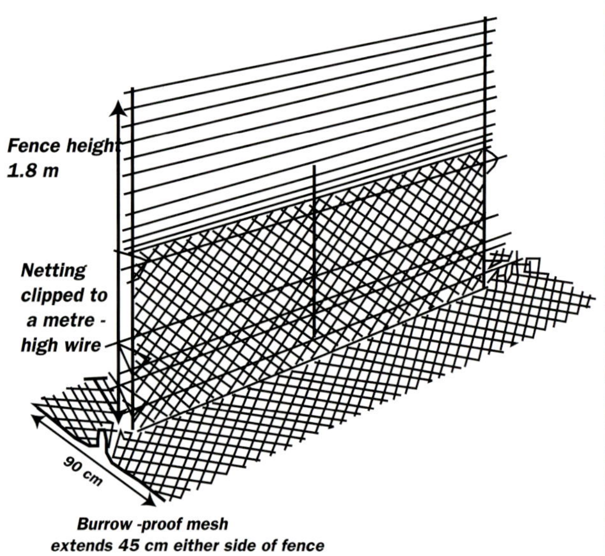

Captive breeding programmes have been underway at Yookamurra for some years. Animal numbers in 2002 were conservatively estimated at: 50 bilbies, 25 burrowing bettongs, 155 brush-tailed bettongs, 30 stick-nest rats and 151 numbats. At Scotia, numbers were estimated at 30 burrowing bettongs, 194 brush-tailed bettongs, 8 mala, 11 plains rats, 30 stick-nest rats, 41 hopping mice, and 82 brindled nail-tail wallabies.

Semi-arid mallee vegetation is widespread on the sandplains and dunefields of the western part of the Murray Basin in the Western Division of New South Wales, and in adjacent parts of South Australia and Victoria—aeolian landforms where the biota is adapted to harsh conditions. The dunefields comprise short, low, linear dunes aligned west–east, rising 2 to 6 metres above the intervening swales and usually between 250 and 1000 metres

The numbat is one of only two Australian marsupials that is active by day. It is a delight to see it fossicking for termites in Yookamurra and Scotia Sanctuaries and in the Dryandra Reserve in Western Australia, where it has been re-introduced and is breeding satisfactorily. *Earth Sanctuaries Ltd*

The lateral root system of a fallen mallee tree shows the contortions that result from trying to find ways through the calcrete layer—like the pot-bound roots of a house plant. Lumps of calcrete are enclosed in the root ball. *John Gitsham*

As mallee grows and its roots find ways through the calcrete layer that lies near the surface at Yookamurra, it pushes up the lumps of calcrete broken by its roots. *M.E.W.*

apart. The south-facing slopes are usually steeper than the north-facing slopes and contain 7 to 20 per cent clay and 5 to 15 per cent calcium carbonate formed as alluvial B horizons. In many places a hard layer of groundwater calcrete lies below the shallow soil A horizon and tree roots have to find cracks to reach the watertable in the sediments of the Woorinen Formation. Soils are essentially poor, the end-product of weathering without renewal through great lengths of geological time, and such nutrients as are present are concentrated in the surface layers where the organic matter is recycled by the biota.

The condition of the upper soil is critical to the germination of seeds, the maintenance of water balance and the cycling of nutrients. The nutrient levels in the upper layer depend on the organic content. Patchy soil disturbance by small mammals can have a profound effect on water harvesting and moisture retention in surface soil, factors that are essential for the establishment of functional microbial and plant communities in such arid and nutrient-poor environments.

Mallee is the arid expression of eucalypt woodland where the trees have a multi-stemmed habit and grow from large underground lignotubers. Old-growth mallee is 'untidy' woodland, with a clutter of old timber and bits and pieces between live and dead standing trees that are often contorted and ragged, with their branches piped by termites. The understorey varies, from sandy areas where spinifex dominates to shrubby or grassy areas with herbs to stretches where much of the ground for most of the time is almost bare, with a dry biotic crust containing lichens.

It is hard to imagine that this sort of country was originally quite productive and was extensively (and excessively) grazed in the early days of European settlement. Greg Martin, project manager for Earth Sanctuaries, has made a search of archives to find out what the region was like then and has used the information as a basis for his interpretation of the factors that created the mallee environment as it exists today. His work is the source of the information presented here, and his are the insights into environmental change and the part played in it by the removal of a major component of the ecosystem—the small mammals whose 'gardening' activities had contributed to the well-being of the mallee in its original condition. I have been privileged to see for myself, on memorable night walks in Yookamurra, the activities of the re-introduced fauna and their beneficial results. The simplistic assumption that the impact of hard-hoofed animals was the cause of soil compaction and hard-capping is challenged by the new view of things that emerges from the Sanctuaries' study, and even the failure of the landscapes to recover when the grazing pressures were removed can be seen in a new light.

The pre-European Murray Plains landscape as reconstructed from archival material[168,169,170]

Early records of soil condition and vegetation in the mallee country leave no doubt that much of the landscape supported productive, species-rich grassy understorey or saltbush, and the ground was soft, the soils were friable. Evidence given to the Royal Commission into the Western Lands in 1901[150,150a], when the crash had come and the plight of farmers in the Western Division was under review, emphasises the changes that European occupation had brought. From soils that were:

> soft, spongy and very absorbent [and where] an inch of rain, then, in spring or autumn, produced a luxurious growth of

fresh green grass, [there had been a] gradual deterioration of the country caused by stock, which had transformed the land from its original soft, spongy, absorbent nature to a hard, clayey smooth surface ... which instead of absorbing the rain runs it off in a sheet as fast as it falls, carrying with it the surface mould, seeds of all kinds of plants, sheep manure, sand, etc., to enrich the lower lying country and plant it with pine, box and other noxious shrubs (James Cotton, stock inspector).

Another farmer giving evidence remembered:

When I got away from the track the horses went up to their fetlocks in loose, friable soil ... it was like a well-tilled field; and the moisture, as it fell, penetrated the soil and fertilised the plants. Now it is scarcely an exaggeration to say that the very ground I am speaking of rings under the horses' hooves ... (George Riddoch).

The hardening of the soil surface happened rapidly. Diaries of early pastoralists indicate that the transformation was, in fact, regarded as necessary for profitable livestock production. Hardened and sealed surfaces allowed run-off during rainfall, and waterholes and dams filled, whereas the original spongy soils had absorbed the rain as it fell. In landscapes without permanent riverine water flow, the filling of reservoirs was all important. The pastoralists were not aware of the impoverishment of the landscape that would follow when surface soil had been washed away with nearly all the nutrients and soil biota. (Numbers of micro-organisms in semi-arid soils are comparatively low, and concentrated in very superficial layers; their activities are restricted to periods when the soil is damp.)

The expansion of pastoralism with its over-grazing that, coupled with drought, had resulted in the crisis in the semi-arid grazing lands, including the mallee, was accompanied by the eradication of the small mammal fauna (as described above and subsequently). Rat-kangaroos, potoroos, bandicoots, bilbies and numbats went into sharp decline and were soon rare or completely absent. All are scratchers and diggers, making pits or holes as they forage for food. They turn the surface of the soil over, incorporating organic matter and aerating it, improving infiltration, and spreading seeds and the spores of fungi. Their activities provide substrate for micro-organisms that improve nutrient and mineral cycling. The scratchings and pits are suitable sites for seed germination and establishment.

The brush-tailed bettong makes between 20 and 100 diggings a night in the course of its foraging. It has been calculated that each animal can churn up more than 6 tonnes of soil in a year—so their activities can constitute major bioturbation. The burrowing bettong and the other scratchers and diggers also disturb and redistribute considerable amounts of soil. Their activities are particularly important in the water-repellant soils that are common in many areas in southern parts of Australia, and where their scratchings occur these soils become easier to infiltrate.

Return of the scratchers and diggers and observed environmental changes

The amount of overnight scratching and digging that is revealed on a morning walk through the Sanctuaries is surprising, once one is tuned-in to notice it.

For those who can monitor the overall situation as it changes through the seasons and after rain, the cumulative and longer-term results of the return of the small native mammals will become clearer. Grasses are already returning where biotic crusts and little else carpeted bare ground, and plants that had become rare are spreading more widely again. Though no scientific experiments have yet been started to monitor changes, it is presumed that soil fungi will return in greater numbers, as it is known from overseas work in deserts that they concentrate in the micro-habitats containing organic matter that result from the diggings of small mammals. Increased microbiological activity, of which the fungi are a part, results in more breakdown of organic matter and more available nutrients for plants. Re-establishing a functional network of organisms, from microscopic upwards, cannot fail to halt and reverse degradation of the ecosystems.

SCRATCHINGS AND DIGGINGS AT YOOKAMURRA

A brush-tailed bettong's digging in the biotic crust. *John Gitsham*

Numbat diggings among the yellow and grey lichens and green mosses of the biotic crust. *John Gitsham*

BIOTIC CRUSTS

An algal crust, like a dark stain, covers the soil and prevents erosion in the Bungle Bungle National Park. *M.E.W.*

Many surfaces of apparently barren, lifeless deserts are in fact colonised by tiny lichens, mosses and liverworts, part of a biological crust that also comprises microscopic fungal hyphae, cyanobacteria and bacteria. In their dormant state, which is most of the time, the crusts may be almost invisible to the untrained eye. It is only after rain that they become active and their macroscopic elements expand. On the hard, crusted soils of Yookamurra Sanctuary in the mallee, these crusts are very visible even in their resting state, covering the open ground that resulted from the years of over-grazing and erosion.

It is the ability of the crusts to be reactivated by very light rainfall, even by the heavy dew that often occurs in arid areas, that makes them so well adapted to arid environmental conditions. Because of this ability, nitrogen-fixing cyanobacteria are able to fix small amounts of nitrogen spasmodically, building up a store of nitrogenous material in the crust. When sufficient rain comes for infiltration through the crust, the nitrogen is available to plants.

The crusts are also a habitat for microscopic invertebrates that digest cyanobacteria and lichens, releasing nitrogen. The larger animals supported by the food chains begun in this way increase filtration through the soil with their burrows.

Severe trampling by sheep destroys crusts and it is very difficult for them to re-establish. In the re-colonisation that follows disturbance, microbiota are the primary colonisers. The return of vascular plants, when it happens, is facilitated by the nutrients supplied by the crust, and by the surface roughness that traps seeds and increases water penetration.

Dry biotic crust with lichens and scratchings at Scotia Sanctuary. *M.E.W.*

Dry biotic crust with lichens at Yookamurra Sanctuary. *M.E.W.*

Microbiotic crusts in semi-arid and arid rangelands[171,172,172a-f,173]

Microbiotic crusts comprise assemblages of mosses, liverworts, algae, lichens, fungi, bacteria and cyanobacteria that form intimate associations with surface soils. They occur widely in Australia's arid and semi-arid rangelands—regions that occupy 70 per cent of the continent and are characterised by highly variable and unpredictable rainfall. In these lands, summers are hot and winters mild to cold; soils are typically ancient, highly sorted and low in nutrients; and the landscapes on which the soils were developed are often flat and highly weathered.

In the semi-arid woodlands of eastern Australia, where the dominant soils are calcareous earths and red earths, crusts and their associated biota may occupy up to 80 per cent of the soil surface in some areas. Their presence is generally thought to be beneficial in maintaining rangeland stability. Land use has been mainly grazing of sheep and cattle on native pastures. During dry periods, the biotic crusts often provide the only biological form of soil cover. When it rains, they prevent raindrop splash, thus reducing the rate of erosion, with loss of fine soil particles, in particular, being prevented. They contribute to small-scale landscape heterogeneity, creating patches with better infiltration and higher available nutrient status.

Microbiotic crusts play a major role in infiltration processes by altering physical and chemical properties of the surface layer and by causing soil surface roughness. The cyanobacteria they contain have the ability to convert atmospheric nitrogen into a form available for absorption by plants, enhancing soil fertility and the chances of seedling establishment. The production of gelatinous sheaths by cyanobacteria promotes aggregation of soil particles, while fungal hyphae physically bind sand grains together, and also produce gels.

Lichens on pebbles and in the crust, with mosses, at Wilpena Pound. *M.E.W.*

In western New South Wales, more than 120 species of mosses, lichens and liverworts are associated with soil crusts.[172a,174] Crusts are most strongly developed in the semi-arid woodlands between Bourke and Cobar, and in the mallee (*Eucalyptus* spp.) and belah (*Casuarina cristata*) plains in the far south-west. A core group of about 15 species dominates most sites. Some species have very limited and localised distribution; some occur only in places where sheep and cattle have been excluded for 50 years, emphasising the vulnerability of certain ecosystems to poor land-use practices. Sites in good condition have more species richness than sites that are degraded. While the extent of crusts (the percentage of soil area that they cover) is not a reliable indicator of the condition of a site, the more degraded the site, the more important the crusts are in increasing infiltration.

Fire affects microbiotic crusts.[172,172b,g,h] It has been shown that in the semi-arid woodlands of eastern Australia, on massive red earth soils, annual fires for seven years resulted in complete elimination of microbiota and also eliminated the cementing gels and mucilaginous sheaths that had accumulated in the soils. A decrease in infiltration resulted. Recovery of crusts took about four years after burning ceased. In the spinifex grasslands of Western Australia there was a marked increase in wind erosion following fire damage to the algal soil crusts that characterise these ecosystems. In the mallee country it was shown that recovery of crusts destroyed by frequent burning was very slow compared with rates on red earths and that a maximum cover of about 40 per cent occurred at 13 years post-fire, which was also the time of maximum fuel availability. This cover level probably represents the upper limit in these landscapes. The cover of crustose lichens increased with length of time since the last fire while cover of algae decreased, so the balance of components of the crust fluctuates, favouring algae in regions burnt more frequently than at 10-year intervals.

Microbiota are commonly pioneering species in re-vegetation of degraded soils. After disturbance by fire or over-grazing, non-vascular plants are the primary colonisers. Depending on soil type, erosion and raindrop splash lead to formation of a non-biological crust that repels water and restricts germination of seeds. The area may later be colonised by biological elements to form a microbiotic crust, or stabilised by filamentous cyanobacteria, and a succession towards re-establishment of vascular plant communities may then result. Lichens only become part of the crust after the surface has been partially stabilised. If a strong grass canopy develops, microbiotic crusts decline. In the arid shrublands, canopies seldom reach a density that eliminates crust organisms, and dense microbiotic crusts can occur under dense stands of mulga (*Acacia aneura*) and under bladder saltbush (*Atriplex vesicaria*).

Green, forked thalli of liverworts, with mosses just beginning to green, in a crust at Wilpena Pound. Dry crusts respond rapidly to water. *M.E.W.*

Lichens, mosses, liverworts and cyanobacteria in soil crusts

Lichens are the product of a complex and highly successful symbiotic relationship between a fungus, and an alga or a cyanobacterium. They are living examples of the sort of cooperation between members of different Kingdoms that characterised the early evolution of eukaryote cells, and of the complex organisms that followed that fundamental breakthrough. The fungus creates the body of the lichen from matted hyphae, but cannot synthesise its own food; the alga or cyanobacterium lives within the fungal body, photosynthesising and producing the food for the host.

About 75 per cent of Australia's lichens are crustose or scaly, those of arid-land crusts being almost exclusively of this type.[174] Crustose lichens are almost two-dimensional, lacking a lower surface and tightly bound to the soil. Some look like large bird droppings many centimetres across. Scaly lichens have a thicker thallus and are often attached to the soil by root-like structures. Some have the appearance of an aggregation of pyramid-shaped crystals, an arrangement believed to minimise heat absorption and to channel water into the centre of the thallus. The other 25 per cent are leafy or shrubby and are found mainly in less arid areas.

Lichens are only abundant in winter rainfall areas as they are damaged by heat when wet and active. If global warming means change in rainfall patterns and more summer rainfall in the arid zone of southern Australia, as has been predicted, lichens may be lost from biotic crusts and cyanobacteria may become more important, though more fragile, as land stabilisers. *Chondropsis semiviridis* is an amazing lichen that spends most of its time rolled into a ball and blowing across the soil surface.[172f] When wet it unfurls into a series of lobes. It is a 'vagrant' lichen (like those of the Middle East that were the Bible's 'manna from heaven').

The colours of lichens result from the different chemical substances in their tissues. Differences in these chemicals can be used to classify lichens. Certain chemicals manufactured, particularly when under stress, have commercial uses—for instance, in litmus paper, dyes, medical products, and poisons.

Mosses and liverworts in biotic crusts[172a] in arid regions have survival techniques enabling then to desiccate and then to reactivate when water is available. *Asterella drummondii*, the Y-shaped liverwort thallus seen in many arid crusts, has plate-like scales on its lower surface; as the plant dries out the sides of the thallus curl up and inwards, and the scales protect it until the next wetting.

Cyanobacteria in the crusts and hyphae from the lichens exude gel-like substances that cause aggregation of grains of sand, making the soil beneath less prone to erosion.

Biotic crusts cover the ground and the surface of pebbles between spinifex clumps at Wilpena Pound in South Australia. *M.E.W.*

CASE STUDY: *HOMO SAPIENS*, THE HUMAN VERTEBRATE—IMPACT ON LIVING SOILS AND BIODIVERSITY

(This section is based on a contribution by David Marsh, Allendale, Boorowa, in New South Wales. He suggested the inclusion of an account of the changes in attitude and practices by farmers and the dilemmas they face at this time, when scientific knowledge about the microscopic world is becoming generally available.)

Preceding pages of this book have documented some of the complexities of the webs of interacting organisms of all sizes, from sub-microscopic to the largest macroscopic, that comprise the living Biosphere. The links between sustainable natural ecosystems and complexity—due to biodiversity and the worlds-within-worlds that they encompass—are obvious. Having had such glimpses of how ecosystems function, even though full understanding is beyond us and the realm of the microscopic, in particular, is so little understood, equips us better to examine current land-use practices and see why they are unsustainable. Most agricultural practices involve simplification of systems and loss of biodiversity, with usually little recognition of the fundamental role of micro-organisms.

Since the beginning of settled agriculture, the aim has been to increase productivity per unit area. As human populations have grown, this intensification has become increasingly necessary. Many agricultural practices have short-term economic gains for humans but long-term negative consequences for the environment. Most involve simplification and biodiversity loss in the ecosystems that replace natural ones—mainly monocultures that need every sort of artificial support because they do not have biological supports from complex interactions within the ecosystem that is created.

Many of the products applied to landscapes have suppressing effects on soil biota; many pollute soil and water, surface and groundwaters. Many have very high energy-consumption levels in their manufacture. Modern high-input food production systems in highly industrialised countries use more energy than they produce when transport, processing and distribution are taken into account. Though such systems obviously cannot be supported into the long term, huge amounts of money are poured into research to continue this unbalanced agriculture of comfort and convenience. It is deemed profitable because the social and environmental costs associated with it are not factored into the prices of the commodities that it churns out. The belief that productivity increases can go on indefinitely and that there is a technological fix for every problem that arises still holds sway, and no one has thought about the limitations imposed by the law of diminishing returns.

One has only to look at the continuing clearance of marginal land for monoculture cropping that will involve high inputs and technological fixes before it peters out to see what is ahead. Worldwide, these marginal landscapes are the fastest desertifying regions.

In the end, human survival will depend on smarter food production. One has to wonder what might be achieved if the research funds that are currently used in trying to make the unsustainable sustainable were channelled into research to develop systems that mimic nature in their complexity and resilience.

Australian farmers and sustainable land use

Farmers are becoming more aware of the importance of soil biota. Many are realising that farming methods that foster species diversity above and below ground are more resilient. The general aim has to be to achieve high groundcover levels. Already enough information is available to help farmers to decide what options they might try in the grazing industry as well as in cropping. Trial and error will always be involved, because there can never be any fixed rules when every bit of land has special characteristics, when no two seasons are alike and when change and evolution within any ecosystem are ongoing. They will need to be aware of what ecosystems need for survival within the context of farming systems—not as a separate issue. For too long they have put conservation of ecosystems in a separate category to agriculture. The two are linked and it is the as-yet-unvalued parts of ecosystems that underpin agricultural production. Therefore, each farmer will have to develop an awareness of how much intensive production is prudent in a particular landscape before the negative trade-offs begin to dismantle the ecosystem of which they are an integral part, and which ultimately makes their productivity possible. There are opportunities for research here. Farmers need simple, cheap and effective methods of assessing the optimum matrix of land use for their particular landscape.

The opportunities that exist for **changes in grazing management** are many and diverse. Maintaining set stocking invariably caused species change in the plant cover and often promoted rates of soil loss greater than soil formation rates. Rotational, time-control, cell, holistic, pulse, high intensity/short time, are all grazing systems that limit the time when plants are exposed to animals. Longer periods between grazings promote greater species diversity and more complex age structure. Longer recovery also increases leaf area index, which could be important for returning ecosystems to a more balanced hydrology. Greater leaf area increases the potential for sunlight capture and thus maximises growth potential using a free and renewable energy source—the sun. Greater leaf area means larger root systems, which adds to soil organic matter and root biomass, both major determinants for prolific soil life. Better ground cover results in more effective use of rainfall, reducing run-off and erosion.

Rotational grazing methods have the capacity to swing the successional balance in favour of perennial plants, whereas set stocking generally favours annuals. Increased root biomass with a litter layer at the soil surface creates conditions favourable to soil micro-organisms which recycle nutrients and create stable aggregates. Mycorrhizal fungi enable increased nutrient availability and also provide an extended interface for moisture retention and water uptake.

In places where rainfall and soils can support cropping or high-input pastures, the long-term viability has to be questioned. While short-term economic advantages are loudly advertised, these need to be weighed up against some probable longer term negatives (like intensive production, high capital demand, higher product removal, accelerating acidity, nutrient leaching to groundwater and surface flows, further injections of capital to correct acidity, salinity, etc., etc.).

Many catchments have a solid base of native pasture species that have been pushed into the background by 150 years of less than ideal grazing management. There is a huge opportunity to manage this existing resource more effectively. It is possible that investment in programmes to achieve this may show greater environmental return than some of the expenditure on cost-sharing programmes thus far. Some of the proposed solutions to the salinity issues we face have been simplistic. The landscape is a complex organism and it is likely that more wide-ranging solutions based on changed management will add to any gains achieved so far.

The situations regarding changes to **management of cropping** are more difficult. Technology has enabled spectacular yields to be achieved. The trade-offs are diminishing biodiversity, low levels of soil cover, increasing salinity risk, export of nutrients, high input costs and pollution problems from residues. Although low-disturbance conservation cropping has reduced some of the soil exposure and organic loss problems, it is dependent on herbicides and insecticides which are already developing resistance, and even cross-resistance, to other herbicide groups. Thus a system that looked like an answer to many problems is in danger of breaking down within 40 years of its introduction. Again it is a question of being realistic about how much of this activity can be practised in a particular landscape, and always trying to minimise the time that soils are exposed. A checklist for farmers in a given catchment that could quickly indicate to them if the enterprise they are engaged in is a catchment-friendly activity would be helpful. This concept of catchment-friendly farming could be powerful, but it probably entails some form of accreditation which may be socially unpalatable, and it would probably need a bureaucracy to run it.

A GONDWANAN BEETLE SURVIVES IN THE MALLEE[234]

The omma beetle, *Omma rutherfordi*, recently described from the South Australian mallee and the Stirling Range of Western Australia, is a member of one of the four currently recognised, extant families comprising the suborder Archostemata. This ancient group has a lineage at least as old as the Triassic Period (245–208 million years ago). The omma beetle retains many primitive characteristics seen in its fossil relatives. *D. McClenaghan © CSIRO Entomology*

Rainforest on the Arthur River in north-western Tasmania. The continued logging of the Gondwanan forests in Tasmania is environmental vandalism. They are the largest remaining temperate Gondwanan forests in the world. The fact that they have been here for up to 60 million years and are little changed, with priceless biodiversity, should be enough to ensure their survival, unchanged, in spite of human greed. Their future economic worth as tourist attractions and as ecosystem service providers in a world that is fast losing all its significant forests far outweighs the revenue they bring in as woodchips that are exported for next to nothing to enrich foreign papermakers. *M.E.W.*

Part Four

SOME AUSTRALIAN ECOSYSTEMS:

How they Function as Units of the Symbiotic Planet

We have seen in preceding pages how Symbiosis and Life are inseparable. At every level, from macrocosm—where Life is the Biosphere—to microcosm—where each living cell is a bacterial masterpiece—the fundamental interdependence and interconnectedness of living things and their 'inanimate' global environment is evident. Looking at a few, diverse Australian terrestrial, freshwater and marine ecosystems from this viewpoint provides insights into requirements for their sustainable management or for their rehabilitation where they are suffering degradation.

Terrestrial ecosystems

Most Australian terrestrial ecosystems have to contend with poor soils, with variable and unpredictable climate, and with fire. How some of the different vegetation types are affected in their distribution by the many factors that limit their range or promote their local success is described in this section. Much has still to be learnt, however, for just about nothing is known about the microbiological aspects that undoubtedly have a major influence—any functional ecosystem involves a complete Web of Life that starts with micro-organisms, symbiotic relationships and all the complex interactions between soil and the biota.

Two recent books will hopefully lead to a reassessment of many of our attitudes and assumptions about the Australian environment. In *Australian Rainforests: Islands of Green in a Sea of Fire*, Bowman[175] gives a refreshing insight into the mysteries of rainforest distribution, demolishing most of the pre-conceived ideas that have plagued our understanding of the real, fact-based situation. It has been our habit to believe the 'experts', to accept broad generalisations and to limit our own visions of how things really are by this preconditioning. And when it comes to fire management, grazing practices and the influence of humans on the environment, David Horton's *The Pure State of Nature*[176] has for the first time really looked at our 'sacred cows and destructive myths'—and demolished preconceived ideas about these important subjects.

RAINFORESTS

A widely accepted hypothesis that distribution of vegetation types is controlled by soil fertility only holds good for parts of subtropical eastern Australia.[175] Recent research in the Sydney region provides some support for the concept that there is a relationship between taxonomic richness of rainforest and non-rainforest vegetation and total soil phosphorus concentrations, but topographic position and moisture supply also influence distribution.

Theories about what determines the location and limits of rainforest include:

- **Soil nutrient status, particularly phosphorus** Data collected from 144 rainforest boundaries in north-western Australia, where pockets of monsoon forest are encircled by savanna, showed no correlation between the number of tree species in savanna vegetation and the concentrations of

DISTRIBUTION OF MAJOR RAINFOREST AREAS

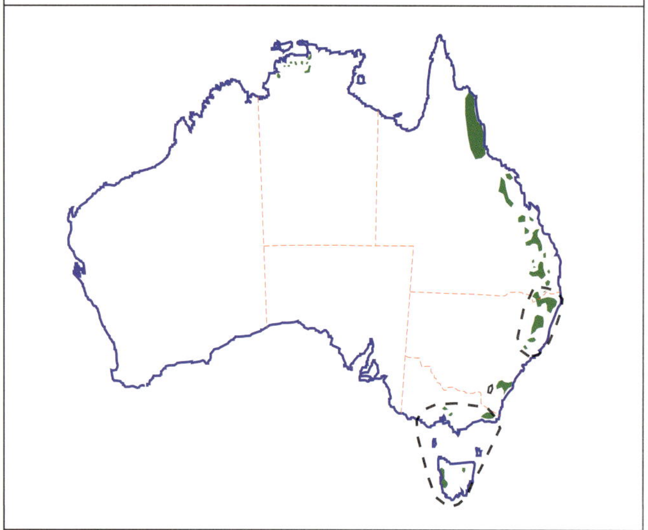

Gondwanan rainforest remnants survive along the eastern margin of the continent and in Tasmania. The original area of rainforest has been reduced by half since European settlement, and a third of what remains has been damaged by logging. The areas ringed are those in which Antarctic beech is found.

Rainforest on Cape York Peninsula, with the golden bouquet tree (*Deplanchea tetraphylla*) in flower. *M.E.W.*

soil phosphorus. Most savannas had soils with less than 500 parts per million of phosphorus, in keeping with expectations, as the northern savannas are known to have impoverished soils. What was surprising about the findings was that some of the most diverse monsoon forests had soils with less than 200 parts per million—thus, according to current reasoning, instead of rainforest the vegetation should be very sclerophyllous. (The evolution of sclerophylly is believed to be related to poor soils and the plants with this attribute that coexisted in the 'mixed' Gondwanan forest, when there was no separating out of those with the ability to cope with water stress, had a great advantage when the continent dried. The characteristics developed in response to nutrient deficiency imparted drought resistance.) Numerous recent studies have shown that rainforests occur across a wide range of soil phosphorus concentrations. For tropical rainforests, the **biological fertility,** which is the rate of circulation of the available nutrients in a closed ecosystem, is the important soil factor. (The invisible, or barely visible, organisms recycling nutrients again running the world!) Biological fertility is distinct from mineralogical fertility, and allows rainforest to flourish on nutrient-poor soils (like the rainforests that grow on pure sand on Fraser Island).

- **The importance of soil parent material** in controlling rainforest distribution has also been over-estimated. There is no evidence that rainforest is restricted to soils with specific textural properties. However, soil drainage is certainly responsible for controlling the spatial distribution of particular species of rainforest trees. On Fraser Island, the largest sand island in the world, subtropical rainforest is restricted to high-fertility sites, and eucalypt forest to low-fertility soils. The fertile soils have developed independently of the siliceous parent material, being created by the recycling of nutrients within the closed rainforest ecosystems.
- Webb and Tracey[177], whose classification of rainforest types remains the authoritative reference, suggested that tropical rainforests form a **sequence along a rainfall gradient** and are replaced by eucalypt savanna at about 600 millimetres annual rainfall. However, there is poor correlation between rainfall and rainforest. Even some species of *Nothofagus*, a genus considered to require perhumid conditions, are tolerant of water stress. It has been suggested that fertile soils allow rainforest to occur in drier climates—an 'edaphic compensation' mechanism—but this is not the case in northern Australia, where rainforest is common on infertile soils in low rainfall areas.
- **Topography** influences rainforest distribution, enhancing moisture supply—as evidenced by the restriction of rainforest to the southern side of ridges on the Great Divide.
- Other scientists proposed that rainforests grew where they were **tapping into aquifers** in seasonally dry situations, but this has been shown not to be the case.
- **Light and climate** were suggested as limiting factors. Many rainforest trees are tolerant of wide variations in light and temperature regimes. A minimum temperature gradient controls the distribution of different species across altitude and latitude gradients, but there is less evidence that minimum temperatures control rainforest boundaries.
- **Fire** Absence of fire on rainforest boundaries permits their expansion, so fireproof localities, topographically controlled, determine their location.

The conclusions reached as a result of consideration of all the evidence are:

- Rainforests are not restricted to sites that are permanently drought-free. Some occur in seasonally arid areas without supplementary water supplies from localised aquifers.
- At a local scale, rainforest boundaries are not necessarily correlated with gradients in soil water availability. Variation in water supply affects species composition, but even this relationship is complicated by the nature of the soils, topography and fire history.

Nothofagus moorei, the warm temperate Antarctic beech, in the rainforest at O'Reillys in the Lamington National Park in the Border Ranges of New South Wales. *M.E.W.*

- Rainforest's tolerance of high light and wide temperature range allows its observed expansion into surrounding eucalypt forests under some conditions.
- It is **biological fertility**, resulting from recycling of nutrients in a closed ecosystem, not mineral fertility, that characterises rainforest habitats—that is, a product of the rainforest maintains the rainforest. Mineral fertility is not the reason the rainforest is there.
- **Fire plays a key role in determining rainforest boundaries. The relationship is not simple, but rainforest trees are unable to survive recurrent fires.**

Surveys in the Northern Territory and inland north-eastern Queensland show a positive relationship between rainforest occurrence and sites with topographic protection from fire. Both dry and wet monsoon rainforest can recover after fire, but repeated burning causes boundary retreat. Some areas of savanna in the north have been considered by some researchers to represent replacement vegetation following rainforest loss as a result of Aboriginal burning practices. However, there is no evidence of return of rainforest when fire is excluded. (The current practice of biennial burning over most of the area makes it virtually impossible to know whether, in time, forest would return. Fires would need to be excluded for 20 years to be sure.) Unfavourable soil biota may be a contributing factor in preventing rainforest trees from establishing in savanna. Fire is known to deplete and change the abundance and diversity of soil micro-organisms, and to destroy some soil organic matter.

In the subtropics, incidence of fire has been shown to control rainforest boundaries. In Tasmania, rainforest can survive infrequent fire, but not repeated fires. The buttongrass plains adjacent to major Tasmanian forests are the product of Aboriginal burning of valley floors.

The softwood scrubs of the Brigalow lands, remnant semi-evergreen vine thickets that may be classified with rainforests (described below), have interesting distribution patterns.

DISTRIBUTION OF BRIGALOW-DOMINANT VEGETATION

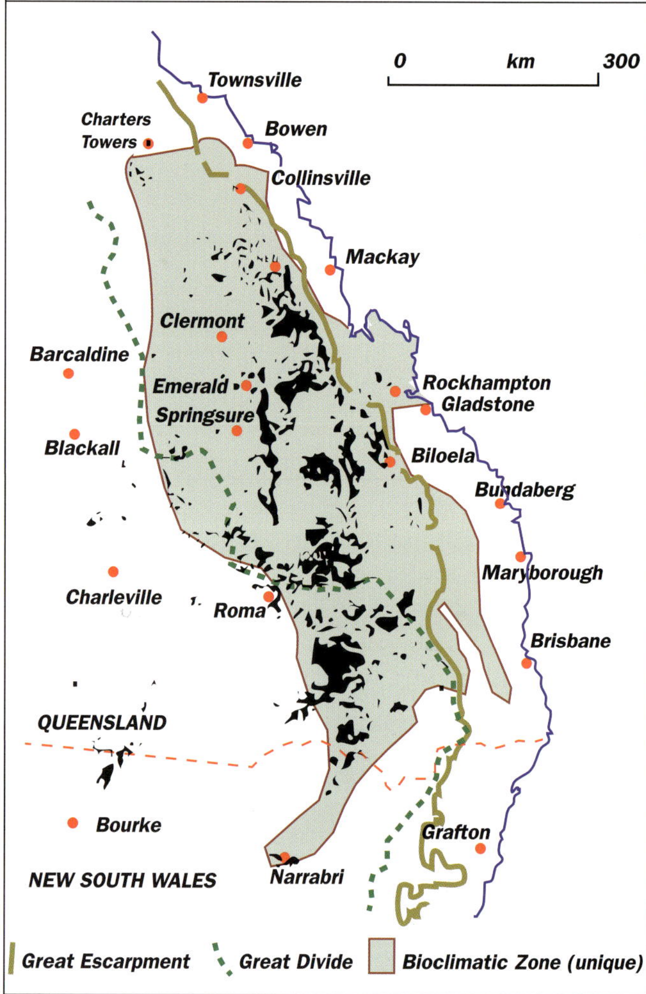

THE BRIGALOW

The Brigalow Belt, extending from just south of Charters Towers in the north to the Queensland–New South Wales border in the south, is a unique bioclimatic zone in which summer rainfall dominates, with 60 to 80 per cent of mean annual rainfall occurring between November and April.[178] The summer peak is separated from the winter peak by autumn and spring troughs. The pattern is bimodal, with a discontinuous period of effective rainfall. In the long hot summers evaporation is high, with high-intensity rainfall and run-off. This is balanced by mild winters with lower evaporation and lower rainfall. Soil water deficits occur in autumn and spring; soil water recharge occurs in summer and winter, with considerable surplus in summer. Flooding rains associated with cyclones saturate the landscape, recharge aquifers and maintain base-flow in the major river systems. Winter rainfall recharges soil moisture and reduces the severity of the driest season, the spring. However, rainfall is variable, due to ENSO, and rivers of the Brigalow domain have some of the most variable flow patterns of any major Australian rivers.

Brigalow grows on clay-based soils, many of which have high sodium content at depth. Half of the brigalow-dominant region occurs on gilgaied cracking clay soils. Rainfall averages between 500–760 millimetres (20–30 inches) per year.

The Brigalow Belt occupies a comparatively narrow discontinuous tract of country lying inland from the eastern coast, aligned north and south of the Central Highlands region of Queensland[179], an elevated, deeply dissected sandstone plateau forming an intermittent divide between northern and southern parts of the belt. The Central Highlands, an area of more than 15 000 square kilometres, is Queensland's most important inland watershed where seven of the state's major rivers have their headwaters. The sandstones comprise a Mesozoic sequence of sediments laid down in the Great Artesian Basin and subsequently overlain by lava flows from the Minerva Hills volcanics in the Springsure–Emerald districts. These underwent deep weathering, erosion and dissection into the 'broken', canyon-type landscape we see today, leaving isolated basalt remnants scattered about on ridge crests and outcrops. Thus, the Central Highlands has diverse topography, soils, moisture regimes, plants and animals in a landscape featuring numerous ranges and tablelands attaining heights of 1000 metres and rainfall of 760 millimetres.

The most highly developed brigalow and brigalow softwood communities occur within this central hub. Extending north and south, the associations become more scattered. The three sectors, south, central and north, have different characteristic associated species. Dry sandstone plant communities comprising mixed eucalypt forest and grassy open woodland with acacia-dominated understorey occupy the higher, uniform sandy regions, while a much greater diversity of plants occur on the heavier clay soils with better moisture ratios of the lowland valley floors and scree slopes. It is in such areas that brigalow scrubs with mixed softwood understorey became established, together with patches of complex pure softwood scrub.

The Brigalow, dominated by *Acacia harpophylla*, occupied at least 6 million hectares of Queensland west of the Great Escarpment; the species was co-dominant over at least another 10 million hectares in the core area before clearing of the landscape began. Rapid occupation by pastoralists saw the whole region occupied by 1861, but 25 million hectares were invaded by prickly pear (*Opuntia* spp.) between then and 1925, when the *Cactoblastis* moth was successfully introduced and its caterpillars cured the problem. In the intervening years the government had been prepared to give away useless, cactus-infected land. The Fitzroy Basin Brigalow Land Development Scheme commenced in 1962, allocating 4.5 million hectares of land north of Roma for closer settlement. Clearing, dense regrowth treated with herbicides, continual burning, and constant root raking cleared the landscape for pastures and cropping. The consequences of the clearing today are erosion, degradation of soils, looming salinisation, biodiversity loss, silting of rivers and their impoverishment downstream of storage dams. Also, the soil that once stored nitrogen resulting from brigalow's nitrogen-fixing ability, has gradually deteriorated under a monoculture of heavily grazed pastures and annual cropping.

The dangers of salinisation as a result of land clearing that affects the hydrology of a region are real. Scientists have been assessing the risks for decades.[116a–d] These four scientific papers referred to, alone, leave no doubt that the situation is real and serious (not only in the brigalow but in Queensland generally). To continue with land management practices that do not take the certainty of future problems with salinity into account is vandalism, and Queensland still has a choice—change now or accept that salinity will desertify the landscape. The Murray Basin salinity problems show what to expect.

Brigalow is adapted to the special conditions that prevail in its domain. Being a legume that has *Rhizobium* nitrogen-fixing bacteria as symbionts is an advantage. (Compared with the red earths of the mulga zone or the gravelly ridge foothills of the Great Divide, brigalow country has heavier, better quality soils, making it good dual-purpose agricultural country by Australian standards.) Brigalow's growth patterns maintain the complex hydrological balance that results from specialised climatic regimes. The plants are drought-resistant to cope with lean times, and their seeds are capable of rapid germination and growth in the infrequent wet spring seasons that occur only a couple of times a century (associated with La Niña episodes of abnormally heavy rainfall). The seeds and seedlings are amazingly salt tolerant.

Brigalow re-sprouts after superficial clearing, producing dense suckers and forming whipstick stands of amazing density in some cases. Poisoning of regrowth was widespread in the 1960s and subsequently, to a point where little of the vegetation type remained intact and efforts had to be made to protect a few remnants and their biodiversity. The endangered status of the remnants of this special ecosystem is a matter for great concern, but of no less concern is the fate of the softwood scrubs that were once widespread within brigalow territory, whose presence and now-threatened status has barely been recognised, let alone publicised.

Of the original 6 million hectares of brigalow-dominated vegetation, by 1985 a total of 30 000 hectares (including 1980 hectares of softwood scrub) had been conserved. This represents 0.5 per cent of the original scrubs. Though several small areas of scrubs have since been added to the conservation network, overall representation still remains grossly inadequate. The very specific adaptation of the brigalow communities to a unique climatic zone with complex hydrology emphasises the risks that now exist in the massive cleared areas—the substituted ecosystems are not in tune with the hydrology of the region and ultimately extensive salinisation is inevitable. (A general account of the Brigalow Belt is given in *Running Down*, pages 201–2.)

With so much of the Brigalow gone it is fortunate that records of the vegetation, the soil types and the ecosystems have been kept by Jim Gasteen, who was involved in early surveys to select areas for National Parks and conservation reserves, and who gathered a great deal of information while studying soils and their distribution, their relationship to plant communities and the effects that 150 years of land development, grazing and fire have had on both. Now retired, he has supplied a comprehensive manuscript, photographs and advice from his unique archives and experience on the little-known softwood scrubs. The account which follows is largely as he wrote it because his was the 'hands on' experience and no one saw more of the territory or understood the implications more fully.[179,179a]

THE SOFTWOOD SCRUBS

(As described by Jim Gasteen.)

Softwood scrubs in the Highlands comprise a complex diversity of shrubs and low trees with a closed canopy height of 5 to 8 metres, often with a tangle of lianas such as *Hoya australis*, *Capparis lasiantha*, *Jasminum* spp. and *Cissus opaca*. Common shrubs and low trees are *Acalypha eremorum*, *Citriobatus spinescens*, *Heterodendrum diversifolium*, *Carissa ovata*, *Alphitonia excelsa*, *Canthium* spp., *Croton insularis*, *Bridelia leichardtii*, *Denhamia obscura*, *Exocarpus latifolius* and a host of others. The following emergent trees were usually present: *Brachychiton rupestre*, *B. australe* and *Flindersia* spp., *Geijera parviflora* (tree form), *Backhousia sciadophora*, *Acacia fasciculifera*. Less common emergents were *Acacia harpophylla* (brigalow), *Macropteranthes leichardtii*, *Cadellia pentastylis* and *Bauhinia carronii*. Rare emergents locally are *Eucalyptus orgadophila*, *E. cambageana* and *Casuarina cristata*. Fringing communities are occasionally present with *Santalum lanceolatum*, *Acalypha eremormum*, *Atalaya hemiglauca* and, rarely, *Flindersia australis*.

Many of the scrubs were moving outwards from previous limits, some new ones were establishing, others were retreating with sharp boundaries. The abrupt transition from scrub to grassy open eucalypt forest in many places is a result of seasonal burning of the mid-height grass communities by cattlemen, and it is likely that the extent of the scrubs was much greater in the past and they now persist where fire has not penetrated.

Soils in the scrubs investigated were acid in the upper parts of the profile with a slight alkaline trend at depths below a metre. The softwood scrubs are completely absent from the sandstone eucalypt communities that predominate on the elevated acid sandy soils throughout the Highlands. Where the sandstone platform has been removed by erosion, exposing the Moolaymber Formation shales with their Triassic plant fossils, soils are friable loamy or silty clays and softwoods tend to come in immediately to form small patches of low (4–6 metre), dense thickets with closed canopy, extending upslope right to the base of the dissecting sandstone cap above—at which point they cut out in a clearly defined sharp boundary on encountering the sandstone proper. Because of the close proximity of retreating sandstone scarps, the scree slopes on which the softwoods have established are paved with sandstone slabs, and fragments of various sizes occur throughout the upper parts of the sandy clay soil profile.

Inside the scrubs, surface soils have a thin sandy or silty layer overlying a tangle of roots, sandstone fragments and small floaters, overlain by leaf litter. Scrub turkeys are abundant and their large mounds are surrounded by bare ground from which they have scratched up all the litter. In the central portion of the Brigalow Belt, the Middle Tertiary Minerva Hills basalts influence soil type and nutrient status. In these conditions softwood scrub species and tree pear (*Opuntia tomentosa*) dominated the understorey in brigalow-blackbutt-bauhinia-wilga layered woodland and forest. Where soils and moisture were favourable, these graded locally into some of the most species-rich softwood scrubs, the blackbutt and brigalow being replaced largely by *Brachychiton*, *Macropteranthes* and *Cadellia* emergent trees.

In the Expedition National Park between Taroom and the Arcadia Valley, differences between soils in the softwood scrubs and adjoining grassy open eucalypt forest lay in the proportion of sand and sandstone fragments in the upper part of the profile. A comparatively rich layer of surface humus was present inside the scrubs. In the lowland plains remote from the influence of weathering sandstone and basalt, deep cracking clay and duplex soils predominate, often with mottled subsoils and with abundant gilgais. The scattered softwood scrubs throughout these areas often lined broad, shallow watercourses with better moisture which formed corridors through the brigalow-dominated vegetation.

In the southwestern sector of the Central Highlands on the headwaters of the Maranoa and Warrego River systems, drier

Two narrow strips of softwood scrubs, separated by grassy woodland of silver-leafed ironbark on the valley floor. In the foreground, the scrub has a fringe community of gum-topped ironbark (*Eucalyptus decorticans*); in the distance, another strip of scrub extends upslope to the foot of the sandstone scarp. *Jim Gasteen*

Pure softwood-vine scrub on deep brown alkaline loam (pH 8.25), with scrub turkey's nest. Closed canopy height 8–10 metres, no emergent trees. *Jim Gasteen*

Softwood scrub on basalt soil being pulled in the Theodore district. *Jim Gasteen*

Partly cleared brigalow–softwood scrub. In the background pure softwood scrubs can be seen spreading up the scree slope to the base of the Precipice Sandstone Formation. These are brown loamy soils with pH 7.75 reaction. *Jim Gasteen*

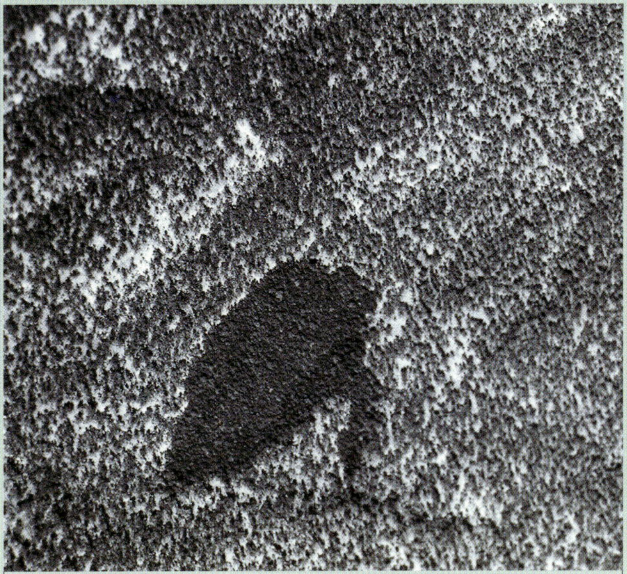

Aerial photo of a small, isolated softwood scrub about 2 hectares in extent, in the centre of several thousand hectares of grassy, open, undulating woodland dominated by narrow-leafed ironbark and bloodwood with sparse *Acacia* species understorey. Soil analyses within the scrub showed that the only difference from the surrounding woodland soil was a grading in the soil profile from acid pH 6.25 to a fairly strongly alkaline pH loamy clay at about a metre depth. *Jim Gasteen*

softwood scrubs with less diversity of plants and animals result from lower rainfall. Those furthest from eroding basalt were poorest because soils are of lower nutrient status and retain less moisture.

So here in the Brigalow we have softwood scrubs restricted in their locations by fire, but primarily excluded from the sandstone plateaux whose free-draining sandy soils are not suitable. The species diversity of the scrubs is related to soil nutrients and to the moisture retention capability of their soils.

GRAZERS IN CLAY CASTLES: KINGS OF THE TROPICAL SAVANNAS

Magnetic termites in savanna in the Northern Territory align their structures, which consist of thin, tapering slabs, north–south. By this means they avoid having most of the surface exposed to direct sunlight during the heat of the day. The termites are tuned in to the Earth's magnetic field to achieve this alignment. *M.E.W.*

THE TROPICAL SAVANNAS

In the savannas of the wet-dry tropics, monsoonal rain and the Big Wet cause rapid understorey growth, and the dry season sees the biota progressively switching into arid mode or inactivity. The eucalypt woodlands with their tall grass understorey have been moulded over time to climatic conditions, and to some extent by fire. Natural fire resulting from lightning strikes has been an environmental force in the progressively drying, increasingly sclerophyll-dominated communities that developed in Australia over at least the last 15 million years. (The northern margin of Australia only entered the tropics 20 million years ago, so tropical savannas are a product of northward drift and global climate change towards more seasonal dryness.) When summer storms herald the monsoon in the tropical savannas, lightning strikes have started fires in the abundant dry fuel, and fire has been one of the evolutionary selective forces.

Co-evolution of the grassy eucalypt woodland savanna vegetation type, its fauna and the way it functions has been a uniquely Australian process, determined by the antiquity of the continent, its worn-out soils, its biota's Gondwanan origins, and the history of climatic change through geological time—the full story that I told in *After the Greening*. In these savanna ecosystems, roles filled by mammals in other continents have been usurped by insects! Who would ever have thought that apart from the few relatively fertile patches on floodplains, where the sedgelands and better quality grassland support most of the mammalian herbivores (including introduced ferals like

Savanna woodland on the Gibb River Road in Western Australia. Eucalypts with tall grass understorey. *M.E.W.*

buffalo and cattle), more than 90 per cent of the grazing is done by insects, as is most of the nutrient cycling?

The supremacy of insects in Australia's tropical savannas is linked to low soil fertility. **Nutrient-impoverished soils produce grasses with low nutritional value, unable to sustain large populations of mammalian herbivores.** When the growing season is short, as in the wet-dry tropics, this is an added disadvantage. (The savannas of Africa that support the huge herds of ungulates are on young and fertile volcanic soils. Those on the old weathered soils that cover much of that continent as well have no large herds. The work done by dung beetles, of which there are 2000 species in Africa, is also significant in increasing fertility of soils wherever animals are present. The paucity of this section of invertebrate fauna in Australia, a consequence of the paucity of mammal fauna, partly explains the low organic content of our soils—native ungulates and their attendant dung disposers go together in nature. One of the most compelling arguments against Tim Flannery's theory of great herds of megabeasts devouring Australian vegetation[180] is the impoverished soils that were already impoverished long before the big fauna evolved. The wide open plains and semi-arid landscapes of Australia have surely *never* supported vast numbers of large animals.)

Because savannas are generally on nutrient-poor soils, rapid recycling of nutrients in the biomass is essential. Soil micro-organisms play a major role in recycling, but only during the wet season.[137] Termites, however, function throughout the year and are responsible for at least 20 per cent of organic matter recycling.

Termites in the Gulf savanna, Gregory River district. *M.E.W.*

Fire is also a recycler of nutrients. The natural fires caused by the lightning that precedes monsoonal rain were randomly spaced in time and location; man-made fires in tropical Australia are now horribly frequent and widespread, particularly in the Top End. (See pages 220–2 in *Listen … Our Land is Crying*.) Fire is not only a recycler but also a destroyer of nutrients, and it is now becoming evident that it is also a destroyer of the precious and unique biodiversity of the region.[150,151a] Recent CSIRO research has advocated a decrease in fire frequency from the mainly biennial, but sometimes annual, current regime to a frequency of one in three or preferably one in five years. Political correctness requires that Aboriginal advice and practice, on which the more frequent burning is based, should not be disregarded, yet the modern Aboriginal Australian has probably little knowledge of the practices of his long-dead elders.

Termites travel up scorched tree trunks, covering them with red mud. *M.E.W.*

Too-frequent burning kills grasstrees and cycads, creates woodland without the mature trees that provide nesting hollows, destroys resident fauna and alters species composition of vegetation. *M.E.W.*

A termite mound, protected by the hard capping of its underground structure, has survived unscorched in the blackened landscape that surrounds it. *M.E.W.*

Nutrients released by fire are mainly deposited as ash, with fine particles going into the atmosphere.[137,137a] Nitrogen and phosphorus are volatilised and lost in hot fires. Studies have shown that tropical grasses move nutrients from above-ground parts to their roots at the end of the summer growing season, a practice that reduces nutrient loss in dry-season fires. At Katherine, research showed that while 90 per cent of the nitrogen in native pastures was lost when burning occurred, the loss was probably offset by inputs from rainfall and the activities of non-symbiotic nitrogen-fixing organisms in the surface soils. Annual burning, even over long periods, has not been shown conclusively to deplete soil nutrient levels. Consequently, it has been suggested that within some tropical savanna systems, nutrient losses by fire and subsequent nutrient gain from rainfall may be closely related and the process might be seen as a sub-cycle functioning within the overall nutrient cycle.

Fire in the Cape York Peninsula

The effect of annual burning—the recommended practice for 'management' of the tropical savannas of Cape York Peninsula—is painfully obvious to anyone travelling the Gulf Country and going all the way to the most northerly tip of the Australian continent. On such a journey in September 2001, the amazing extent of the burning was the first impression. Day after day through blackened, smouldering or actively burning landscapes was enough to evoke a feeling of deep sorrow for a land so mistreated and to wonder what was the fate of the resident fauna. It was also an opportunity to observe the sequence involved in the degradation of the savannas.

The purpose of burning, according to locals, is primarily to refresh and increase the edible grass understorey for the cattle that range over extraordinarily large tracts of country in relatively low numbers, and to 'prevent unexpected natural fires that do so much more damage'. What has not been recognised is the cumulative effects of annual, or even biennial, burning. Perhaps being close to the situation does not enable the land managers to see what is so patently obvious to an outsider who remembers the original savanna landscapes (or has access to archival material that provides a baseline for present observations) and now sees the degraded ecosystems with which they are being replaced. Surely, if the 'managers' opened their eyes and saw beyond a bit of green grass and a tidy landscape for this year, they would realise that the benefits they are seeking are becoming more meagre with each burn and that what they are doing is unsustainable.

Many changes follow too-frequent fires, as seen in the photographs.

- Primary damage to the trunks of some of the older trees becomes more invasive with each burn and termite penetration into heartwood follows, weakening the structure of the trunk. At some subsequent burn, fire takes hold of the trunk and the tree crashes to the ground.

After the fire, a large tree has crashed and lies on the ground. In the next fire it is likely to be burnt up and no trace will remain of its existence. Many of the small saplings in the background that have grown since the last major fire will not survive. *M.E.W.*

Fire damage to the trunk of a mature tree. The next fire will have easier access to the trunk and soon the tree will fall. Some fallen trunks continue to smoulder for weeks after a fire and all that is left is white ash on the black ground. *M.E.W.*

There it continues to burn, as often as not; in newly burnt areas the only record of the fallen tree is an ash diagram of its trunk and branches, white on the blackened earth (like the chalk outline that is drawn round a body at a murder site). By next season all trace of that tree has gone for ever, and any trunks of fallen trees that did not continue to burn when they fell are reduced to ash in their turn when the next fire comes along.

- Many younger trees with less bark thickness and protection than older trees have so much scalding up their trunks that following a burn they are encased in the red mud carried up by invading termites within a matter of weeks. This further weakens them and makes their survival through following fire seasons less certain.
- The re-sprouting of the trees from underground root structures and the growth of understorey species after fire at first leads to whipstick-type and shrubby regeneration in the grassy understorey. Cycads and grasstrees tend to proliferate at first, but too-frequent fire sees them becoming less abundant, and the number of termitaria, or at least the amount of termite activity that is largely sub-surface, appears to increase. Each fire kills off some of the saplings as well as eventually eliminating the more fire-sensitive of the other plants, reducing diversity. Many of the more palatable grasses and forbs are in this category, so the 'better' pasture being sought becomes more elusive.
- Erosion that follows the monsoonal rains when vegetation cover has been destroyed or damaged further degrades the ecosystems and run-off sediment in its turn degrades the river systems.
- In a world that is beginning to show ominous signs of Greenhouse warming, the addition to the problem of the amazingly large amount of atmospheric pollution generated by tropical savanna fires globally has to be considered. (See page 61 for figures.) Australia's contribution in this regard has to be recognised and steps must be taken to reduce it.

The tropical grassy savannas, iconic Australian landscapes of eucalypts in a tall grassy understorey, are being destroyed by their 'management' by fire. It is doubtful whether the changed ecosystems that evolve will be more productive in terms of beef cattle revenue.

Fire and the fate of small mammals in Kakadu

A study of mammal abundance in the Kakadu National Park between 1986 and 1993[181] had recorded substantial declines in numbers, and concluded that this was a 'natural' response to a run of poor wet seasons. When the area was re-sampled in 1999, following a series of unusually good wet seasons[150,151,151a], ongoing downward trends for all mammals combined and for most individual mammal species were recorded—including the northern quoll, the fawn antechinus, the brushtail possum, the northern brown bandicoot, the dusky rat, the black-footed tree rat and the pale field rat. The declines in these mammals are consistent with limited observations elsewhere in northern Australia, suggesting recent and current declines in the kind of animals that have proven susceptible to regional extinctions elsewhere in the continent. (The tropics had previously been thought to be immune to such regional extinction patterns. See Part Three, page 117.)

It was concluded as a result of the recent studies that the serious decline in mammal abundance is a response to some 'management' or other threatening process and is not connected to wet season fluctuations. The most likely explanation is the frequency over vast areas of the tropical savannas of extensive and destructive fires (combined, to a lesser degree, with pastoralism). The hot, frequent fires result in changes to understorey composition; reduction in abundance and productivity of the mid-storey fleshy fruit-producing trees and shrubs; reduction in abundance of large hollow trees, fallen logs and litter; and changes in grass species composition and diversity. The rapid decline in numbers will be followed by regional extinctions if the problems are not immediately addressed. The effects on the ecosystems of the loss of the small mammal component are totally predictable as a result of the observations that have been made, for instance, in the Yookamurra Sanctuary (see pages 124–126).

The widespread decline of a high proportion of mammal species across the monsoonal tropics mirrors the decline in seed-eating birds that has been noted across the same area. (No one has the faintest idea of what the fires are doing to invertebrate and micro-organism populations.)

The dry savanna at Nourlangi Rock in Kakadu National Park burns readily; too-frequent fire is changing its composition and structure and adversely affecting its fauna. *M.E.W.*

Just take a trip across the tropical north in the fire season, just before the monsoon, and you will not need to be told of scientific experiments to prove why Australia's biodiversity is being lost. It is caused by gross fire mismanagement and emphasises the need for a complete reassessment of fire management on a continent-wide scale.

The role of insects in tropical savanna ecosystems

The importance of herbivorous insects in savanna ecosystems[182] has generally been greatly underestimated and under-researched. In Australia in particular, with its species-poor native mammalian fauna, attention has always focused on cattle production, not insect activities. Research on tropical savannas has been mainly concentrated in the Kakadu region.

Herbivorous insects are of three types—grazers, folivores and seed predators. Grazers feed predominantly on grasses and forbs; folivores eat the leaves of trees and shrubs; seed predators eat seeds either before or after they are dispersed. The most important grazing insects are grasshoppers and harvester termites, the latter being mainly detritivores (like other termite types) because they clean up dead material rather than fresh. **In Australian tropical savannas, grasshoppers are probably the major grazing animals.** It is known that consumption rates for grasshoppers in African savannas can be comparable to those of large populations of native ungulates and cattle. Little is known about folivores in Australian savannas but some work has been done on their significance to eucalypts and acacias in southern parts of the continent. Harvester ants are the main post-dispersal seed predators—mainly omnivorous species of *Monomorium* and *Pheidole*, but including an endemic radiation of seed-eating *Meranoplus*.

The low nutritional value of tropical grasses keeps kangaroos and other macropod numbers low. Grasshoppers are claimed to be more important as grazers than all the native mammals combined, and are major competition for domestic stock. An inverse relationship has been shown to exist between the importance of grazing insects on the one hand, and the importance of mammalian grazing on the other, associated with differences in soil fertility. Thirty sites were studied within the Kakadu National Park, with termite species richness found to be positively correlated with grasshopper abundance and richness, but negatively correlated with both soil fertility and the abundance of macropods and rodents. The increasing dominance of insects with decreasing soil fertility is not just a direct function of reduced mammalian presence, but involves a positive response by the insects. Though poorly understood, it seems possible that low-fertility sites result in a reduction of plant chemical defences—it is likely that plants' commitment to defence depends on the nutrient budget, and those on infertile soils are less well defended.

The role of termites in soil processes in tropical savannas

Termites occur in large numbers over the savannas of much of northern Australia, particularly in the more arid regions. The only areas of semi-arid tropical Australia that do not support large numbers of mound-building termites are the cracking clay soils. (Africa has termites functioning in the same way in the same sort of environments.)

While microbial activity is restricted to the four wet months of the year, termite activity is continuous throughout the year in the northern savannas. The release of nutrients by their breaking down of organic matter, and the fixed nitrogen that they make available by virtue of their symbiotic gut bacteria, are important in these impoverished environments. In a few instances, amounts of carbon recycling have been investigated. Near Townsville, for example, two mound-building species were shown to mineralise about 250 kilograms of carbon per year, representing about 10 per cent of the annual carbon turnover at the site. However, the contribution of subterranean termites at the site could not be assessed; they may be responsible for as much as 20 per cent of the carbon mineralised in the system. Termites were able to exploit the niche left open by the inability of soil micro-organisms to perform the function during the dry season. At sites near Charters Towers similar figures were obtained.

The role of termites in formation of topsoil in the red and yellow earths of the tropical north has been discussed in Part Three (page 97). In the Kakadu National Park, the decay of abandoned termitaria, and therefore the release of accumulated nutrients, is considerably hastened by the activities of other fauna, particularly burrowing vertebrates. The termites are replaced in termitaria by other insects, small lizards, rodents, snakes, native cats and bandicoots, in about that order.

Ants in the tropical savannas

The ant fauna in eucalypt-dominated savanna in the Kakadu National Park is extremely diverse.[183] Because of the great importance of ants in the savanna ecosystem, their reactions to frequent burning are important, affecting nutrient cycling and other soil functions and characteristics.

Experiments to determine how ground-foraging ants were affected by different fire regimes involved 81 species in 24 genera. Plots were burnt annually, biennially and not at all, over 14 years. The ants were classified into functional groups based on habitat requirements and competitive interactions. The most important groups were:

- Dominant species of *Iridomyrmex* (11 species)
- Generalised myrmicines: *Monomorium*, *Pheidole* (22 species)
- Hot climate specialists: *Melophorus*, *Monomorium*, *Meranoplus* (14 species)
- Cryptic species, many genera (13 species)
- Opportunists, mostly *Rhytidoponera* and *Tetramorium* (11 species)

In experimental plots burned annually, high numbers of dominant *Iridomyrmex*, hot-climate specialists and opportunists, were found, and low numbers of the generalised myrmicines and cryptic species. The reverse was true for unburnt plots, and intermediate results were obtained under biennial fire regimes. Many species were common under one regime, but rarely or never recorded under the others. Structural changes in the habitat, levels of litter accumulation and insolation on the ground affected the ants directly while the abundance of the dominant *Iridomyrmex* increased competition for resources after fire.

It is likely that ants play a significant role in seed collection in the savannas. Many of species of *Acacia* in the Kakadu region have white arils (fleshy appendages), attractive to ants.

THE KWONGAN: SANDPLAIN FLORA OF WESTERN AUSTRALIA

The Kwongan sandplain flora of the South-West Province of Western Australia is of particular interest as it shows high rates of adaptation to the extremely poor soils of the region. The South-West Province is a mega-diverse region with 5710 native species, of which 3000 (53 per cent) are provincial endemics and another 1524 (26 per cent) endemic to Western Australia.[184] Most of the species are from relatively few plant families and as a result, many rather similar plants coexist. The Kwongan is a 'fynbos' mixed shrubland assemblage in which sclerophylly is a dominant characteristic—this, along with habitat specialisation, is a response to the low-nutrient status of the soil. Specialised mechanisms for enhancing nutrient uptake are found in 88 per cent of the species that grow on these poorest sandplain soils. (The same sort of conditions and the same adaptations to them occur in the flora of the south-western Cape in South Africa, largely involving members of the same Gondwanan families that occur in the Kwongan.)

The Kwongan can be described as a 'democratic' community where there are no dominants and where each species becomes a specialist, exploiting subtle differences in the levels of the critical resources such as water and nutrients in different areas.

The South-West Province has a Mediterranean climate with winter rainfall and long, hot and dry summers. Thus, temperatures are low when water is readily available, and water is scarce when heat and evaporation are at their greatest, presenting problems for vegetation.

The Kwongan plant community has evolved a number of adaptations to the climatic and low nutrient situations.[185]

- **Increased efficiency of water uptake** This is achieved by increased fineness of the root system and proliferation of long root hairs. The greatest development of this trend is seen in proteoid, capillaroid and dauciform roots. Most new season's rootlets and specialised roots are confined to

View across the sandplain to the sea. The diversity of plant species in the Kwongan exceeds that in rainforest. The whole ecosystem is under threat because of urban sprawl and spread of agriculture along the sandplains. *M.E.W.*

BOTANICAL PROVINCES
OF WESTERN AUSTRALIA

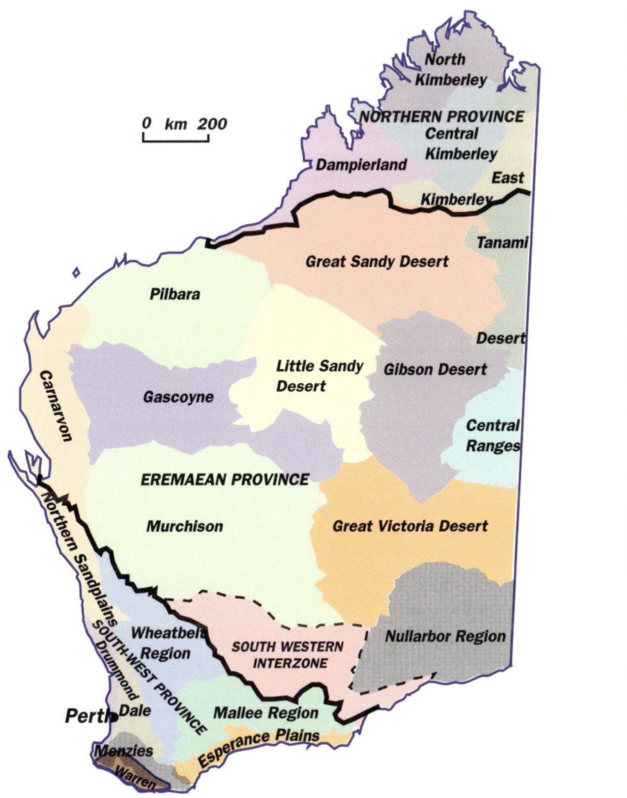

After Beard et al.[184]

the upper 15 centimetres of the soil, especially in or near the decomposing litter zone. Such nutrients as occur in the poor sandy soils are concentrated in the top few centimetres. Substances exuded by the root systems attract and feed the micro-organisms of all sorts that are involved in nutrient cycling. Symbiotic relationships between mycorrhizal fungi and roots also serve to increase absorptive area and make for more efficient connections between roots and pore water. Mycorrhizae are also enormously important in nutrient availability, by solubilising the phosphorus that is in short supply and virtually unavailable without their intervention. The selective uptake of nitrogen by host-specific mycorrhizas is also of great significance.

- **Tapping novel sources of nutrient** Carnivorous plants are well represented in the flora, obtaining their nitrogen and other nutrients by catching small invertebrates in sticky glands on their leaves and digesting them with enzymes produced by the glands and by micro-organisms that live in them. Nitrogen-fixing symbioses (in legumes and casuarinas and a number of other plants) are important in nutrient cycling, while the molybdenum, cobalt and other trace elements that they require are more readily available in the poor soils after the frequent fires that occur in the region. A number of root parasite plant species, whose haustoria (root structures that attach them to their host's root system) absorb phosphorus and

Wildflowers in the Kwongan are a tourist attraction, persisting along roadsides and in reserves. *M.E.W.*

potassium preferentially, are also prominent members of the Kwongan community. The grasstree *Kingia australis* has a novel adaptation—aerial roots that penetrate its own persistent leaf bases and extract nutrients from them, recycling them through the plant's vascular system—using the plant's own old tissue as food.

- **Prolonging water uptake** This is facilitated by the tendency for rootlets to cluster, trapping water by capillary action and prolonging nutrient release.

Proteoid roots consist of hundreds of densely packed and extremely hairy rootlets arising in longitudinal rows from a parent lateral. The root hairs account for as much as 90 per cent of the surface area of the proteoid root, yet only about 10 per cent by volume. Proteoid roots are characteristic of the family Proteaceae, and perform the functions that mycorrhizal fungi perform in most other plants. They are short-lived, developing and functioning during the wet season.[186] Root hair production increases as the soil dries and they act as links across gaps between soil particles, plugging the gaps with mucigels that maintain connection with pore water and which also supply a substrate for mineral-solubilising micro-organisms. There is evidence that a bacterial co-factor is involved in the production of the proteoid roots in the first place, though how this works or which organisms are involved have yet to be discovered. When nutritional levels, particularly of phosphorus, drop below a certain threshold, the production of extra long root hairs escalates rapidly.

Studies on *Leucadendron laureolum*, a South African species[187], found that proteoid roots amounted to 40 per cent of the root mass in a 9-month-old plant growing in sand and increased the total surface area of roots by 15 times. Because rates of chemical exchange between roots and their environment depend largely on the surface area of the ultimate roots, nutrient uptake is greatly increased by the proliferation of fine rootlets and root hairs. Potassium uptake is at least twice that in normal roots, and increases up to 13 times have been shown experimentally. Energy requirements for acquisition of nutrients via roots are high and phosphorus moves preferentially to young proteoid roots. Some storage of inorganic polyphosphates also occurs in them, ready to be mobilised later for shoot growth. Proteoid roots are now known to exude water from their surface at night.[222]

The legume *Viminaria juncei*, which grows in flooded habitats or on heavier soils that become waterlogged in winter, produces proteoid roots and also pneumatophores (aerial roots to assist in oxygen supply to the root system).[187] These are connected to aerenchyma (aerating tissue) within the lower stem, roots and nodules, enabling plants to fix nitrogen under waterlogged conditions.

Dauciform roots are carrot-shaped rootlets, densely covered in long root hairs and also forming clusters, present in seven of the nine Cyperaceous genera (sedges) in the South-West Province of Western Australia and in the Mediterranean climatic region of South Africa. **Capillaroid rootlets**, characteristic of the Restionaceae, operate in the same way as the finely branching cluster rootlets of the other types described above. The clusters increase capillarity, keeping the soil damp and prolonging water and nutrient intake. Restios have dual root systems, with sparsely branched tap roots penetrating deeply into the soil and the thin, highly branched cluster roots abundant near the soil surface.

Mycorrhizal symbioses are ubiquitous in the Kwongan flora, except in those plants that have proteoid roots carrying out the same functions. The fungal symbionts are always primitive genera from the Endogonaceae that are obligate symbionts (unable to function as free organisms), and *Glomus* is most common.[185] There appears to be a wide host range for each species. Spores survive in the soil, but most infection of plants is by the fungal hyphae that spread everywhere. A network of fungal hyphae makes it possible for the plants in a community to be epiparasitic and divert nutrients away from less efficient, connected, host plants. This sort of **nutrient sharing** is an important feature in the Kwongan communities where there are no aggressively dominant species. The incidence of mycorrhizal symbiosis is greatest where soils are least fertile and where interspecific competition is greatest. Zinc, potassium and sulphur are available to plants in greatest amounts where symbionts are concentrated.

Nitrogen fixing is carried out by the many legumes that are present in the flora. Abundant shrub legumes are a feature of Kwongan communities and are estimated to add between 2 and 15 kilograms of nitrogen per hectare each year. In such nutrient-deficient ecosystems, the mineral nutrient requirements for nodule formation and nitrogen fixation are more likely to be limiting than the carbohydrate required. (Calcium and probably zinc and copper are needed for formation of the infection thread that penetrates a root hair.

Boron is necessary to ensure sugar transport and vascular continuity between the nodule and the plant root.)

Macrozamia reidlei, a cycad that grows in the banksia woodland of the Perth Coastal Sandplain, contributes as much as 19 kilograms of nitrogen per hectare to the ecosystem, courtesy of its nitrogen-fixing symbionts. Some banksias have nitrogen-fixing cyanobacteria in the nectar that their flowers produce. (Nectar production on a large scale, and flowering patterns that result in the Kwongan being a place where something is in flower no matter what time of year, are characteristic of the flora. It has been suggested[188] that because the nectar is available throughout the year for the small mammal pollinators that have co-evolved with the flora, the advantages outweigh the energy costs of making nectar in this nutrient-impoverished environment. Less energy is required in making nectar than in making tissue, and it is essential to attract pollinators when individuals can be widely scattered in a democratic society like the Kwongan.)

Carnivorous plants are well represented in the Kwongan flora and there is a high rate of endemism, with 93 per cent of the 70 species that occur in the South-West Province and in similar South African ecosystems being endemic. *Drosera* species are common, occurring in damp places but also on deep dry sands or lateritic surfaces. Their leaves have stalked glands that secrete mucilage to attract and capture small invertebrates. The glands produce six different enzymes as well as organic acids, and the symbiotic bacteria that live in the mucilage add other enzymes, especially chitinase. All act to break down the prey and release nutrients into the mucilage. Nitrates, phosphorus, sulphur, calcium, amino acids and carbohydrates become available, ready to be absorbed by the plant. Rates of metabolism in the glands are high.

A dense mat of proteoid roots of *Banksia* under leaf litter. *Michelle Watt*

The endemic pitcher plant *Cephalotus* has glands in its pitcher which produce enzymes to digest prey that have drowned in the water it holds. *Pseudomonas* and other bacteria are also active in the nutrient cycling process, and the breakdown products are absorbed in the pitcher solution.

Species of *Utricularia*, bladderworts, have substituted nodule-like bladders for roots. Sensitive hairs on these trigger opening to catch insects, and enzymes in the bladders digest the prey. Such nutrients as are in the water in which the plants grow and which fills the bladders, plus those from the digested prey, are absorbed into the vascular tissue.

Parasitic plants are unusually well represented in the Kwongan. Ten of the 14 families of parasitic Angiosperms in the world occur in Mediterranean Western Australia and South Africa. The South-West Province has 161 species, of which 75 per cent are endemics, and four endemic genera.

The structures that connect the parasite to the host are known as haustoria. The root haustoria of hemiparasites are borne on long lateral branches that lack root hairs. Tracheids in the vascular core fuse with the xylem in the host root, making a physical connection. *Nuytsia floribunda*, the West Australian Christmas tree, fixes to any root, even those of grasses and delicate herbs, yet the tree is large and has a large lignotuber. The haustoria are attracted to the polysaccharide-type root exudates (a chemotrophic response).

Stem parasites are comparatively few, comprising the mistletoes. Some are host-specific like *Amyema linophyllum* that grows only on casuarinas, *A. miquelii* only on eucalypts and *Viscum continuum* only on acacias. Studies on *Amyema linophyllum* on *Casuarina obesa*[189] have shown that its rate of transpiration is greater than that of its host but its assimilation rate is the same—that is, it is less water-efficient.

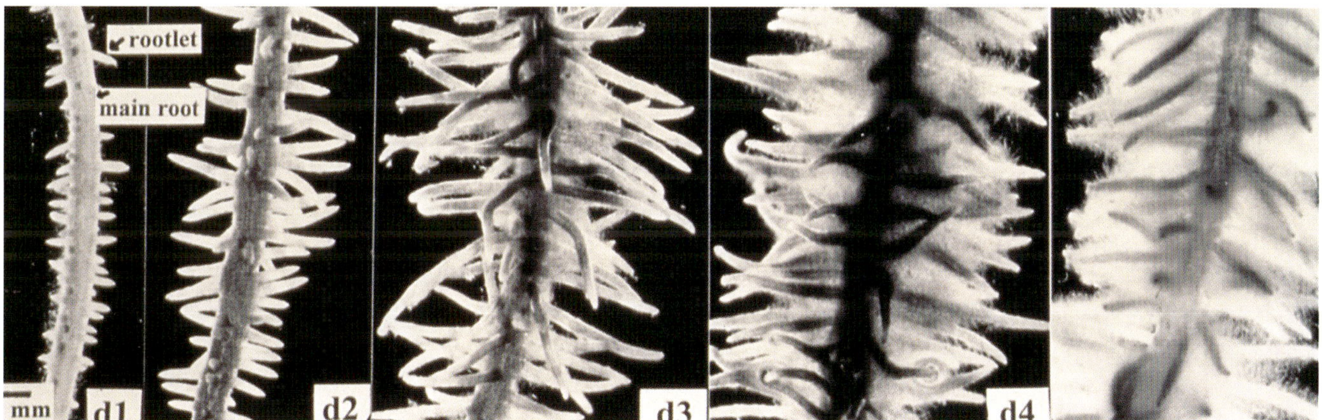

Stages in development of proteoid rootlets on a lateral root of white lupin at days 1–4 and day 8. *Michelle Watt*

Some other adaptations to the poor soils and Mediterranean climate in the Kwongan are noteworthy.

- In nutrient- and water-limited habitats with abundant light, the cotyledons of the germinating seeds of some plants maximise supply of mineral nutrients to the shoot, enabling the seedling to maximise carbon supply for rapid root elongation. The shoot remains a very small, highly efficient food factory while the root grows rapidly down to access groundwater. Investigations of *Hakea* species[190] have shown that a first-year seedling can have a shoot a few centimetres high but a taproot system that goes down a metre or more. In the next growing season an extensive system of lateral roots and hairy root clusters develops, and the shoot grows rapidly.
- Investigation of eight species in Proteaceae has shown[191] that the plants channel nutrients into their seeds so that some have 30 to 500 times as much phosphorus and 8 to 100 times as much nitrogen, copper and zinc than normal plant tissues. Seedlings need to establish deep roots fast in order to survive and without such a reservoir of nutrients in the seeds would be unable to compete in the nutrient-poor environment.
- Some plants growing in the shallow pockets of soil in granite outcrops where desiccation and heat are at their most severe have a short winter growing season, at the end of which their aerial parts die and they aestivate by virtue of their underground corms or stem tubers.[192] These specialised organs have high levels of protein and carbohydrate, like seeds, and can survive desiccation. On wetting, adventitious roots appear and a new shoot system starts to grow. Such plants are known as geophytes. The resurrection plant *Borya nitida*, widespread on the exposed granite outcrops, has the ability to withstand desiccation of the complete plant, not only the root system. When winter rain falls, the leaves and stems rehydrate, the plant resumes full activity and eventually flowers.
- A 'micro-stilt' habit, unique to plants in the Kwongan, is seen in 70 species of small herbs, involving nine genera in three families of monocotyledons and two families of dicotyledons.[193] Most of the micro-stilt plants are species of *Stylidium* that are very vulnerable to fire and grow mostly in open spaces, on gravels or fringes of rocky outcrops. In this adaptation to severe drought and high surface soil temperatures the mature shoot system is partially or wholly raised above ground by aerial prop roots. When the seeds of these plants germinate early in the winter, the early development is the same as in other plants, with a rootlet and a shoot. Then aerial roots develop from the shoot and the seedling root is lost. By summer the aerial roots are heavily lignified, with their internal tissues insulated against heat and drought. The elevation of the shoot system even two centimetres above the surface of the ground has been found to lower temperature by 6° to 10°C.

The very high degree of specialisation and adaptation of the Kwongan flora in response to the poor soils and the dry Mediterranean climate is illustrated by all the examples described above. The poor soils have been a feature of the region, and of much of the rest of the continent, through great lengths of geological time, although exacerbated by glacial stages of the Pleistocene ice age during the last 2.6 million years. The Mediterranean climatic regimes became established in the south-west of Western Australia (and in the south-western province of South Africa) only in the last 5 million years. A great deal of adaptation to poor soils had been occurring throughout the evolution of the flowering plants (in the last 125 million years) and the development of sclerophylly was one such response. Plants that already had that adaptation were suited to habitats in which there was increased water stress, and also had features and strategies that imparted some tolerance of fire. (A similar suite of adaptations is seen in the South African Cape flora, which has closely related genera in the same families that are numerically abundant in the Kwongan—testimony to the antiquity of the basic modifications.) The progressive drying of the Australian continent since it became an island continent 45 million years ago and started its northward drifting saw the expansion of such pre-adapted plants into ever-larger areas, while broadleaf forest was increasingly restricted to ever-decreasing locations where water was not a limiting factor.

Geographic isolation of the flora of the South-West Province of Western Australia, of which the Kwongan is a part, has occurred during the last 5 or 6 million years. Combined with the intensification of seasonal aridity, isolation helps to account for the very high rates of endemism and also for the proliferation of adaptations and the development of some novel responses to the unique conditions.

When one realises how much adaptation has been involved

Borya nitida and lichens growing on an exposed granite outcrop have the ability to withstand desiccation and very high temperatures. *M.E.W.*

in creating a sustainable natural ecosystem that is in balance with its environment, and admits that one knows nothing of the micro-biological realm that supports the macroscopic assemblages with which one is familiar, perhaps one can see clearly why monocultures of introduced plants, like wheat, are not sustainable substitutes.

(The Mallee ecosystem is described in the Case Study on Yookamurra Sanctuary, pages 124–125; Mulga ecosystems on page 112.)

Fresh water—rivers and groundwater

About 96 per cent of the Earth's water budget is found in the oceans. Fresh water comprises only 3.5 per cent—and much of this small fraction is tied up in polar ice and virtually unavailable. When one realises that lakes and rivers comprise only about 2 per cent of the world's *available* fresh water and about 97 per cent of that fresh water resource lies underground, largely unseen and much less considered in terms of conservation, one gets a new perspective. The global situation of limited and finite supply and rapidly expanding human population is becoming clearly apparent. The need to understand the hydrology and ecology of subsurface waters is becoming ever more urgent, but up until now most research and interest has focused on surface water resources.

Australia, the driest vegetated continent, has few large rivers with high discharge and few deep permanent lakes. Its surface water deficiency is more than compensated by a rich array of groundwater systems. Many surface water systems have varying reliance on groundwater.[194] In regions with reasonable rainfall, rivers derive baseflow from subsurface waters. Depending on the nature of the riverbed and the porosity of sediments, there may be considerable exchange of stream and groundwater along the course of the river channel and below the banks in what is known as the **hyporheic zone** (HZ).

Subsurface flow is also common in many arid-zone rivers. Rivers may exist for long periods as a series of pools connected by hyporheic flow, or as completely dry surface sediments for much of their length. The river red gums that line dry watercourses far into almost desert country are surviving because their deep root systems have access to this underground source, even when surface flows are often separated by many completely dry years.

Groundwater-dependent ecosystems fall into four broad categories: river baseflow systems, aquifer and cave ecosystems, terrestrial vegetation and wetlands. Many of these ecosystems are threatened either directly by groundwater abstraction or indirectly through disruption of hydrological exchange. Local groundwater pollution as a result of human activities can affect groundwater-dependent ecosystems—and even a city like Perth becomes one of these threatened ecosystems, as it is more than 60 per cent dependent on groundwater from unconfined aquifers of the Swan Coastal Plain for its urban water supply.

GROUNDWATER CONTAMINATION BY HYDROCARBONS

In 1994 there were 144 'known incidents' of groundwater contamination in Australia.[195] Industrial effluent, sewage contamination and petroleum spills or leakages (11 per cent of total), in that order, were the most clearly defined. Eight years later the recorded number of contaminated sites was about 80 000, about 30 per cent of them associated with petroleum storage and distribution. Australian industry spends over $300 million per year on contaminated site clean-up. Where the contamination is of organic origin, as in the vast majority of cases (including fossil fuels), there are fortunately bacteria in the natural environment capable of dealing with the problem.

Scientists at the CSIRO Centre for Groundwater Studies in Western Australia have made a particular study of groundwater contamination because Perth relies on groundwater for 60 per cent of its urban water use. The unconfined aquifers of the Swan Coastal Plain are particularly vulnerable and their management is crucial to the viability of the city and its surrounding areas. Contamination by pesticides and other organic compounds that are regularly applied to the soil surface, either purposefully or through accidental release, has potentially serious consequences.

Problems with petroleum contamination are now global and even Antarctica has had fuel spills. The extent of this sort of environmental pollution is so great worldwide that active remediation is virtually impossible. Examination of sites of leakage of petrol from service station tanks on the Swan Coastal Plain has shown that usually most of the contaminant is transported to the surface of the groundwater where it floats as a 'light non-aqueous phase liquid'.[195a] The aromatic hydrocarbons either dissolve slowly in the groundwater and move in solution in the direction of groundwater flow, forming a plume, or they volatilise and become a gaseous phase in the soil profile between the watertable and the surface.

The different components of gasoline have different qualities and their solubility or volatility determines the degree to which they are likely to be transported off-site. Benzene is the most persistent, travelling in groundwater furthest from the leakage site, broken down slowly by aerobic organisms. It is relatively soluble in water and volatile, and it is a known carcinogen so it is of greatest concern. The toluene component is dealt with by sulphur bacteria that function in the anaerobic conditions that apply in much of the groundwater and does not travel as far. Overall, naturally occurring soil bacteria effect the clean-up, and it is only the residual benzene contamination that requires oxygen, which is in short supply at depth below service-station tank installations, that may require attention.

It has been suggested that supplying oxygen to the plume would enable the aerobic bacteria to speed up destruction of benzene and lessen its contamination of groundwater. Bioremediation by addition of microbes at the same time would be beneficial.

GROUNDWATER CONTAMINATION BY CHEMICALS

Pesticide contamination of groundwater has been found at a number of sites on the Swan Coastal Plain.[196] Atrazine, the most commonly and widely used herbicide, has been detected here as elsewhere in the world, usually in low concentrations. In 1993, a serious contamination of groundwater by atrazine and fenamiphos, a widely used pesticide, occurred in the suburb of Dianella.[197] Extensive plant death occurred in gardens that were watered with groundwater, emphasising the vulnerability of unconfined groundwater aquifers to chemical contamination.

CSIRO scientists have found that the rate of mineralisation of organic pesticide and herbicide chemicals decreased rapidly with depth below the thin, relatively 'microbially rich' surface layers of the soil.[196] Atrazine degraded slowly, disappearing from 'fertile' topsoil in about a year, but taking 3 to 5 years at depth of a metre. As Perth receives 94 per cent of its annual 890 millimetre rainfall between April and October, it would be expected that application of chemicals during this period would result in rapid flow through the soil towards the watertable, with little time for microbial activity in the surface layers. Their application in the dry season would allow more time for mineralisation and reduce the risk of groundwater contamination.

Laboratory experiments were undertaken to determine under what conditions bioremediation (addition of mineralising bacteria) would be most effective. It was calculated that under anaerobic conditions, in the absence of microbial activity, half-lives for atrazine and fenamiphos were 730 and 1000 years respectively. Oxygenation and adding atrazine-mineralising bacteria from the contaminated site, the *Pseudomonas* sp. strain ADP, resulted in the half-life of atrazine being decreased to less than 20 days. The best result for fenamiphos under the experimental conditions was 6.8 years, which is exceedingly long compared with its rate of decay under natural conditions in surface soils.

GROUNDWATERS AS LIVING ECOSYSTEMS

A range of subsurface waters exists in Australia on a scale from massive aquifers like the Great Artesian Basin, 1.7 million square kilometres in extent, to cave streams and pools, water giving rise to springs; and the fluctuating interface zones between rivers and groundwater—the saturated hyporheic zone below and alongside rivers.[194,198] Until recently, exploitation of groundwaters has involved no recognition of the fact that they are living ecosystems, and we need to assess the contribution made to water quality and health by the amazingly biodiverse communities that are being discovered in subsurface freshwater habitats. Most of these habitats harbour invertebrates, many from ancient groups long extinct in surface waters and whose ancient lineages make them organisms that deserve to be protected.

Subterranean fauna may occur in any size void below ground, ranging from the pores between grains forming alluvial deposits, particularly gravels, and in boulder deposits such as screes, but particularly in the crevices, caves, sinkholes and conduits that develop on soluble rock terrains, principally limestones.[199]

Groundwater is one of those extreme environments where only highly adapted organisms would be expected to exist, and biodiversity might not be expected to be great. However, biodiverse relictual biotas have been discovered in many other parts of the world and are now coming to light in Australia, even in the limited sampling that has recently occurred. There is less biodiversity than in surface waters, but Australia is predicted to be a 'storehouse' of subsurface biodiversity. The subsurface habitats in an old, very stable, continent have proved to be ideal refugia for surface forms able to colonise these habitats when climatic conditions made their original habitat inhospitable, and for forms that were already adapted and living underground and thus able to survive the climatic changes that were affecting surface-dwellers. Most of the groundwater animals are believed to have evolved from surface ancestors already adapted for life in a subterranean environment, like those described below, in the hyporheic zone. Such creatures are frequently small, elongated, with reduced eyes and pigment; those that are true groundwater animals are collectively known as **stygobites**. When they colonise subterranean voids they become genetically isolated, undergo adaptive changes that further suit them to life in the dark underground, and eventually become separate species.

Subterranean faunas with ancient lineages have not only survived climate changes and physical changes in

An amphipod, resident of the hyporheic zone. *Andrew Boulton*

THE TETHYS SITUATION

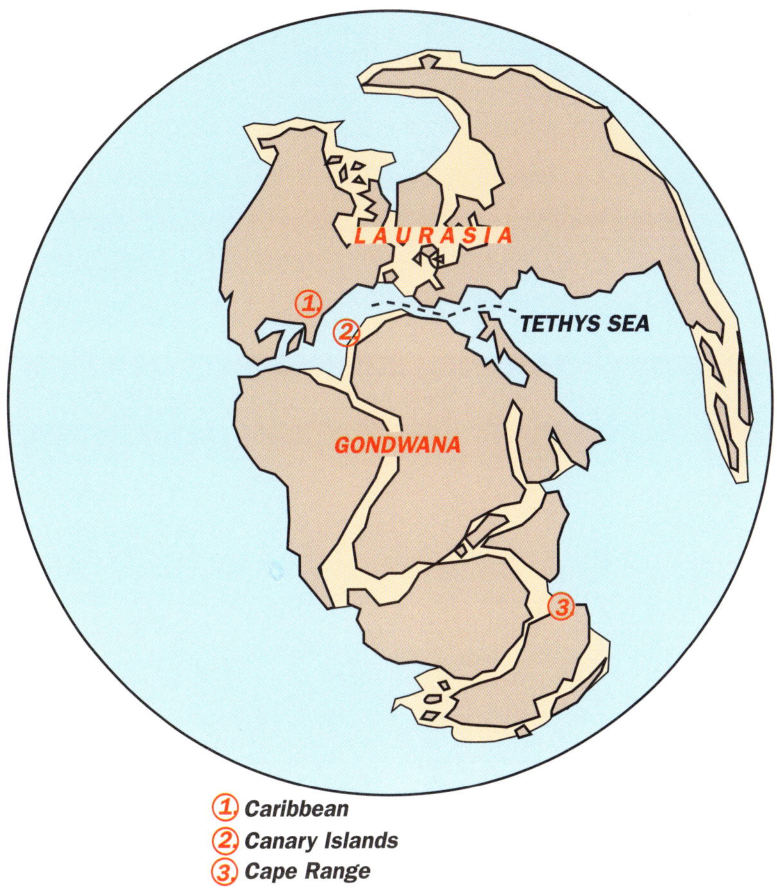

Location of landmasses during the Jurassic (165 million years ago).
1. Caribbean; 2: Canary Islands; 3: Cape Range, lying adjacent to the ancient Pilbara Craton that has been land for the last 600 million years.

THE CAPE RANGE PENINSULA

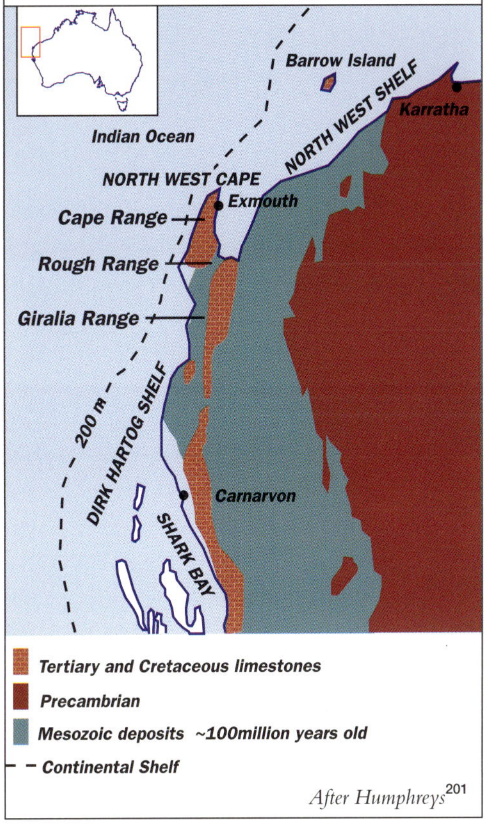

After Humphreys[201]

geomorphology, but have also survived great changes in geographic position when the supercontinents in which they originated have fragmented. In this context, close affinities have been established between ancient-lineage species found in the Cape Range in north-west Western Australia and species that are now resident in the North Atlantic region, in the Caribbean and Canary Islands, suggesting an ancestry shared with species found in regions around the opening Tethys Sea.[199–202] (Pangaea, the 'one Earth' of Triassic times 200 million years ago, began to rift into northern and southern supercontinents during Jurassic times with the opening of the North Atlantic and eastern Pacific.)

It is now known that arid north-western Australia contains one of the world's most diverse subterranean faunas in limestone caves and karst and calcrete aquifers. Even two amazing blind fish occur in the limestone systems—a cave gudgeon in the groundwaters of the Cape Range and Barrow Island, in both fresh water and salt/fresh water affected by marine tides, and an eel.[203] The entry of these fish into groundwater and their adaptation to the new environment may possibly be explained by the dramatic changes in sea level that occurred during glacial stages of the Pleistocene ice age.

The Cape Range Peninsula and Barrow Island: Limestone caves and groundwaters[200,201,202]

The Cape Range Peninsula lies just within the tropics, in the arid zone that occupies most of Central and Western Australia. The region has been arid for a long time, supporting vegetation mainly of *Triodia* hummock grasses (spinifex) and *Acacia*, and is adjacent to the Great Sandy Desert.

Rainfall in the region is low and unpredictable, at the whim of tropical cyclones, monsoons, north-westerly epitropical flows and frontal events that occur spasmodically, interrupting complete dryness with rainfall events in which about 1 in 10 is very large. Annual evaporation of 3219 millimetres exceeds the median annual rainfall of 280 millimetres. The episodic and unpredictable rainfall pattern greatly influences the lives of the limestone cave faunas (**troglobites**) that rely in part on organic matter and water entering the caves following heavy rain. The intermittent flooding rejuvenates the cave fauna, permitting it to occupy larger areas and to commence breeding.

Marine limestones of different ages, from Eocene to Late Pliocene, comprise the Cape Range Group, with the Tulki Limestone that dominates on the peninsula riddled with caves, subterranean tunnels and sinkholes that support a unique fauna which displays a high level of endemism from cave to cave. The Cape Range attains an altitude of 314 metres and is deeply dissected. (The folding of the limestone rock that elevated the

The Cape Range, looking west towards the Ningaloo fringing reef. The deeper gorges penetrate a marly (harder) limestone that is not cavernous. This harder limestone forms a barrier between adjacent cave areas and has resulted in the segregation of individual cave faunas, leading to their genetic differences. *Bill Humphreys, WA Museum*

Cape Range and exposed it to erosion occurred in the Miocene.) The caves above sea level are freshwater habitats. Seawater intrudes far inland in the tunnels and voids below sea level, below the groundwater which remains as a freshwater lens floating on the seawater. On the coastal plain, several caves and sinkholes and a number of wells and bores permit access to groundwater. Marine tides are apparent in bores as far as 3.5 kilometres inland, indicating open-conduit flow of the groundwater within the coastal plain. (The cavernous nature of the limestone, to depths of more than 100 metres below sea level, is attributable to the sea-level fluctuations during the Pleistocene ice age when rainwater eroded the soft limestone down to a deeper watertable during glacial stages.) Some sinkholes originating in higher parts of the limestone have a freshwater zone above, a mixed zone and a saline underground zone, with distinct faunal zoning according to depth.

The terrestrial fauna in humid caves in the Cape Range includes cave-adapted invertebrates showing affinities with the litter fauna of wet temperate and tropical Gondwanan rainforest. The nearest wet temperate forest to the south is 1200 kilometres away, while the remnant Gondwanan forests of eastern Australia are more than 3000 kilometres away. (Caves that dry rapidly after rainfall events do not have a cave fauna; those that take inflow only in heavy rainfall events, which may occur only every few years, and dry slowly over many years tend to have a diverse fauna.) Tree roots, particularly of *Ficus* and *Brachychiton*, penetrate into some caves on the coastal plain. Though they usually do not reach the floor, their fine rootlets serve as foci for condensation, providing water for the trees and food for cave-dwellers at the same time. On the coastal plain, root mats form in the groundwater—an advantage to plants that live in an arid region—and may provide food for cave fauna.

Palaeoclimatic information supplied by the fossil pollen record[201] indicates that complex rainforest occurred in north-western Australia in the Eocene (58 to 37 million years ago). Rapid contraction of forests occurred in the Miocene, most forests being confined to riverine and deep gully locations by 6 million years ago, followed by fragmentation of remnants as the continent dried with the approach of the Pleistocene ice age. Evidence from cave faunas suggest that aquatic and terrestrial components of the fauna were adapted to and inhabiting the subterranean realm before the fragmentation of habitats occurred, and already speciating. The genetic provincialisation that results in different caves having different variants results from the isolation of habitats during the severe dry conditions that prevailed in glacial stages of the ice age.

It has been realised that while the loss of rainforest is tied to the onset of aridity, the assumption that relict cave faunas also owe their origin directly to the drying may not be valid.[201] The fauna may have already colonised the underground realm and undergone a great deal of speciation while forest still thrived locally. The cliché of nature abhorring a vacuum is clearly illustrated by extremophiles in general—if there is a niche to be filled, some organism will fill it. If the niche offers protection from the changes that threaten the populations from

THE BUBBLER MOUND-SPRING, LAKE EYRE

Mound-springs, which occur mainly around the margins of the Great Artesian Basin where the Jurassic aquifers abut against older rock formations and water comes to the surface, are of particular interest in the context of ancient lineages and relict faunas. Not only is much of the water that emerges fossil water up to 2 million years old, but some of the inhabitants that have come up with the groundwater have Gondwanan connections; others have used the springs as refuges during times of severe climatic change much more recently when other freshwater habitats have disappeared. Mound-springs have highly endemic faunas that are only beginning to be studied and their complete protection must be guaranteed—especially as in the arid zone surface water forms a magnet for stock and humans. *M.E.W.*

which the troglobites have evolved, then they will survive in an area that is no longer suitable for their progenitors.

Bundera Sinkhole

The Bundera Sinkhole[204] is located on the Cape Range Peninsula, about 1.7 kilometres inland from the Indian Ocean on the wide coastal plain where the watertable lies about 5 metres below the surface. The cave has a single opening and a flooded passage extends downwards for 70 metres, inclined at 30 degrees to the horizontal. The sinkhole is an example of an **anchialine ecosystem**—an inland, underground system that is affected by marine tides. In other parts of the world such systems are noted for their relict faunas and for their species richness. Bundera Sinkhole is the only deep anchialine system known in Australia, and the only one so far known in the Southern Hemisphere.

Anchialine systems in drowned sinkholes typically have a freshwater to brackish layer overlying seawater and are stratified by availability of light, temperature and density differences, and degrees of salinity. The marine section typically consists of seawater with low oxygen content, and one or more layers of fresh, brackish to increasingly saline water meet the seawater at a density interface. Below this interface, such ecosystems support a wide range of relict life forms that are clearly related to ancient faunas in the Caribbean and in the Canary Islands. In particular, remipeds, a type of crustacean only recently discovered and confined to these sorts of environments, occur below the interface, at which level sulphur bacteria are also found.

Bundera Sinkhole has surface water half as salty as seawater; the concentration of salt increases with depth towards the interface with seawater. At the entrance, the water is mostly covered with a deep scum of algae that form the basis of food chains for a fauna that is dominated by crustaceans of many types, some with connections to Tethyan species. Because of the small surface opening, the arid conditions at the surface and the penetration of light right down to the interface, the zonation and layering in the sinkhole water is remarkably stable. The water has a strong density interface, with a sharp thermocline (zone where temperature change occurs) at 8 metres with the temperature increasing about 5°C.[205] Light penetration goes only as far as the interface, so photosynthesis fuels activities above it and chemosynthesis is the basis for food chains below it. Salinity increases from half to full seawater between 8 and 16 metres. Sulphur bacteria, forming veil-like colonies suspended in the water, are active in the thermocline zone, releasing hydrogen sulphide gas (H_2S) as a by-product of their chemosynthesis.

The remipeds that live below the interface are endemic to the Bundera Sinkhole, although their affinities lie with the other known remipeds found on either side of the North Atlantic. Long isolation in widely separated habitats has resulted in speciation. The blind gudgeon and eel that live in the sinkhole are found throughout the water profile.

Ecosystems in groundwater calcretes in north-western Australia

Groundwater calcrete (limestone) zones are a common feature in arid regions, where they form at or near the watertable as a result of deposition of carbonates from groundwater flow. Their formation is in part due to micro-biological activities in the fluctuating saturated zone and often results in creation of karstic features like sinkholes and connected subsurface conduits like those in limestone cave environments. Such calcretes become valuable freshwater reservoirs as well as suitable habitats for the specialised groundwater biota of stygobites.

Extensive calcrete deposits have developed in palaeorivers carved into the ancient Western Shield that comprises the Pilbara and Yilgarn Cratons and associated basins in Western Australia. These drainages were incised 300 million years ago when Australia and Antarctica were still connected, and today contain few active rivers but many chains of salt lakes. (See *Running Down*, pages 12–29, on palaeodrainages.) As rainfall recharges groundwater and moves down-gradient towards a salt lake, it gradually approaches the surface where evaporation occurs and carbonates are deposited. The process involves major changes in groundwater chemistry and the water progressively changes from fresh to hypersaline. These gradients occur at intervals down the length of a palaeochannel, resulting in the formation of a series of isolated calcrete deposits, each immediately 'upstream' of a salt lake.

DISTRIBUTION OF CALCRETE AQUIFERS IN WESTERN AUSTRALIA

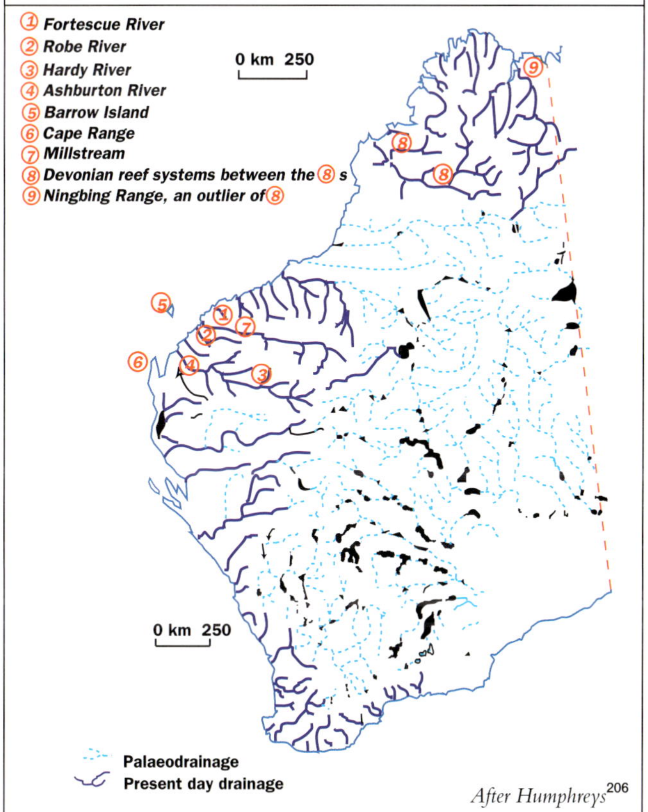

After Humphreys[206]

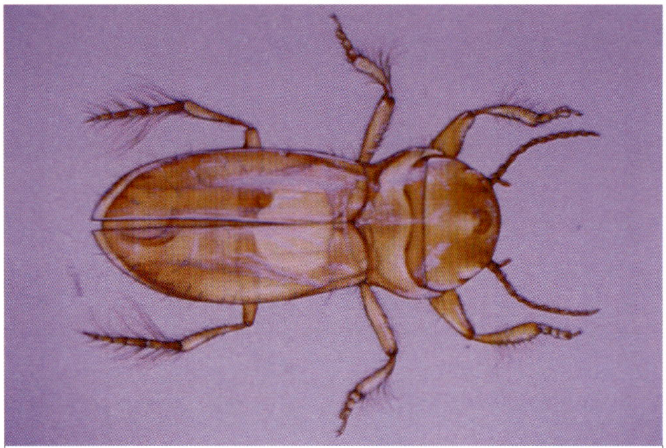

The eyeless and translucent diving beetle *Tjirtudessus eberhardti*, one of nearly 50 species inhabiting groundwater calcrete in the arid zone of Western Australia and the Northern Territory. *Painting by Elyse O'Grady*

Examination of a number of these deposits has so far shown that separate and unique assemblages of stygofauna occur in the different calcrete bodies.[194] Each is a localised biodiversity 'hotspot', and some of the species involved have Gondwanan connections, with related species found only in South Africa or Brazil. Study of the fauna is in its infancy and is urgently needed because increasing water extraction from palaeovalley reservoirs puts their survival at risk.

Originally, Aboriginal peoples used the underground water that the calcretes contain, but only in a very limited way; then pastoralists increased water extraction; and now, very large-scale development of mining below the watertable and on-site mineral processing has led to enormously increased extraction of water. The calcrete itself is being removed and used in construction and for neutralising acids in the mineral-processing activities. It is ancient, fossil water that is being mined from the deep palaeochannel aquifers below the calcrete layers, and the coupling arrangements between the two water systems are not understood. How such activities affect the stygofauna underground and the groundwater-dependent ecosystems on the surface is unknown. The chances of the well-being of either being a priority, when short-term financial gain is driving the water-mining, are slim indeed.

Recent investigations into a groundwater calcrete aquifer in the Murchison District have revealed a remarkable community of animals specialised for subterranean life.[206] Most of them are crustaceans of different kinds, but the main predators are blind diving beetles (Dytiscidae) of which more than 50 species inhabit calcretes of Western Australia and the Northern Territory.

Ecosystems of the limestone cave systems at Ida Bay, Tasmania

The Ida Bay limestone cave system of south-eastern Tasmanian harbours a diverse cave fauna of about 100 invertebrate species, of which 15 are obligate cave inhabitants and highly endemic. Commercial mining of the limestone has created far-reaching effects on the biota. The influx of sediment and acid run-off from mining operations, and physical alteration of conditions by human activities, have had a very noticeable impact on numbers of organisms and on the composition of invertebrate faunas. (The acid waters that dissolve the limestone and cause new cave features result from the exposure of pyrite in the mined limestone which weathers to release sulfate.[235]) The enormous limestone system comprises caves and tunnels of all sizes, ranging from the minute to the spacious caves that tourists and cavers visit.

Since mining has stopped and rehabilitation has begun, some understanding of the biotas and their requirements has been achieved. The importance of linkages between the underground and surface components has been recognised. Streams are crucial, importing food and organic matter for terrestrial and aquatic organisms within the system. For example, glow worms in the Exit Cave depend for their survival on the aquatic insect larvae that streams bring in for their food; and water percolating into caves through tiny fissures brings nutrients from the photosynthesising world outside to the creatures that live in the dark. The physical and biological interconnectedness of systems above and below ground is again emphasised, and a further connection exists to the hyporheic zone and the groundwaters below.

'THE LITTLE THINGS THAT RUN THE WORLD'[1] IN RIVERINE ECOSYSTEMS

Australia is a continent with unusual drainage patterns. Being in large part relatively flat, with vast areas draining towards the Centre, it has become a land of floodplains. A large proportion of its rivers do not drain to the sea, and most in the arid three-quarters of the continent are ephemeral. The combination of these factors results in a great deal of river flow being subsurface. It is now realised that floodplains are an essential part of the river systems, and particularly so in arid and semi-arid regions, depending to a greater or lesser degree for their life-giving water on irregular surface flooding and less irregular subsurface supply to shallow watertables.

Many present-day rivers lie in the beds of palaeorivers that facilitate underground flow. One has only to look at the Pleistocene ice age history of the continent, when in glacial stages Australia was twice as dry and twice as windy as it is now, and when about 80 per cent was subjected to build-up of wind-blown sand, to realise why so much flow is subsurface. In this context, rivers that flow into progressively drier terrain on their journeys westward from the Great Divide did not form the fully connected river systems that they do today when the first European explorers and settlers arrived in the continent. Archival records tell us that in many cases they terminated in swamps or billabongs, to join the 'system' only at times of great floods. (See page 212 in *Running Down*.) This can be attributed partly to the Pleistocene situation, when dry river valleys

Cotton growers at Dalby in Queensland rely on overland flow, when the nearby Condamine River floods most years, to fill their gigantic earth dams, and on an allocated allowance of alluvial aquifer water to subsidise their irrigation programmes. Big operators admit that their extraction of the shallow alluvial water in years when there are no floods exceeds recharge of the aquifer; and they also admit that the losses to evaporation and seepage in the dams and channels are enormous. This is an example of the sort of physical alteration to overall hydrology of an area that results from human intervention. The effects on riverine and groundwater biota are largely unstudied. *M.E.W.*

would have filled to some depth with aeolian material and floodplains had deep sand cover. The archives also tell us that it was possible for the Aboriginal guides who accompanied explorers to find fresh water on the plains many miles from the nearest river simply by digging down a very short distance, to what we now know was alluvial aquifer water on floodplains that were connected to river water underground and regularly topped up by overland flood flows.

The impact of human activities that caused erosion and alteration of headwater valleys and whole catchments has resulted in the present-day situation where there is incision of river beds and restriction of most flow to river channels. Many groundwater-dependent ecosystems in Australia have been affected by the changes to hydrology that result from our land management.

It has been customary to study groundwater and river ecology as separate entities. Hydrologists have focused on the former, biologists on the latter, and most ecological studies have until recently dealt only with visible animals and plants in the rivers. Only lately has attention turned to the interface between groundwater and riverine environments, and only still more recently has attention been paid to the micro-organisms involved in the webs of life that ensure river health. With increasing evidence being found worldwide for diverse life in groundwaters, and some microbial life being found at surprising depth in aquifers, the interconnectedness of the two realms can be seen. Just as there is an awareness emerging of the importance of micro-organisms in the regolith and their penetration even into the lithosphere, so also is a new world opening up in freshwater research.

Rivers do not act simply as pipes or drains conducting water and solutes from the catchment downstream. There are multiple flow paths above and below the ground where constant exchanges of water and materials occur, and living organisms—bacteria, micro-organisms and invertebrates—make them a very important part of the Biosphere.

Australian research in the hyporheic zone[207–213]

Recent riverine research in Australia has concentrated on a largely neglected area—the **hyporheic zone**.[208] This is the zone of saturated sediments (from fine-grained muds to the coarsest river gravels) that lies below many rivers, extending laterally beneath their banks and beyond. Streams that run over bedrock or extremely fine sediments largely lack this zone, but most rivers have at least some reaches where the bed is sand, gravel or cobbles, where water exchange occurs potentially in three dimensions: vertically and longitudinally as sequences of upwelling and downwelling along a stream reach, and laterally as exchange across gravel bars and alluvial aquifers.

The hyporheic zone is the interface between riverine and groundwater systems. In this zone, dynamic exchanges of water and materials occur between the groundwater below, lateral alluvial aquifers and the river flowing above. Like transitional zones between terrestrial ecosystems, it is an ecotone—a

A gravel-bed stretch of the Never Never River in northern New South Wales. *Andrew Boulton*

A rich growth of algae along a stretch of Bungeroo Creek in the Flinders Ranges. *Andrew Boulton*

region where the biota of the different realms meet and intermingle, where special conditions apply and specialisation is seen in some inhabitants. The biological processes that occur in this zone affect the quality of water in the systems. This vital zone is threatened or compromised in many rivers by siltation, toxicants, increasing salinity, physical extraction of gravel, and altered groundwater inputs resulting from pumping of the alluvial aquifers.

As water flows through the hyporheic zone, microbial and chemical processes alter the water chemistry by transforming nutrients, consuming oxygen, decomposing organic matter and other activities. The water that emerges from the zone and mixes with the river flow is of different quality as a result and is capable of altering surface processes like algal growth and leaf-litter breakdown, and the composition of invertebrate faunas. Upwelling zones are often marked by swathes of algae that benefit from the nutrient-enriched upwelling water. In some streams in the Flinders Ranges, a longitudinal transition in algae can be seen along the gradient of declining nutrient concentration below an upwelling zone.

In some rivers, the volume of water in the hyporheic zone may be three or four times that of the surface flow. The zone may extend laterally for several kilometres in rivers with large alluvial floodplains, but only where sediments are not so fine as to affect porosity.

Boundaries of the hyporheic zone are elastic, fluctuating

FLOW PATHS IN A RIVER REACH

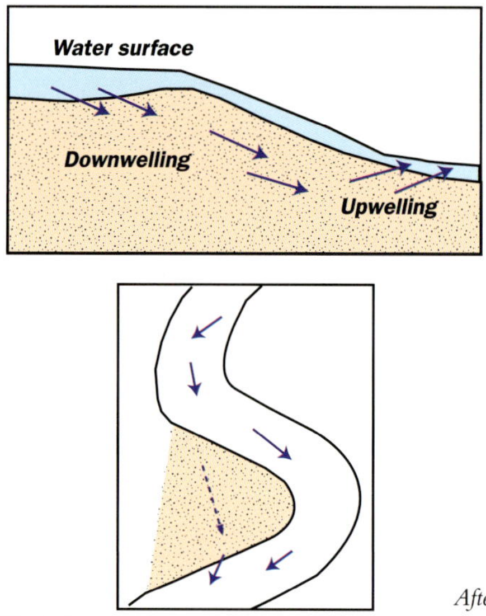

After Boulton[208]

Hyporheic water travels vertically, horizontally, and laterally in the river.

with changes in surface discharge, watertable elevation and variations in permeability. During low flows, little surface water enters the zone, most of the contribution coming from deeper groundwater or the alluvial aquifers. At high-flow times, surface water can flood the zone. Downwelling water carries oxygen, nutrients, organic matter and tiny invertebrates into the zone, and may also carry unwanted substances like silt, toxic substances and acid run-off. As the water percolates through the layers of sediment they act as a biological filter (like gravel does in a fish tank). The biofilms coating the sediment particles perform the filtering, often enhancing water quality during the exchange. They also act as a 'lawn' for minute grazing invertebrates. The finer the sediment particles, the greater the surface area available to be colonised by biofilms, so coarse river gravels have less filtering ability than finer sediments, although a gravel bed is more permeable.

A diverse fauna of tiny invertebrate creatures has been found in the hyporheic zone of unpolluted rivers, mainly undescribed and new to science. Many that are specific to groundwater habitats are giving exciting clues about past climates and continental movements as they act as 'living fossils'.[207] Many that live permanently in interstitial spaces between grains of sediment are physically altered, having lost their eyes and developed elongated sensory appendages—the same sort of adaptations as are seen in creatures living in the dark in caves above ground. The zone is also thought to provide refuge for small surface invertebrates during times of river drying or flooding.

Hyporheos, residents of the hyporheic zone

Hyporheos can be divided into two major categories—occasional and permanent. Occasional hyporheos, as their

THE HYPORHEIC ZONE AT THREE SPATIAL SCALES

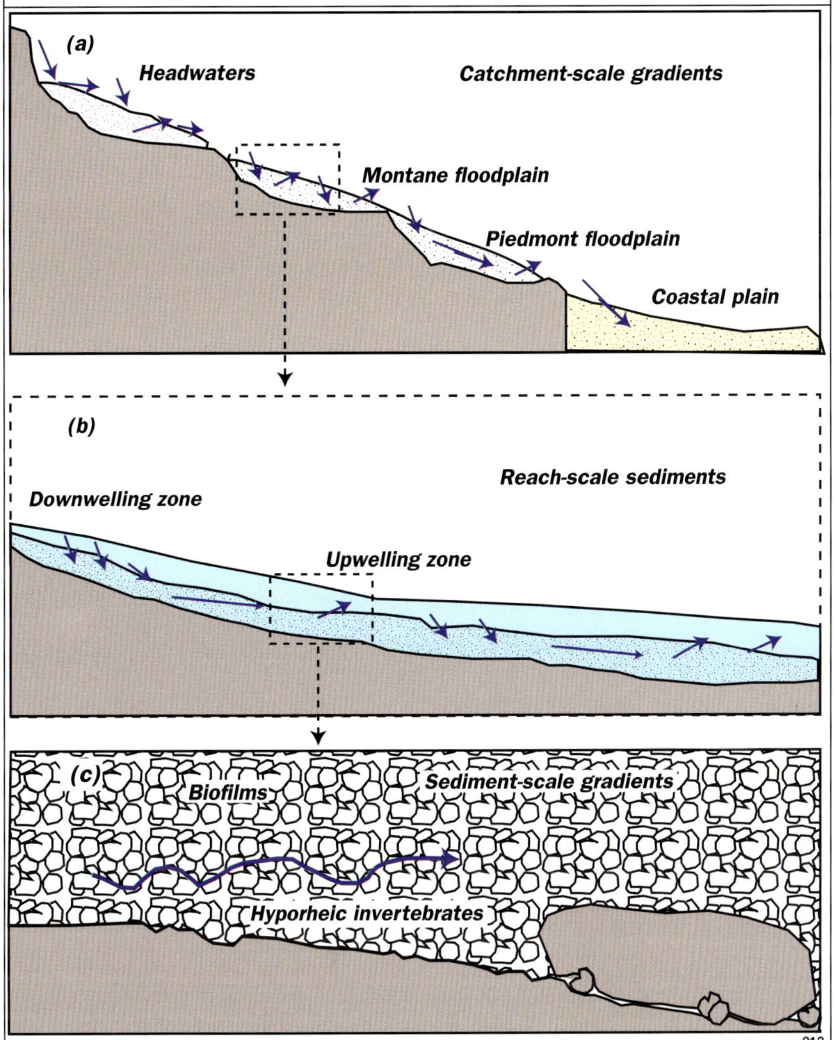

After Boulton et al.[213]

A: At catchment scale, gradients are seen in relative size of the hyporheic zone; amount of water retained; sediment size. B: At reach scale, upwelling and downwelling zones alternate, generating gradients in nutrients, dissolved gases and sub-surface fauna. C: At sediment scale, microbial and chemical processes occur on particle surfaces, creating microscale gradients. Arrows indicate water-flow paths.

name suggests, are surface-dwelling invertebrates such as insect larvae that occasionally enter the hyporheic zone. The permanent hyporheos comprise many types of crustaceans, segmented worms, flatworms, rotifers and water mites that are specialised for subterranean lifestyles and seldom occur at the surface. Biofilms provide an important food source for the hyporheos, whose grazing activities in turn are believed to enhance biofilm productivity. The breaking down by invertebrates of leaves and twigs trapped in sediments increases surface area for microbial activity, and prevents clogging of the system. Very little is known about food webs in the zone.

The physical effects of the small invertebrates and micro-organisms on the zone parallel those of soil organisms on soil structure. They alter pore size through their activities (movement, feeding, pellet formation); act as vectors for microbes; graze biofilms; are prey for surface predators; and generate nutrients through their excretion and decay. Some representatives of the permanent hyporheos are relatively large crustaceans, reaching up to 15 millimetres in body length. These invertebrates can act like tiny bulldozers in fine sediments, moving rapidly through interstitial spaces, their activities ventilating the spaces and generating tiny water currents through them.

Studies of the effects of human activities on biodiversity in the hyporheic zone are in their infancy. However, interesting results were obtained in a study in New Zealand that looked at the relative abundance of species in the hyporheic zone in rivers that ran through native forest, through land that had been cleared and converted to pasture, and through pine forest.[210] Biodiversity was greatest in the natural forest river, decreased in the river running through pasture, and was much decreased in the pine forest stream. Thus it can be seen that biodiversity in river systems suffers the same fluctuations as in the terrestrial systems from which their run-off water contribution is derived. This is another graphic illustration of the interconnectedness that characterises the Biosphere.

Studies in Australia in the Never Never River that arises in the Dorrigo National Park showed that close to the pristine headwaters the hyporheic zone supported a diverse biota. (In a 'hotspot' near Bellingen, 28 species of water mite, at least 11 of them new to science, were identified.) Downstream, where the river runs through valleys that have been heavily logged and where river flats are subject to cattle grazing or intensive cropping, samples contained few invertebrates and great quantities of interstitial silt. Clogging of the sediment spaces appears to have eliminated most of the hyporheos; dissolved oxygen levels were found to be low in the downwelling zones, implying disruption of hydrological exchange of subsurface water with that of the stream. In the lower reaches of the river, gravel extraction has been extensive, as it is in so many gravel-bed rivers, further disrupting the connections between surface and subsurface waters.

Siltation of rivers, a major consequence of so many land-use practices, is obviously having serious effects on the health of complete hydrological systems. It is no longer enough to legislate for control of river water abstractions and environmental flows or allocation of groundwater entitlements. The big picture that involves recognition of the interdependence of the systems has to be the basis for management.

Marine ecosystems

We started our study of the Living Biosphere and the Symbiotic Planet with an account of microbial life in the oceans, and we journeyed on to see Life-as-we-know-it evolving from bacterial beginnings. We have seen how microscopic organisms are still essential links in the Webs of Life and we have marvelled at the interconnectedness of everything, animate and inanimate. Now we have come full circle, back to the oceans that gave us life (for we are part of Life's continuum), and the final pages of this book give glimpses of two major marine ecosystems.

Microbes and micro-organisms in the marine realm are functioning as their ancestors did in the primaeval seas, reinforcing our understanding of the course of co-evolution of Life and the Biosphere. If you are like me, you will find yourself amazed by the detail of ornamentation, and perfection of detail in structure, of some of the micro-organisms that scanning electron microscopy has revealed, and will join me in thanking the Australian Antarctic Division in Hobart for providing the images that take one's breath away in the final study on the Southern Ocean.

CORAL REEFS AND SYMBIOTIC ALGAE, EPITOMISING SYMBIOSIS

Major coral reefs in the oceans today are formed by stony corals, the Scleractinia, whose growth and survival would not be possible without their symbiotic dinoflagellates (generally called zooxanthellae) which are algae, now classified as Protists.[51]

Only the Scleractinia, with their very large numbers of small polyps, are capable of producing the skeletal architecture that will withstand the immense force of ocean waves. The photosynthetic activities of the symbionts enhance the processes of calcification, enabling the reefs to grow quickly enough to overcome climatic and biological forces of change and destruction. Much of the carbon fixed by the endosymbionts is exported to the coral host, helping it to grow and respire. In addition, the symbiosis allows the coral to take up inorganic nutrients such as nitrogen (as ammonium) from the environment, an advantage not shared by corals that do not have symbionts. Because the tropical oceans in which corals live are 'nutrient deserts' this is an enormous advantage.

The symbiotic dinoflagellates that colonise reef-building corals all belong to the one genus, *Symbiodinium*. Until the advent of DNA analysis (and of Small Sub-unit RNA in

Aerial view of a coral atoll and its fringing reefs near Heron Island, Great Barrier Reef. *Photo and caption Dr Peter Harrison*

Detail of *Favia* brain coral polyps, showing the zooxanthellae. *Photo and caption Dr Peter Harrison*

Brain coral polyps (*Favites pentagona*) expanded at night. Note the dark brown polyp tissues, coloured by masses of zooxanthellae, contrasting with the animal pigments seen in the oral disks and tentacles. *Photo and caption Dr Peter Harrison*

particular) it was impossible to separate the different strains within the genus, but it is now known that there is great diversity, and four major clades, A–D, are identified.[214] For the algae, the advantage of the symbiotic relationship is living in a nutrient-rich environment instead of in the nutrient-poor seawater outside, where concentrations of phosphate and ammonium are very low. Ammonium in the host's cytoplasm has been shown to be between 5 and 50 times as high as in the surrounding water; concentrations of phosphate can be a thousand times as high. Additionally, the nutrient levels remain fairly constant within the organisms, unlike conditions outside that are subject to large fluctuations.

Marine organisms in many other important groups also have symbiotic relationships with algae: soft corals and other coelenterates, sponges and many molluscs, including giant clams. The clams have been shown to have the same genus of dinoflagellates as endosymbionts as the stony corals—*Symbiodinium*—and the same clades have been identified, often with several different types as endosymbionts in the host clam at the same time.[215] Some sponges also have the same genus of endosymbionts.

(Other, free-living dinoflagellates can form huge blooms and create the poisonous 'red tides' in the oceans. Some species are bioluminescent and create phosphorescence on the sea at night; some are not photosynthesisers and eat other organisms.)

Coral bleaching on the Great Barrier Reef

Bleaching today threatens coral reefs globally. It involves the expulsion of the symbiotic organisms that are responsible for colour as well as nutrition, which results in the death of the coral after it has remained white for a fairly short period. If bleaching is transitory, algae can return and the coral can recover.[215a] During the 1980s, repeated episodes of bleaching occurred on reefs around the world, including the Great Barrier Reef. The causes of bleaching are still unclear. While stress of many origins is involved, temperature increase appears to be the most serious, causing irreversible bleaching. Because global warming is known to increase extreme fluctuations of temperature, and because coral reefs are essentially tropical and algae are known to desert their coral hosts above 30°C, this is a global crisis. Whether the corals produce substances toxic to algae when they are under stress, or whether they simply do not produce enough food for their symbionts, is not known. Some bleaching is related to pollutants reaching the reefs in run-off from the land.

Recent research into the different clades of *Symbiodinium* endosymbionts in *Acropora* and other stony corals has shown

Schools of fish swim through stands of partly bleached *Acropora* staghorn corals. Note the upper sections of each branch exposed to light are bleached, while the lower shaded sections contain masses of zooxanthellae and are brown. Heron Island reef, April 1998, during the massive coral-bleaching event which led to significant mortality in some areas of the Great Barrier Reef, and catastrophic mortality on some reefs worldwide. *Photo and caption Dr Peter Harrison*

AUSTRALIA'S GREAT BARRIER REEF UNDER THREAT

View of masses of bleached *Acropora* corals on exposed reef flat at Heron Island, April 1998.
Photo and caption Dr Peter Harrison

Partly bleached colony of *Favia* brain coral showing healthy zooxanthellae population density in polyps on left, and bleached polyps on the right. Magnetic Island, central Great Barrier Reef.
Photo and caption Dr Peter Harrison

Brain coral colonies bleached and starting to die at Orpheus Reef, May 1998. *Photo and caption Dr Peter Harrison*

Bleached soft coral colony at Orpheus Island reef, May 1998. *Photo and caption Dr Peter Harrison*

that zoning of the various clades occurs according to high and low light penetration and temperature. Members of some clades are much more sensitive than others; varieties that like bright light and tolerate temperature fluctuations occupy higher zones on the reef while less-tolerant kinds choose polyps in shaded situations or lower on the structure. In some cases, bleaching has been seen to start in the transition zones where levels of irradiance are critical to residents who are either at the bottom of their favoured zone or at the top of the suitable zone below—so it is individuals already under stress that are most vulnerable when other factors threaten viability. Because of the different tolerances of different strains to higher or fluctuating temperatures there is hope that some corals might be able to survive global warming by changing their symbionts. Those better adapted to warmer water would naturally be favoured by the decreased competition for host space if warming was gradual enough for adaptation to take place.

Reef-building corals and their symbionts do indeed epitomise symbiosis, as the heading of this section suggested. It is amazing just how much genetic, and therefore metabolic, difference can exist between single-celled organisms that appear identical and are quite unable to be separated as different varieties except by their cellular chemistry. Each individual is the product of its own endosymbiogenic history; each coral host is characterised and suited for its niche by its endosymbionts.

A number of real threats to the Great Barrier Reef have recently been recognised.[216] Scientists warn that the reef is being slowly poisoned by actions on the mainland. Areas of reef close to the shore are at risk from agriculture, especially cane farming on the coast and the cattle industry in the headwaters of streams that drain the mainland. A fundamental change to practices is necessary to improve water quality on the Reef. Large-scale vegetation clearing has to cease; responsible use of pesticides, herbicides and fertilisers is essential.

Recent sampling has shown that crabs from localities from Brisbane to Cairns are contaminated by dieldrin; dugongs contain dioxins in high concentrations; diuron, a herbicide,

DDT and atrazine are damaging reef ecosystems; heavy metal pollution, seen in high mercury levels found in Bowling Green Bay and high cadmium levels on Hinchinbrook, comes from fertiliser use on the mainland. In a huge, symbiotic system such as the Reef represents, imbalances caused by all these sorts of pollutants are extremely serious.

Trace metals affect tropical marine ecosystems. It has been shown that copper, in particular, is toxic and even relatively low concentrations significantly affect fertilisation rates.[227] Copper also impairs or inhibits settlement of coral larvae.[228] Copper is present in sewage, in fungicides and herbicides, and in antifoulant paints used on boats, and is seen particularly in reef lagoons. In the Magnetic Island region, the marine sediments contain high concentrations of copper, and disturbance by dredging, and even disturbance caused by cyclones, can liberate it. Cadmium and zinc (in laboratory experiments) were shown to be less toxic overall than copper, but to affect different species at different developmental stages to some extent.

Concern about increasing nutrient levels affecting the Great Barrier Reef led to the ENCORE experiment, commenced in 1993. The effects of controlled additions of dissolved inorganic nitrogen and/or phosphorus on patch reefs on One Tree Island at the southern end of the Reef were investigated over a two-year period. The results of the increased nutrient levels on reef corals were:

- Organisms containing zooxanthellae (corals and giant clams) assimilated dissolved nutrients rapidly, responding initially to increases.
- Coral mortality increased with increase in nutrient loading.
- Nitrogen additions stunted coral growth.
- Corals exposed to elevated nitrogen levels produced fewer and smaller eggs and less testes material.
- Phosphorus additions increased the rate of calcification and linear extension of corals but skeletal density was reduced, making them more easily broken.
- Elevated phosphorus levels resulted in more and smaller eggs and more testes material.
- Settlement of coral larvae was reduced in nitrogen treatments but enhanced initially in phosphorus treatments, decreasing at higher levels.
- No significant changes in rates of bio-erosion and grazing were detected.

Other organisms were affected in various ways:

- Benthic crustaceans (stromatopods) living in coral rubble suffered recruitment decline in nitrogen and nitrogen + phosphorus treatments.
- Grazing rates and reproductive effort of various fish were not visibly affected. Because fish remove large amounts of epilithic algae, their grazing masks the effect of nutrient increases on algal communities. Increases in algae no doubt promote changes in fish densities, grazing rates, etc.
- Nitrogen fixing by sediment micro-organisms increased in the phosphorus treatments; denitrification increased in the nitrogen treatments.

Conclusions reached in the ENCORE study were that impacts of increased nutrients were subtle and sublethal—but this was a limited experiment, and the Great Barrier Reef is under threat. Since about 1850, when European settlement began in the region, 83 per cent of coastal catchments on the adjacent mainland have suffered urban or agricultural development. The flux of nitrogen and phosphorus into the Great Barrier Reef lagoon has increased fourfold, from 2500 tonnes of phosphorus in 1850 to 10 000 tonnes in 1991; from 17 000 tonnes of nitrogen in 1850 to 70 000 tonnes in 1991 (and how much more since?). While inshore reefs are most impacted by run-off from the continent, river plumes also reach the outer reef on occasion.

Coral reefs: The global situation

A new estimate of the extent of the world's coral reefs by United Nations experts has revealed that they are not nearly as extensive as previously thought. Only 284 000 square kilometres exist—in 0.01 per cent of the global ocean—and widespread degradation has occurred, particularly in the last five years. Reefs are 'over-fished, bombed and poisoned', smothered by sediment and choked with algae that thrive on nutrient-rich sewage and fertiliser run-off.

Indonesia has the most extensive coral reef systems—with an areal extent of 51 000 square kilometres (18 per cent of the world's coral). Australia comes next with 49 000 square kilometres (17 per cent). The Philippines, Malaysia and New Guinea between them account for about 45 000 square kilometres (about 16 per cent) and the greatest biodiversity is found in this region, with 500 to 600 coral species in each nation's sector. It is disastrous that reefs in this most diverse region are the most threatened by human activities.

In Indonesia, 82 per cent of the reefs are at serious risk, mainly from the blasting method of fishing, regularly used, that stuns the fish so that they float on the sea surface and are easily collected, while structural damage to the reef and its ecology are ignored.

Australia cannot afford to be complacent about its reefs—32 per cent are threatened, mainly by pollution and sedimentation. Global warming is threat enough without those recognised by a report by the Great Barrier Reef Marine Park Authority that finds that pollution levels have increased by 1500 per cent in the last 150 years. Pollution discharged into the lagoonal waters that lie between the coast and the reef show sediment loads increased by 300 to 900 per cent; phosphate discharges up by 300 to 1500 per cent; total nitrogen up by 200 to 400 per cent; and pesticides and heavy metals are now detectable in tidal sediments. Tourist activities also cause damage.

CASE STUDY: MICROBIAL RESEARCH IN THE AUSTRALIAN ANTARCTIC DIVISION

(Provided by Programme Leader, Biology, Professor Harvey Marchant)

The Australian Antarctic Division, located south of Hobart, Tasmania, is part of the Federal Government department, Environment Australia. The AAD is the lead agency for Australia's Antarctic programme. Various scientific research programmes are conducted by scientists employed by the AAD, including investigations on micro-organisms in marine and other aquatic environments. Research is conducted near stations on the Antarctic continent, on board the icebreaker *Aurora Australis* and in Australia on material returned from Antarctica. The research being undertaken is part of national and international collaborative programmes.

The Southern Ocean is a major site for the exchange of carbon dioxide between atmosphere and the sea. The activities of micro-organisms (autotrophic and heterotrophic protists, bacteria and viruses) play a significant role in determining when, when, where and how much carbon dioxide is taken up by the sea or released from it. In addition, marine micro-organisms produce dimethyl sulphide (DMS), a compound which when ventilated to the atmosphere is oxidised to sulphate aerosol particles which can act as nucleation sites on which water vapour forms droplets, promoting the development of clouds.

Microbiological research in the AAD is directed to ascertain how microbial communities in the Southern Ocean are influenced by global changes and to ascertain their role in marine processes that modulate the exchange of carbon dioxide and other chemically active gases between ocean and atmosphere. New marine micro-organisms are frequently being found and described. Their community structure and activities, including growth rates and grazing rates and selectivity, are being explored, as is why particular communities are found at particular locations at certain times of the year. The responses of phytoplankton and bacteria to elevated ultraviolet radiation as a result of the springtime 'ozone hole' continue to be investigated. This research is designed to better understand the microbial activities that drive the whole marine food web as well as the role of these organisms in global change processes.

BIOLOGICAL ZONES IN THE SOUTHERN OCEAN[20]

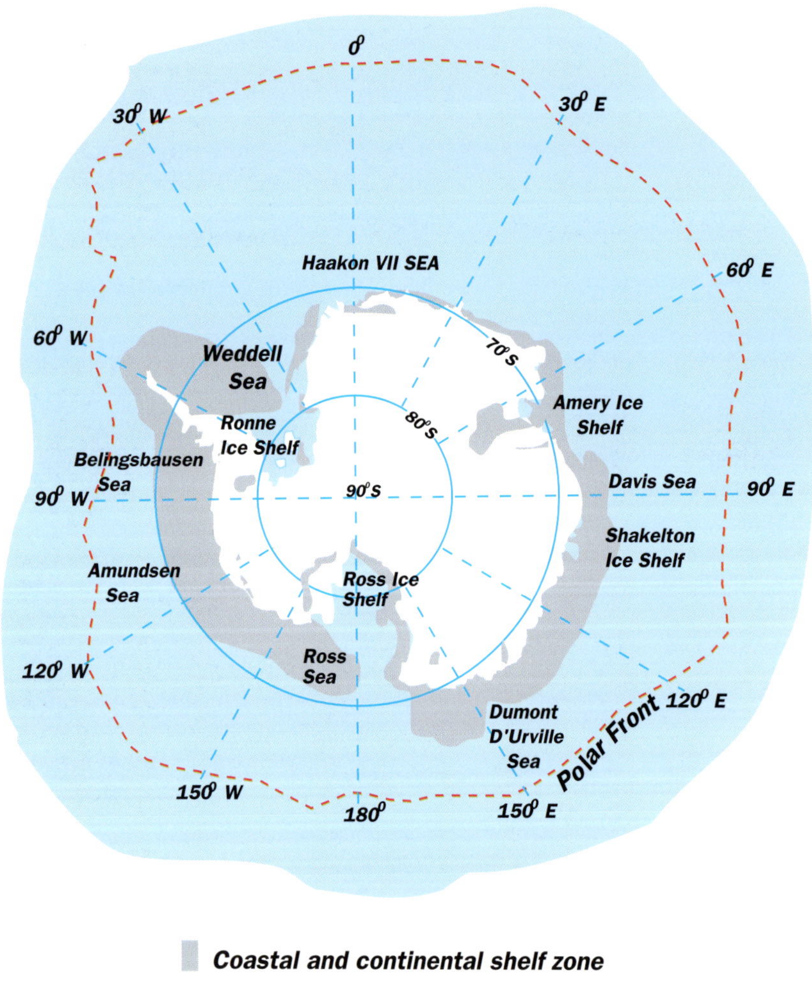

CASE STUDY: CORAL STRESS RESEARCH AT SOUTHERN CROSS UNIVERSITY

Associate Professor Peter Harrison and his postgraduate students at the School of Environmental Science and Management at Southern Cross University have been studying the effects of natural and human-induced stress on reef-building corals and their reproductive success, over the past two decades. Their research has revealed that critically important sexual reproductive processes of reef corals, including fertilisation, larval development and settlement, are very sensitive to a wide range of environmental stressors. Reproductive success is impaired or blocked by relatively small changes in environmental conditions and water quality, including elevated temperatures, reduced salinity, increased nutrient levels, oil pollutants and elevated levels of trace metals. In many cases, reproductive success is impaired or blocked at levels of disturbance or concentrations of pollutants that are much lower than those that are known to affect adult corals. Hence, coral reproductive success can provide very sensitive indicators of stress for coral reef environments. The ultimate aim of the research is to determine ecologically meaningful guidelines for maintaining suitable water quality and environmental conditions on coral reefs, to enable successful reproduction by the corals that build these magnificent reef structures, and provide habitats for thousands of other reef species.

PROTISTS OF THE SOUTHERN OCEAN

A ciliate lorica covered in *Emiliania* coccoliths. The organism that lives inside this lorica is not photosynthetic. It eats other organisms and decorates its house with the scales of its prey. *Geraldine Nash, AAD*

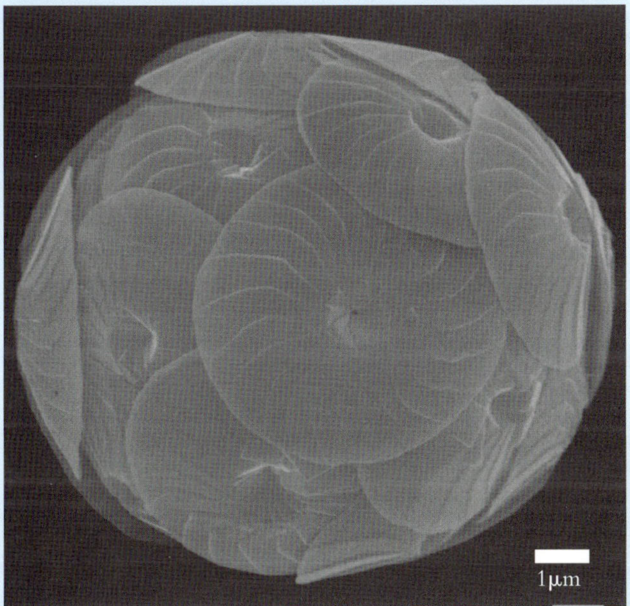

A photosynthetic coccolithophore, *Oolithotus* sp. The scales that cover this organism are more disc-like than the open spoke pattern seen in *Emiliania*. *Geraldine Nash, AAD*

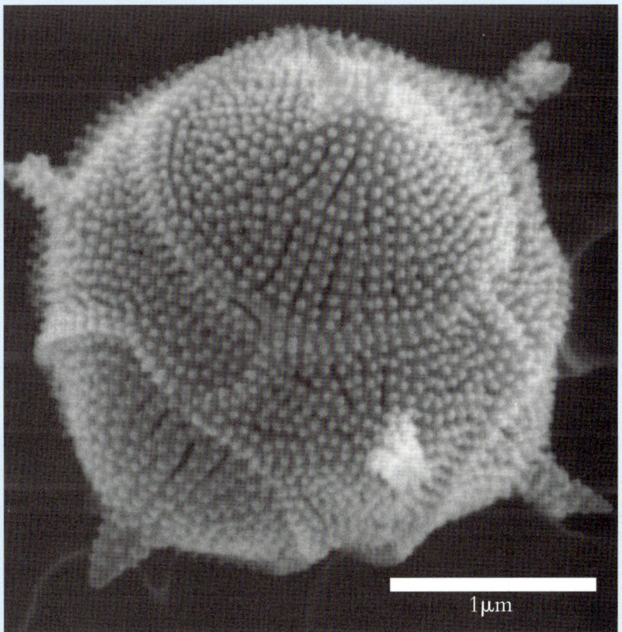

A photosynthetic organism of the algal group the Parmales, *Tetraparma pelagia*. *Geraldine Nash, AAD*

Disc-like cells of the diatom *Porosira pseudodenticulata* are more than 50μm in diameter. The short chains that are formed during division of diatoms are easily broken up. The glassy walls, made of silica, accumulate as thick deposits on the seafloor. Diatoms are photosynthetic. *Fiona Scott, AAD*

LIFE IN THE SOUTHERN OCEAN

The Southern Ocean around Antarctica was thought to be one of the richest marine ecosystems on Earth, with a simple food web in which diatoms fed shrimp-sized crustaceans called krill, which were in turn consumed by fish, whales, seals and birds. It is now known that this is a gross over-simplification and that most of the Southern Ocean is no more productive than more nutrient-poor parts of the world's oceans, and its biology is as complex as that of warmer waters. The Southern Ocean comprises several connected ecosystems, and cannot be regarded as a single ecological unit.

The eastward-flowing Antarctic Circumpolar Current dominates in the Southern Ocean. Largest of the Earth's currents, it developed when lands that had been attached to Antarctica in Gondwana had moved away sufficiently for it to start in the widening proto-Southern Ocean in response to the spinning of the Earth. Fossil evidence indicates that it was fully established by 27 million years ago and that by 22 million years ago the Polar Front had formed.[20] Here warm water from the north meets the cold water of the circumpolar zone and the water temperature abruptly drops by about 2°C. The Polar Front is one of the Earth's major oceanic boundaries and is a significant biological barrier.

Near the Antarctic continent, the coastal currents flow in the opposite direction to the circumpolar current, travelling westwards. This is usually attributed to the shape of the continent with its large indentations, the Ross and Weddell seas, the northward-projecting Antarctic Peninsula and the varying contours of the seafloor—all of which change the breadth and speed of the main eastward-flowing current and give rise to localised eddies. Contributing factors are believed to include wind shear from the southern edges of subantarctic cyclonic winds and geostrophic currents from northward-flowing fresher coastal waters. The patchy distribution of marine life around the continent results from the diverse conditions that local currents and conditions provide. South of the Polar Front, an ice-free zone surrounds the seasonal ice zone, which is ice-covered for most of the year. Close to the coast are permanently ice-covered regions.

The Permanent Open Ocean Zone (POOZ) is nutrient rich but has relatively low levels of primary production. Dominated by the strong easterly current, it is a zone of high winds and rough seas that cause deep mixing of surface waters. Phytoplankton carried below the photic zone do not have sufficient light to photosynthesise. (See also page 36 for factors limiting photosynthesis and productivity in this zone.) The phytoplankton in this zone are usually tiny nanoplankton, mainly diatoms and flagellates. They are grazed by several groups of organisms, but generally not directly by Antarctic krill, *Euphausia superba*. Coccolithophores are involved in phytoplankton blooms as far as 60° South in this zone. Zooplankton in this zone and elsewhere in the Southern Ocean include many sorts of invertebrate grazers and larval fish, all much less studied than krill but forming a high percentage of the zooplankton biomass. Other euphausiid crustaceans, copepods, salps, amphipods and arrow worms play key roles in the marine ecosystem.[20] Salps (planktonic sea squirts) and copepods are involved in seasonal blooms. The salps are filter feeders and can effectively remove all sizes of plankton from the water. They dominate to the exclusion of other organisms, including krill.

The Seasonal Ice Zone (SIZ) is ice-covered in winter, but is essentially open water in the summer. It is the most productive zone of the Southern Ocean, contributing at least 40 per cent of the phytoplankton biomass[22] and around 60 per cent of annual pelagic primary production.[61] Blooms of phytoplankton are common in the shallow, less saline water that is produced as the melting sea ice retreats southward in spring and early summer. Large diatoms and blooms of the alga *Phaeocystis antarctica* are dominant, grazed by abundant krill and other creatures. Baleen whales, crab-eater seals, penguins and other birds feed on the krill.

The Coastal and Continental Shelf Zone (CCSZ) retains most of its ice even at times of maximum retreat. Phytoplankton blooms may be intense but are short lived. Antarctic krill are uncommon, replaced by smaller *Euphausia* species, and mammals are less abundant than in the SIZ.

The coastal and continental shelf areas of Antarctica have a rich benthic (bottom-dwelling) fauna and patchy seaweed and algal microflora, except in the zone that is scoured by ice and glacial debris. Most communities are dominated by animals that feed on particles of organic matter that rain down from the surface waters.

Studies of benthic communities of the CCSZ in McMurdo Sound have shown that ice-scouring takes place regularly to a depth of 15 metres, so the only algae in this zone are those capable of growing fast in the short summer, and the animals are mobile foragers—sea urchins, starfish, worms, crustaceans and fish. A rich flora and fauna exists below this level. Filter-feeders, including anemones, soft corals, molluscs, sea squirts, sponges and tube worms, are attached to the sea floor. Some of the sponges have algal endosymbionts to help with their nutrition, and the glassy spicules that are developed in sponge tissues act as optical fibres (light tubes), transmitting light to their symbionts and enabling them to function. Mobile scavengers are abundant. Where the seafloor is sandy or muddy, burrowing animals are found in greater concentrations than elsewhere. Large proboscis worms (*Parborlasia corrugatus*) can grow to several metres in length. They are carnivores that feed by enveloping their prey in their proboscis. They protect themselves against other predators by exuding acidic mucus, the ever-useful slime.

Micro-organisms, bacterial, algal and faunal, are abundant on the seafloor and in the sediments. Micro-algae also occur in profusion, attached to seaweeds or other organisms. They make a substantial contribution to primary production. Seaweeds are sparse below 40 metres and absent below 100 metres, restricted by light availability.

For most of the year, the sea surface is frozen around

MARINE SNOW[218]

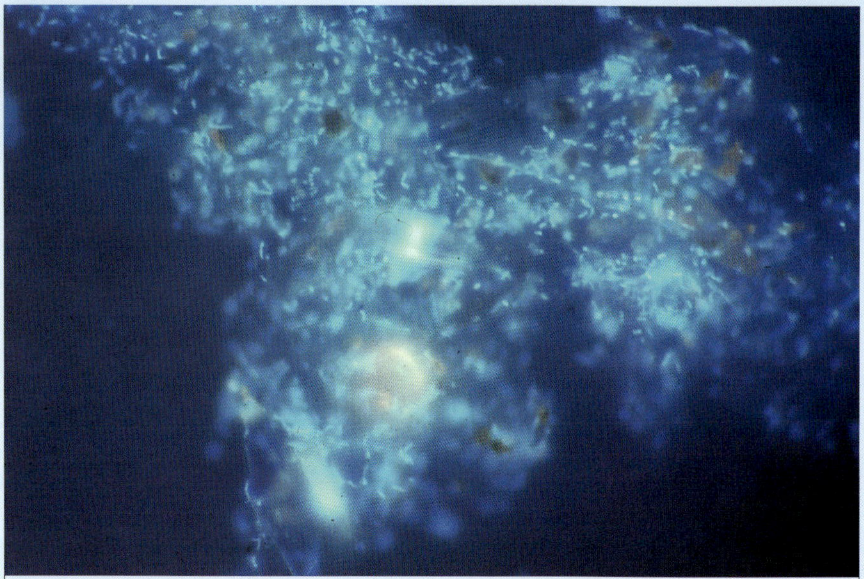

Marine microbial communities often exist as aggregates called marine snow. This marine snowflake has been treated with a stain that makes the DNA of the constituent organisms fluoresce blue-white when exposed to UV light with a fluorescence microscope. The small stained rods are bacteria, the large diffuse stained areas are the nuclei of protozoa, and the red-stained cells are algae. *Harvey Marchant, AAD*

Marine snow is the name given to fragile aggregates, visible to the naked eye, that occur in the world's oceans in drifts like a snowstorm. Marine snow particles are formed from aggregations of zooplanktonic faecal matter, mucilage from phytoplankton and zooplankton, and non-living detritus in the form of vast numbers of fragments of the outer coverings of microscopic organisms—the silica frustrules of diatoms, calcium carbonate scales from coccolithophores, for example—and organic compounds, polysaccharide-rich, produced by phytoplankton. The snow particles are rapidly colonised by phytoplankton, bacteria and protists, forming thriving communities where close proximity aids recycling of nutrients within the microbial loop. Bacteria are frequently concentrated on the transparent exopolymer particles (TEP). Main producers of the TEP are diatoms and *Phaeocystis*, both of which dominate in major blooms as the sea-ice melts in spring, and the bacteria themselves.

Zooplankton feed on the marine snow. Particles and organisms that are too small to be individually captured by filter-feeding and predatory invertebrates when dispersed in seawater can be ingested when aggregated in marine snow. The abundant nanoplankton that have been reported from gut and faecal material of Antarctic krill[219] suggest that krill also feed on the aggregates. So eating marine snow effectively short-cuts the food chain, if the eaters do not mind a lot of 'roughage' in their food.

Aggregation leads to increased sedimentation, as the larger aggregates sink to the seafloor more readily than the individual detritus and micro-organisms would.

Antarctica, the freeze beginning in March and reaching maximum extent in September, when sea ice covers 20 million square kilometres (an area nearly three times that of Australia). The sea ice is a major habitat for microscopic organisms. Diatoms proliferate on its underside, as well as at the interface between the ice and the snow that settles on it. Concentrations of diatoms create a rich ochre-coloured layer on the lower surface of the ice, clearly visible, promoting the original concept of diatoms as the basis for a simple food chain in Antarctic waters. Small creatures that graze on the micro-organisms find refuges in the cracks and other features in the ice. Seals and penguins breed on the ice. Limited areas where enough light is available support some species of seaweed, and these patches are food for grazers. Their patchy distribution results in the patchy distribution of creatures that feed on or live in seaweed.

In the autumn and winter, when the sea freezes, highly saline water is excluded from the forming ice. (Ice has lower salinity than the water from which it forms.) The cold, very salt water sinks to the seafloor and flows northwards as Antarctic Bottom Water, starting the deep abyssal currents that create the vertical circulation and mixing of the global ocean. During spring and summer, the release of less saline water from the melting ice produces a shallow, nutrient-rich layer within which phytoplankton bloom in the 24 hours of daylight. Localised areas of high output briefly support large biomass and such events gave rise to the idea of highly productive Antarctic waters.

Though most of Antarctica is permanently ice-covered, it has been found[217] that soils can develop and support communities of terrestrial plants, algae, lichens and animals. Each gram can contain between tens of thousands and billions of bacterial cells (Archaea and Eubacteria) as well as many minute invertebrates.

Greenhouse warming and the Southern Ocean

There is serious concern that global warming will reduce the extent of sea ice and diminish the production of Antarctic Bottom Water. Global ocean circulation would be affected and movement of major currents from their present flow paths could have serious repercussions. (See pages 174–5 in *Running Down*.)

In Antarctic waters, local greenhouse effects from reduced production of sea ice would be: stagnation of the bottom water, depleting its oxygen content and making it uninhabitable by marine organisms; reduction in amount of less-saline meltwater, diminishing ice-edge blooms; reduction in habitat for organisms and thus lower productivity in the

CASE STUDY: AUSTRALIAN RESEARCH INTO ANTARCTIC MARINE LIVING RESOURCES

(Stephen Nicol)

Australian research itno Antarctic marine biology began in the late 1970s, spurred on by the establishment of the international BIOMASS (Biological Investigations into Marine Antarctic Sustems and Stocks) Program which had as its focus the distribution and abundance of Antarctic krill. This interest was itself driven by the development of a fishery for krill, and the BIOMASS Program sought to find out more about Antarctic krill and the interrelationships between krill and the other elements of the marine ecosystem. Australian Antarctic marine research was made feasible by the conversion of the supply ship, the *Nella Dan,* into a functional research vessel and it was with this ship that the Australian Antarctic Division made its first concerted forays into open ocean research. Australia was a participant in the highly successful First International BIOMASS experiment (FIBEX) in 1981 and was fraught with logistic difficulties. FIBEX was the first attempt to survey the distribution and abundance of Antarctic krill using acoustic techniques and the results from this huge effort, using 13 ships from 11 nations, were translated 10 years later into catch limits for the krill fishery in the Atlantic and South West Indian sectors of the Southern Ocean.

Krill from the Southern Ocean

In parallel with scientific focus on the waters around Antarctica, diplomatic efforts in the late 1970s and early 1980s concentrated on developing a regime to manage the harvesting of krill, fish and squid in the Southern Ocean. These efforts resulted in the signing of the Convention on the Conservation of Antarctic Marine Living Resources (CCAMLR) which came into force in 1982. CCAMLR was a groundbreaking convention because it took an 'ecosystem approach' to management, an approach adopted because of a recognition of the central importance of krill in the Antarctic food web and also in light of the history of over-exploitation of seals and whales whose recovery should not be jeopardised by future harvesting activities. The Convention is overseen by the Commission for the Conservation of Antarctic Marine Living Resources (also known as CCAMLR) which first met in Hobart in 1981 and has since met there annually. The Commission uses the best scientific advice (provided by its Scientific Committee) to set catch regulations on the fisheries in Antarctic and subantarctic waters. The major fisheries in the region have been those for krill in Antarctic waters and for a number of fish species, such as icefish, Antarctic rockcod and Patagonian toothfish in the subantarctic. Since the coming into force of CCAMLR Australia has had an active research programme which has become the prime source of information to CCAMLR on the harvested species and on the wider marine ecosystem off East Antarctica. A CCAMLR Ecosystem Monitoring Program was established near Mawson Station in 1989. A krill research program was established at the Australian Antarctic Division in the early 1980s and a series of krill surveys were conducted in the Prydz Bay region in the early 1980s to the early 1990s and a single large-scale krill survey was carried out off the coast of East Antarctica in 1996; these surveys were used by CCAMLR to set precautionary catch limits in the waters off the Australian Antarctic Territory. Fish surveys were also carried out in the Prydz Bay area in the early 1990s and then the focus for fish and fisheries research switched to Heard Island where scientific surveys in the early 1990s were used by CCAMLR to establish management procedures for this region. In 1997 an Australia-based fishery for Patagonian toothfish began in the Heard Island area and this led to an increased focus on fisheries research by the Australian Antarctic Division. In 1999 the Antarctic Marine Living Resources (AMLR) Program was established specifically to provide information that can be used by CCAMLR, and by the Australian Fisheries Management Authority (AFMA), to regulate fisheries in Antarctic and subantarctic waters. This science programme conducts strategic research into issues such as: stock assessment of fish and krill, incidental mortality of seabirds in long-line fisheries, ecosystem monitoring, the development of novel techniques to examine ecosystem interactions, ecosystem modelling and research into the life history parameters of harvested and dependent species. Although harvesting of fish is unlikely to increase into the future because of lack of further significant stocks to exploit, fishing remains controversial because of reported illegal, unregulated poaching of valuable populations of Patagonian and Antarctic toothfish. The krill fishery is unique amongst the worlds' fisheries in that the current catch (~100 000 tonnes per year) is so much lower than the conservatively estimated precautionary catch limit (~5 million tonnes per year). It is highly likely that this fishery will increase as new uses and processes are developed and this increase will require careful management. Squid are also thought to be abundant in Antarctic waters but no viable fishery has yet been established. The existence of these fisheries and the continued demand for quality scientific information on harvested species, dependent species and on the Southern Ocean ecosystem will ensure that Australia will require a scientific programme focused on Antarctic fisheries into the foreseeable future.

INTERACTIONS IN THE SOUTHERN OCEAN

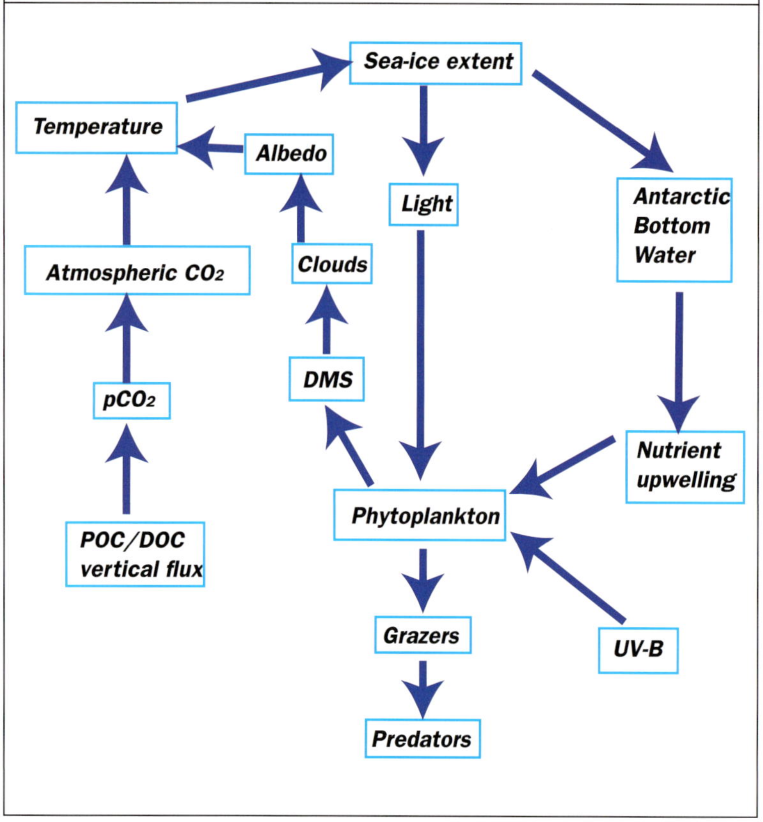

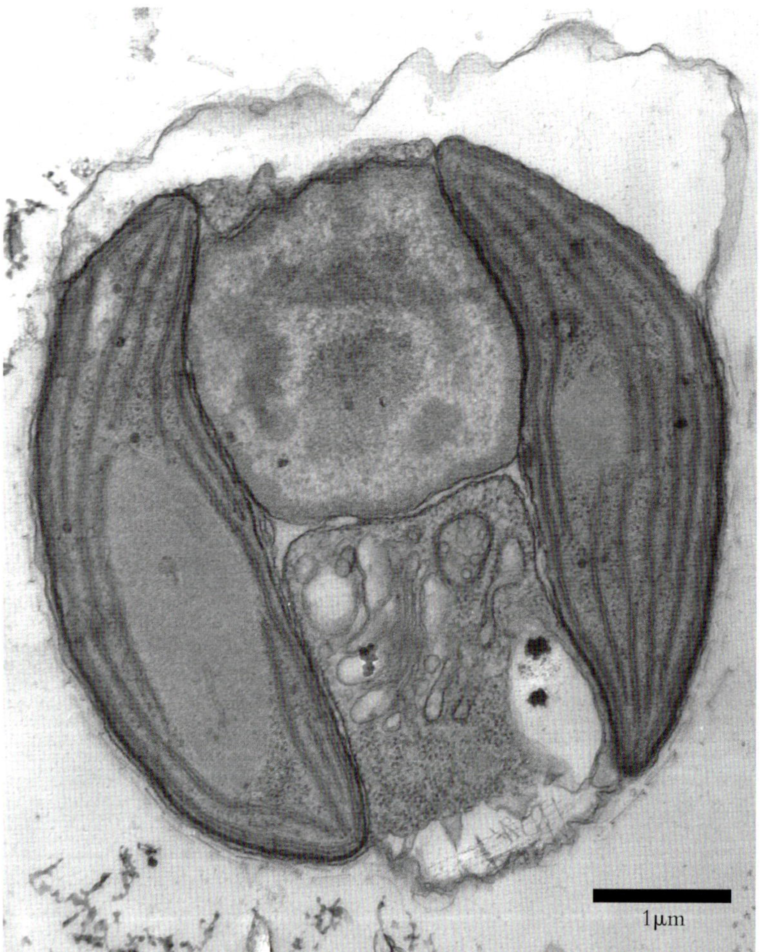

Chloroplast of *Phaeocystis*. *Geraldine Nash, AAD*

food chains. Increased carbon dioxide in the atmosphere would not be expected to increase photosynthetic activity because available light and micronutrients limit productivity. Changes in relative abundance of species would result from temperature changes, with increase in the picoplanktonic cyanobacteria, for instance, and therefore in the zooplankton that graze them.[22] Coccolithophores are also temperature dependent, being absent from water below 4°C. An increase in their abundance beyond their 60° South limit today would have implications for increased global warming as they are significant producers of carbon dioxide.

One likely effect of global climate change is the production of airborne dust through the spread of deserts. Ice-core analysis has shown a correlation between historic atmospheric CO_2 concentrations and the input of iron and silica in dust, particularly during the last glacial maximum. Because the productivity of large areas of the Southern Ocean is constrained by shortages of iron and silica, it is likely that a change in ecosystem structure would occur with an increase in nutrient concentrations downwind of continental dust sources. The added impact of temperature changes, ozone hole effects and other factors that currently influence distribution of organisms makes for a complex picture.

While increased CO_2 on the one hand is expected to increase phytoplankton growth and increase species diversity due to invasion by organisms presently unable to grow at the current low temperature regime, on the other hand the effects of declining sea ice and increased ultraviolet-B radiation are expected to have the opposite outcome.

According to marine scientists[22], the overall impact of these competing influences is likely to be changes in microbial species composition with increased species diversity, and a net decline in biological production in the Southern Ocean.

The ozone hole and life in the Southern Ocean[47]

Since the mid-1970s there has been a marked reduction in the concentration of stratospheric ozone in south polar regions. The Antarctic ozone hole develops in September and dissipates in November, when ozone-depleted stratospheric air is transported to mid-latitudes, including Tasmania and the southern Australia mainland. Increased ultraviolet radiation impacts on marine and terrestrial environments, having deleterious effects on all living things.

The terrestrial flora of Antarctica comprises mosses

A magnificent mite, microscopic but perfect in its decoration and structure, lives under rocks in the Stilwell Hills at Mawson Station. *Geraldine Nash, AAD*

(true Plants), and algae and lichens (protists and fungal symbioses). When not covered by snow, they are exposed to ultraviolet radiation. The mosses *Grimmia* and *Bryum* are known to contain substances that are effective ultraviolet filters, with high levels in early summer when irradiance is high, decreasing as summer progresses. The same sort of substances are found in the terrestrial alga *Prasiola* and the cyanobacterium *Nostoc* that grow on the poorly developed soil in regions (usually with a northerly aspect) where nutrient enrichment results from penguin and other bird nesting sites. Antarctic terrestrial vegetation supports invertebrate communities of nematodes, tardigrades, rotifers and arthropods. Preliminary studies of the impact of ultraviolet radiation on phytoplankton in freshwater lakes on South Georgia found that exposure to ultraviolet-B reduced motility of some species while others seemed unaffected.

A survey of 57 mostly intertidal, marine Antarctic organisms, comprising fish, algae and invertebrates, including krill, found that most contained mycosporine-like amino acids, which are believed to provide ultraviolet screening. However, phytoplankton suffer reduced primary productivity and exhibit reduced amino acid, protein and pigment concentrations, nitrogenase activity and nitrogen metabolism under ultraviolet irradiation. While different species have different degrees of response, little is known of the full picture. Wavelength and many other factors influence response, but there is no doubt that Antarctic marine phytoplankton are ultraviolet-stressed. Some, like Antarctic colonial stages of *Phaeocystis*, produce 5 to 10 times as much ultraviolet protection as strains examined from other parts of the world. Diatoms apparently have little ultraviolet protection. It would seem, therefore, that under increased ultraviolet levels, the relative abundance of *Phaeocystis* to diatoms in the marginal ice zone would increase unless diatoms' more efficient repair mechanisms compensate.

KRILL IN THE SOUTHERN OCEAN

Krill are a group of about 80 different species of crustaceans, the euphausiids, which resemble shrimps and live in the open ocean. They are food for fish, birds and mammals. There is a greater mass of Antarctic krill, *Euphausia superba*, than any other single species on Earth. Recent estimates of krill abundance, based on echo-sounder records and re-examination of historical information on their range, is between 60 million and 155 million tonnes. It is difficult to estimate accurately because of their patchy distribution and their ability to avoid capture in research-sized nets.

Krill have a complex life cycle.[20] They are believed to spawn close to the surface between late December and March, with females producing several broods of up to 10 000 eggs of about 0.5 millimetres diameter. The eggs sink to depths of about 1000 metres, where those that haven't been eaten hatch. The embryos go through a series of non-feeding larval stages before rising to the surface to become feeding larvae. This descent and ascent is believed to take three to five months. In their second summer they develop into juveniles and swarm. By the start of their second winter they are about 25 millimetres long. They undergo many moults before becoming adult. Because they feed on phytoplankton that is scarce during long dark winters, they have developed the ability to survive for long periods with no food, during which times they may shrink in size because they use up muscle tissue rather than depleting their fat reserves. They also regress to earlier developmental stages.

Krill swarms characteristically can contain 30 000 individuals per cubic metre—easy prey for whales, seals, penguins and fishing trawlers. Often swarms are all males, or all females, or all juveniles. One swarm of sexually immature males near Prydz Bay, estimated to weigh 57 000 tonnes, coloured the sea brown. A superswarm off the Antarctic Peninsula in 1981 covered 450 square kilometres and contained more than 2 million tonnes of krill

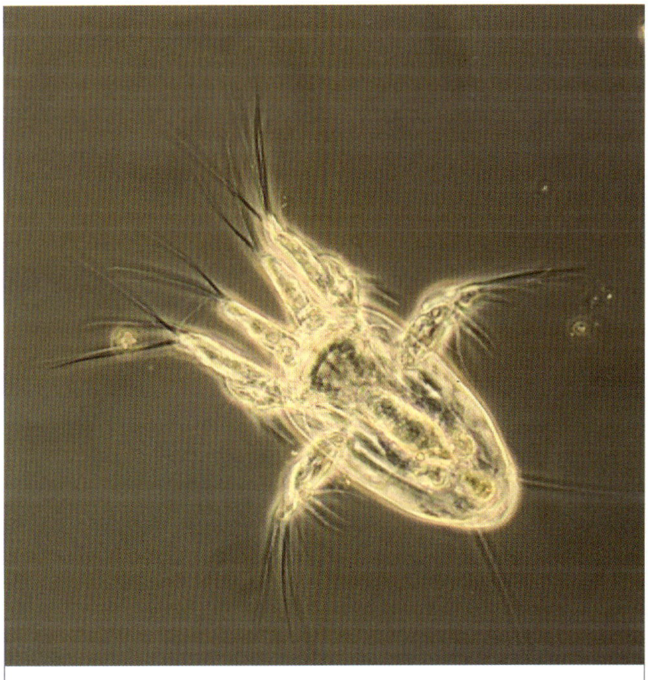

A juvenile stage in the life cycle of a copepod, a small marine invertebrate from sea ice. Magnified many thousand times. More than 70 species of copepod are recognised in Antarctic waters. They are usually microscopic, few larger than a millimetre. Some graze on phytoplankton, others are carnivores, others are omnivores. Copepods are the chosen food of some fish. *Harvey Marchant, AAD*

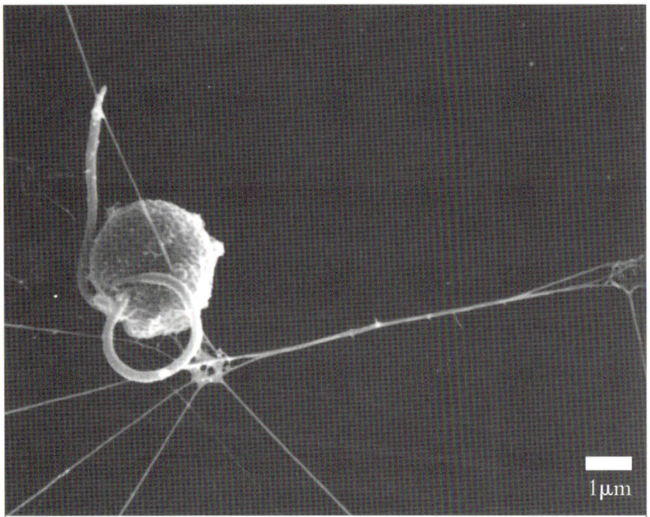

Phaeocystis sp. A motile cell showing the features common to all such cells—two flagellae, haptonema, pentagonal star and five threads. The body of the cell is covered in scales. *Anna McEldowney*

A culture flask containing *Phaeocystis antarctica*. The round colonies can be over a centimetre in diameter and contain thousands of cells. *Harvey Marchant, AAD*

Phaeocystis, *a major contributor to algal blooms in polar waters*

A type of phytoplanktonic algae, the haptophytes, are important and abundant in blooms in the marginal ice zone in polar waters in spring.[59,61] (The haptophytes, which include coccolithophores, are a major component of plankton throughout the world's oceans.) Several coccolithophores are now known to be part-time heterotrophs, consuming other organisms when they cannot photosynthesise.[20]

The organism that dominates in both North Polar and Antarctic blooms is *Phaeocystis*. (*P. antarctica*, found only in Antarctic waters, is now recognised as genetically different.) The alga has two life forms, occurring in colonies and also as separate motile unicellular individuals. The alga in blooms occurs mainly in its colonial form, where numerous individuals are aggregated in a jelly-like sphere. The free-swimming phase of *Phaeocystis* has been found attaching itself to diatoms and becoming part of the community that lives under the ice during dark winters. Much has yet to be learned about the extremophiles that survive in such hostile places as the frozen seas.

(A life history like that of *Phaeocystis*, with both colonial and independent stages, reminds us of how multicellularity evolved. The aggregated organisms later took other kinds of organisms with specialised attributes into their colony and different functions were taken over by them for the benefit of the whole group. Then loss of individual motility followed and members of the colony became cells of the multicellular organism.)

Phaeocystis is usually the first alga to bloom as the ice starts to retreat. Diatoms start to increase in abundance at the same time but their rate of proliferation declines as the *Phaeocystis* peak approaches, and they reach their peak after *Phaeocystis* has started to decline. *Phaeocystis* is believed to play an important role in the cycles of trace metals in Antarctic waters, monopolising manganese during its dominant times and making zinc, iron and copper in short supply for other

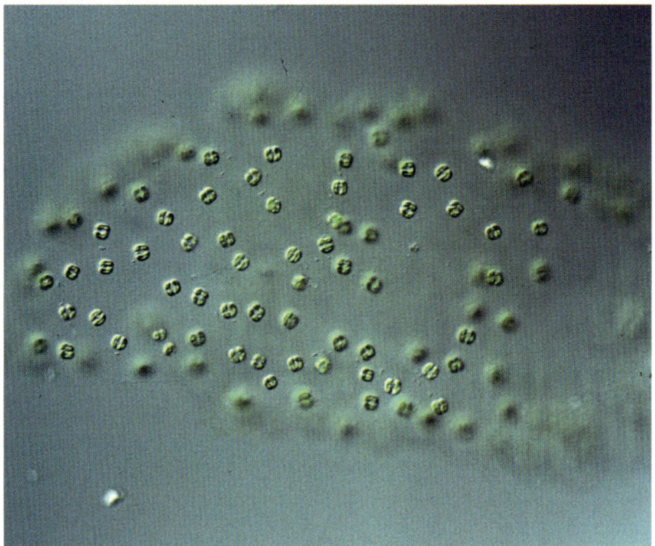

Phaeocystis antarctica, a haptophyte, one of the most important algae in the sea. A colony of about 100 cells encased in mucilage. *Harvey Marchant, AAD*

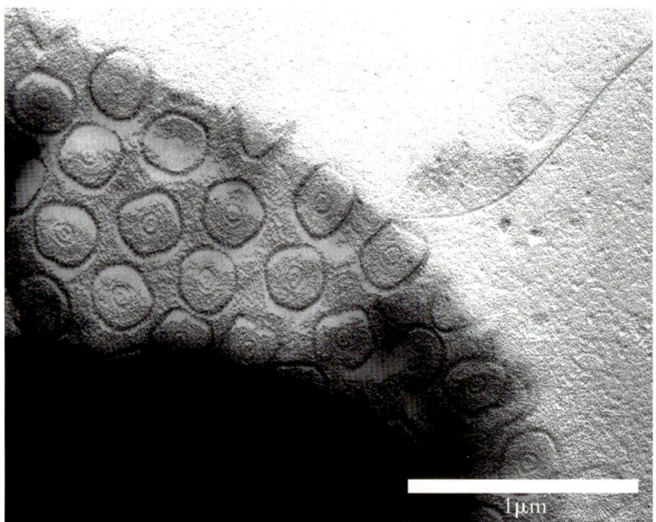

Phaeocystis scales, shadowcast. *Anna McEldowney*

The head of the mite pictured on page 167. *Geraldine Nash, AAD*

organisms. Its secretion of much of the carbon that it has photosynthesised as particulate and dissolved organic matter is of vital importance to the heterotrophic organisms that proliferate following blooms. This matter is also implicated in marine snow.

The *Phaeocystis* blooms are a major source of particulate and dissolved organic carbon, fuelling the food chain in the photic zone and making some contribution to the store of carbon that sinks to the bottom of the ocean. *Phaeocystis* is also a major source of the precursor of dimethyl sulphide, which is rapidly oxidised and produces the sulphate particles that act as cloud condensation nuclei.

Grazers in clay castles reign in the savannas of tropical Australia. *M.E.W.*

LOOKING BACK, LOOKING FORWARD...

I said in the Introduction to this book that as a result of the research that went into its making I was a 'proudly bacterial Gaian'. Now, as I come to write the last pages, after what has been for me an exciting and often mind-blowing journey, along byways that I had never wandered into before, I would re-phrase and change pride to humility—because it is impossible to comprehend the full implications of the Continuum of Life. All one's human perspectives of a world evolving to suit us, and our importance in the overall scheme of things, change. I find deep 'spiritual' contentment in this new perspective, knowing that the wonders that I have glimpsed are more awesomely complicated than I could ever have imagined, and are largely beyond the grasp of my human mind.

Here I am, my cellular endosymbionts performing their housekeeping duties within every individual bacterial cell of the 100 billion, of more than 200 different types, that contain my DNA; knowing what to do and when to do it; enabling the metabolism that keeps me alive; providing the neurones that enable me to think and begin to understand a little the meaning of Life. What makes me me are the ghosts of endosymbionts in my bacterial cells now functioning to give me my brief instant in the river of Time. Every one of my ancestors since the beginning of Life 'went forth and multiplied' successfully, and endogenous retroviruses in my humble genome enabled me, in my turn, to leave progeny to carry on the living. How fortunate to be part of the Greatest Mystery ...

So, what have we learned from the synthesis of information about the world in which we live as presented in this book?

- That Life and the Biosphere are indivisible—Life is the Biosphere?
- That the Biosphere is solar-powered and Life-as-we-know-it depends on photosynthesis—this is a green planet, and the photosynthetic bacteria, protists and plants (in that order) have been responsible for making a Life-friendly world through geological time and keeping it that way?
- That bacteria and microscopic life forms still control the Biosphere, recycling the finite materials that are essential for Life and making the biospheric organs—the soil, atmosphere and waters—alive?
- That, in fact, the total amount of carbon in prokaryotes, invisible though they be, is 60–100 per cent of the estimated total carbon in plants (doubling estimates of the amount of carbon stored in living organisms)?
- And that Archaea and Bacteria contain tenfold more nitrogen and phosphorus than plants, and represent the largest pool of these nutrients in living organisms?
- That fungi have a vital role and soil can be described as a fungal phenomenon?
- That invertebrates are enormously important in the scheme of things?

  ... and so much more about the interconnectedness of all living things in ecosystems and about the complexity of interactions at all levels, from the microcosmic to the largest macrocosmic—the Biosphere itself.
- (And, perhaps, that the 'highly evolved and complex vertebrates at the top of the evolutionary tree'—as we see ourselves—are superfluous to the functioning or well-being of the Biosphere, and in fact are a threat to its health ... ?)

The evolution of Life and its bacterial beginnings, and the co-evolution of the Biosphere, seen as a coherent picture alerts us to many concepts that we have tended to overlook in our search for answers to environmental problems.

Water and Life in a desert landscape. Davenport Springs in Australia's Centre. *M.E.W.*

Firstly, complexity and biodiversity are essential for sustainability of ecosystems, and natural ecosystems are maintained in balance by their complex webs of life. Then, because the Biosphere is alive and metabolising, everything is forever changing, renewing, evolving, and it is the interconnectedness and symbiotic functioning of living matter at all levels that maintains the checks and balances. The natural laws that apply in the natural world, particularly those that balance populations and nutrients, maintain equilibrium; the only species that has been able to function outside these laws is mankind—and only by virtue of technology. The result is gross global over-population and damage to the Life-supporting systems of the living Biosphere. In order to have a future, humans will have to accept their responsibilities to the rest of Life and find ways to co-exist, in far fewer numbers and having far less impact. The days of technological fixes that enable us to go on blindly without regard for the health of the Biosphere are numbered.

Earth, Air, Life and Water, the living Biosphere. *M.E.W.*

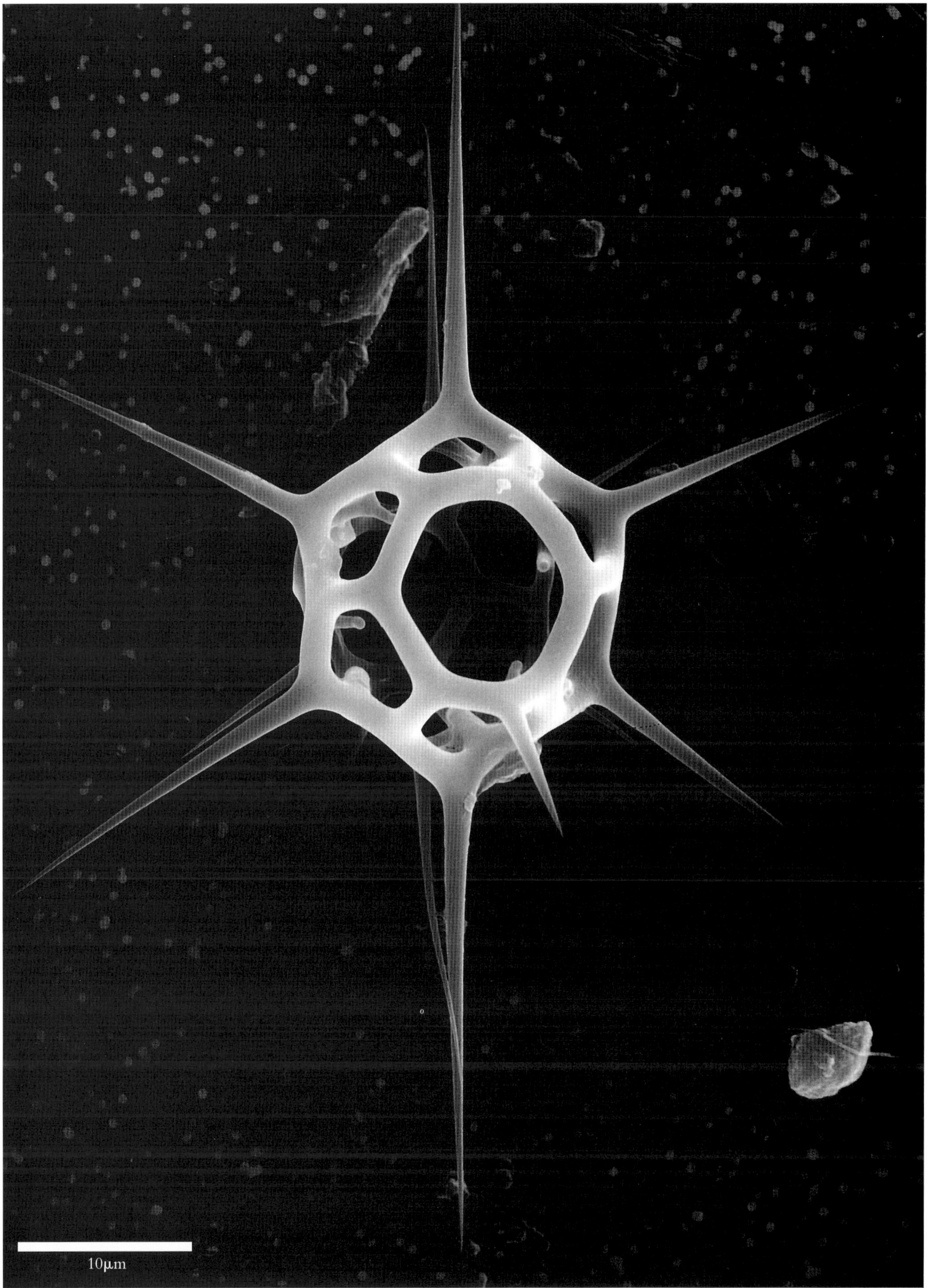

The dramatic skeleton of the silicoflagellate *Dictyocha speculum*, a photosynthetic organism. *Geraldine Nash, AAD*

Glossary

actinomycetes a type of bacteria

aeolian blown by the wind

aerobic living or active in the presence of free oxygen

aerosol particle small enough to remain suspended in the atmosphere

albedo fraction of incoming radiation which is reflected back without being absorbed

algae protists: single or many celled, aquatic, reproducing by spores

algal blooms periods when algae proliferate within a river or water body

alluvial deposits accumulation of sediment transported by water

alluvium unconsolidated sediments, silts, sands, gravels, carried by streams, deposited on land

ammonification the reduction of nitrates and nitrites to ammonium compounds by soil bacteria

anaerobic living or active in the absence of free oxygen

anabranch part of a river that branches off the main stem, then rejoins or reaches a wetland

anchialine (or **anchihaline**) bodies of haline waters with subterranean connection to the sea, showing marine and terrestrial influences; typically occur in volcanic or limestone bedrock

anion a negatively charged ion

aquifer a porous layer of rock that is saturated and can supply water to springs or wells

aquitard a layer that retards but does not prevent the flow of water to or from an adjacent aquifer

Archaea Super-Kingdom of prokaryotes, ancestral and now still important

archipelago chain of islands

arid zone region with annual rainfall usually less than 250 mm

artesian water underground water that is stored under pressure

atrazine herbicide used to control weeds

autotrophs organisms that synthesise organic substances from inorganic material

barrage construction across the mouth of a river to prevent entry of seawater

basalt fine-grained volcanic rock formed from a lava flow

base flow river flow sourced from dry weather groundwater inflows, not surface run-off

basement the undifferentiated complex of rocks that underlies the rocks of interest in an area

basin an area of continued subsidence of the earth's crust where sediment accumulates

bed-load the sediment that moves by sliding or rolling along the bed of a stream or river

benthic of animals or plants that live on or are attached to the bottom in a water body

billabong waterhole which fills during floods, usually formed from an old river channel

biodegradable readily decomposed by micro-organisms

biodiversity full range of organisms, including their genetic diversity

biofilm accumulation of bacteria attached to solid surfaces

biomass a measure of the total mass of living matter in a particular habitat or area

biota all living things in an area—plants, vertebrates, invertebrates, fungi, bacteria

bioturbation disturbance of soil by biological activity

blue-green algae cyanobacteria, prokaryotic, not algal (an inappropriate term)

bore a deep drill-hole tapping an aquifer

calcrete carbonate deposit formed in soil or in vicinity of groundwater table as a result of evaporation; typically forms in arid climates with rainfall <200mm, evaporation >3000mm

Carboniferous Period 360 to 286 m.y.a.

carrying capacity population size at which some limiting factor prevents further increase; or size of population beyond which environmental degradation occurs.

chemo-autotrophs organisms deriving nutrient from chemical reactions of inorganic substances, as in sulphur and iron bacteria

clay sediment particles 0.002—0.004mm in diameter

coarse particulate organic matter large fragments of decomposed organic matter (>1mm)

colloidal particles material in a very fine state of division that will not settle out of solution

colluvial deposit material transported to a site by gravity

community group of organisms of different species occurring in the same place or habitat

competition organisms or species competing for a common resource (food, nesting place, etc.)

craton part of the Earth's crust little deformed for a prolonged period; a shield region

Cretaceous Period between the Jurassic and the Tertiary, about 144 to 65 m.y.a.

crustaceans animals, mostly aquatic, with a hard outer shell—includes lobsters and crabs

cyanobacteria 'blue-green algae'; photosynthetic bacteria important in functioning of the Biosphere

decomposition biochemical processes that break down organic matter into simple compounds

deflation wind erosion

delta a fan-shaped mass consisting of the deposited load of a river where it enters the sea

detritus decomposed plant or animal material
Devonian Period 408 to 360 m.y.a.
diatom microscopic unicellular protist with shell made of silica
dissolved oxygen measure of the amount of oxygen in solution
diversity number of species in a community or habitat
DNA deoxyribonucleic acid; the molecule that is the key component of a living cell; it contains genetic information as well as instructions for the biochemical functioning of the cell
DOM dissolved organic matter
duricrust an iron or silica-cemented crust capping landscape features and slowing erosion
ecology scientific study of living organisms and their relationship to one another and the environment
ecologically sustainable ensures protection of biodiversity, ecological processes, soil and water
ecosystem organisms in a community and their environment interacting as an ecological unit
ecotone region where two distinct ecosystems meet
endemic organism found only in a particular area
endogenous being part of the internal space
endo- prefix that indicates containment
endosymbionts symbionts within a cell or organism
endosymbiosis the process of acquiring endosymbionts
endorheic rivers that do not reach the sea but run into wetlands or dry river beds
ENSO the El Niño–Southern Oscillation; climatic oscillation producing drought and floods
environmental flow providing water for the needs of the environment
enzyme a protein that acts as a catalyst
Eocene an epoch of the Tertiary Period from 58 to 36 m.y.a.
ephemeral pools and rivers temporary wetland areas and rivers that only run some of the time
epigean surface environment as opposed to the sub-surface (subterranean, hypogean) environment
erosion the wearing away of the Earth's surface by running water, ice, wind and sea
escarpment a steep slope forming a linear feature in a landscape
estuary the broad mouth of a river where it enters the sea
eukaryote cell with a nucleus and organelles, or organism composed of such cells; all living things except bacteria are eukaryotic
euphotic zone open water zone of an ocean with sufficient sunlight to support photosynthesis
eustatic related to world-wide changes in sea level
eutrophication increasing nutrient levels
evapo-transpiration total amount of water lost from soil, water bodies and vegetation
extremophile organism that lives in an extreme environment
feral introduced organism which has become naturalised
ferricrete iron-cemented sediments
fine particulate organic matter fragments of detritus <1mm
floodplains areas beside rivers that flood when the river overflows its banks
fluvial sediment sediment that is washed by the action of a river
food web network of food pathways within an ecological community
genome total gene complement of an individual organism
geomorphology study of the origin, character and development of land and rivers
geothermal pertaining to heat of the interior of the Earth
gibbers small scattered stones forming a surface cover on deserts
gigalitre (GL) 1 000 million litres
groundwater water below the surface of the soil
halocline more or less abrupt change in salinity with depth of water column
halophyte plant that tolerates very salty soil
hardpan extremely hard to cemented soils, claypan areas
headwater streams streams in the upper catchment of a river
heterotrophs organisms unable to synthesise organic substances from inorganic
Holocene geological epoch within the Quaternary period; the last 10 000 years
hot-spot volcanoes volcanoes which result from local weaknesses in the earth's crust
hydrological cycle the continuous interchange of water between land, sea, other water and air
hydrology study of the behaviour of water above or below the ground
hypersaline saltier than seawater
hypogean the subsurface or subterranean environment as opposed to the surface (epigean) environment
hyporheic zone interstitial spaces within the sediments of a stream bed; a transition zone between surface water and ground water
interstitial occupying spaces between particles, especially in alluvia
ion an electrically charged atom
Jurassic period between the Triassic and Cretaceous in the Mesozoic era, 208 to 144 m.y.a.
karst soluble rock landscape developed in limestone terrain
lacustrine sediments sediment laid down in a lake
laterite sediment impregnated with iron
leaching process by which materials are removed by percolating water
lignotuber woody underground rootstock
limno- prefix, restricted to fresh water
lipid a naturally occurring fat or fat-like material consisting of fatty acids, etc.
lithosphere Earth's crust of solid rock
littoral refers to shoreline
local run-off reliant on flows derived from nearby sources
macro-invertebrate aquatic invertebrate, usually between 1 and 100mm
mafic igneous rock composed of one or more ferromagnesian minerals

mallee habit multi-stemmed growth of a eucalypt from a lignotuber
megalitre (ML) 1 million litres
meiosis cell division in which daughter cells receive half the number of chromosomes in the parent cell; sexual cell division producing gametes that have to fuse with other half-chromosome-number cells to initiate a new individual
mesophyte plant that requires an average amount of water
Mesozoic Era the Triassic, Jurassic and Cretaceous periods
Metazoa multicellular animals
microclimate humidity, temperature, moisture prevailing in a small area
Miocene an epoch of the Tertiary Period, from 23 to 5 m.y.a.
mitochondria organelles that supply the energy needed for metabolism in a cell
mitosis cell division in which each daughter cell receives the same number of chromosomes as the parent cell had
morphological relates to the form and structure of organisms
mound-spring a natural spring where water in a pressurised aquifer comes to the surface
nitrification microbial conversion of ammonium and nitrite compounds to nitrates, generally by soil-nitrifying bacteria
nutrient material eaten or taken in for maintaining an organism or allowing it to grow
nutrient cycling path of nutrients through the ecosystem
Oligocene epoch of the Tertiary Period, from 35 to 15 m.y.a.
oligotrophic pertaining to low amount of available organic matter, low energy density
organelle a membrane-enclosed entity within a eukaryotic cell, developed from an endosymbiont
organic carbon carbon derived from living things
palaeo- prefix which denotes situation in geological time
Palaeocene the first epoch of the Tertiary Period, from about 65 to 58 m.y.a.
pelagic marine organisms that belong to the open sea
Permian Period a geological period of the Palaeozoic Era, from 286 to 245 m.y.a.
pH a measure of acidity or alkalinity (a pH of 7 indicates neutrality)
photosynthesis process by which organisms synthesise carbohydrate using solar energy
photo-autotroph organism that obtains metabolic energy from light
phreatic part of a cave system that lies below the local groundwater table; below the unsaturated or vadose zone
phytoplankton photosynthetic plankton, either algal or cyanobacterial
pisolites spherical, pea-like balls of ferricrete
plankton floating plants and animals, usually <1mm
Pleistocene epoch of the Quaternary, from 1.6 m.y.a. to 10 000 years ago
Pleistocene ice age the last 2 million years, when glacials and interglacials alternated
Pliocene the last epoch of the Tertiary Period, from 5.3 to 1.6 m.y.a.
pluvial period period with high rainfall, active rivers
prokaryote cell without a nucleus or inclusions, bacterial or archaean
pycnocline more or less abrupt change in water density with depth, typically at the halocline or thermocline in the water column
Quaternary Period geological period comprising the Pleistocene and the Holocene
recharge the movement of surface water into the groundwater or an aquifer
refuge place where the biota can shelter from predators or from extreme environments
refugia regions where certain types or suites of organisms can persist during a perion when most of their original geographic range becomes uninhabitable because of climatic change
regolith the mantle of weathered rock and soil overlying the bedrock on the earth's surface
rhizosphere interface between roots and soil
riparian associated or living on banks of rivers or streams
RNA ribonucleic acid; the companion molecule to DNA that carries instructions from it to the place of execution
run-off water that flows into rivers and streams across the land
salinity the amount of sodium chloride or dissolved salts in a unit of water
salinisation the accumulation of salts in the soil or water body to a level that causes degradation
saltbush Chenopod shrubs
saprolite weathered and altered rock at the interface between lithosphere and watertable/soil
saprophyte organism that lives on decaying organic matter
savanna grassland with sparse trees and shrubs
scarp steep face on the side of a hill
scarp retreat the process by which the edge of a plateau migrates inwards due to erosion
sclerophyll vegetation plants with hard leathery leaves and other modifications for coping with drought and poor soils
sediment particulate matter that has been transported by water and subsequently deposited
sedimentation deposition of particulate matter by water
seed bank seeds that lie dormant in the soil
semi-arid area with 250–350mm annual rainfall
SET the Serial Endosymbiosis Theory of evolution
silcrete sedimentary rocks impregnated with silica
silt coarse mud
southern beech *Nothofagus* spp.
stygo- prefix referring to groundwater habitats
stygobite species that are specialised for subterranean living
stygofauna groundwater-inhabiting fauna
suspension loads fine insoluble particulate matter carried by water
swale shallow depression in undulating ground; low area between adjacent dunes

symbiosis an association of dissimilar organisms, whether to their mutual advantage or otherwise
tectonic relating to processes in the Earth's crust which create landscape features
Tertiary Period from 65 m.y.a. to the Pleistocene, which starts 1.6 m.y.a.
thermocline more or less abrupt change in temperature with depth, in air and water columns
topographic describes geographical features
transpiration loss of water vapour by plants
Triassic Period geological period between Permian and Jurassic, from 245 to 208 m.y.a.
troglobite species which exist only within caves
trophic pertaining to nutrition
troposphere lower part of the atmosphere, up to about 11 kilometres above the Earth's surface
turbid water which is opaque with suspended matter
vadose zone unsaturated zone; a zone of rock above the water table, mostly free-draining; lies above the phreatic zone
vascular plant plant with specialised internal transport for nutrients
virus a filterable micro-organism that cannot metabolise and is too small to be considered alive; it must enter another organism in order to replicate itself
water-mound aquifer a sand body which has a saturated zone acting as a water reservoir
water table the groundwater-saturated zone in the regolith
wetland area that is either temporarily or permanently flooded
xeromorphic applied to plants that are adapted to water shortages
xerophyte plant adapted to arid conditions
yabby freshwater crayfish
zooplankton small animals living in water, usually drifting

Coccoliths (scales) of *Emiliania huxleyi*. Astronomical numbers of these beautiful scales contribute to seafloor deposits that tie up carbon in sediments. *Geraldine Nash, AAD*

Bibliography

1: Wilson EO. 1987: The little things that run the world. *Conserv. Biol.* 1, 344–346

1a: Vernadsky VI. 1926: *The Biosphere*. Synergetic Pr. Inc. Arizona

1b: Carroll SB. 2001: Chance and necessity: the evolution of morphological complexity and diversity. *Nature* 409, 1102–1109

2: Roper MM & Gupta VVSR. 1995: Management practices and soil biota. *Aust. J. Soil Res.* 33, 321–339

2a: Roper MM. 1998: Rotations for living soils. *Proc. WCFA Conf. Feb. Sharing the Wealth.*

3: Seuss E. 1875: *The Origin of the Alps*. Wilhelm Braumuller. Vienna

3a: Cancela da Fonseca JP. 2000: On Vernadsky's biosphere. *Agora Web Ecology* 1, 86–96

4: Lovelock J. 1979: *Gaia: A New Look at Life on Earth*. Oxford University Press

5: Lovelock J. 1988: *The Ages of Gaia*. Oxford University Press

6: Hutton J. 1788: Theory of the Earth. *Trans. Roy. Soc. Edinb.* 1, 209–304

7: Margulis L & Sagan D. 1995: *What Is Life?* Simon & Schuster, New York

8: Margulis L. 1998: *The Symbiotic Planet. A New Look at Evolution.* Phoenix

9: Uwins P. 2000: How small can you get? A story of genuine low-lifes. *Proc. 15th Aust. Geol. Conf.* Sydney. Public Symp. 22

10: Highfield R. 2001: Meet the one-cell wonders all animals can relate to. *Telegraph*, London (SMH 22.12.01)

11: Woese CR et al. 1990: Towards a natural system of organisms: Proposal for the domains Archaea, Bacteria and Eucarya. *Proc. Nat. Acad. Sci.* 87, 4576–457

12 : Smith D. 2000: Extremism. *SMH* June 10

13: Jones S. 2000: *Almost Like a Whale*. Doubleday

14: Cowan DA. 1995: Hypothermophilic enzymes: Biochemistry and biotechnology. In *Hydrothermal Vents and Processes*. Geol. Soc. Spec. Publ. 87, 351–363

15: Canfield DE. 1999: A breath of fresh air. *Nature* 400, August, 503–505

16: Brocks JJ *et al.* 1999: Archean molecular fossils and the early rise of the Eukaryotes. *Science* 285, 1033–1036

17: Nisbet E. 2000: The realms of Archaean life. *Nature* 405, 625–626

18: Summons RE et al. 1999: 2-methylhopanoids as biomarkers for cyanobacterial oxygenic photosynthesis. *Nature* 400, August, 554–557

19: Smith GD et al. 1990: Cyanobacterial nitrogen fixation in arid soils of Central Australia. *FEMS Microbiol. Ecol.* 74, 79–90

19a: Rogers RW. 1989: Blue-green algae in southern Australian rangeland soils. *Aust. Rangel. J.* 11(2), 67–73

20: Marchant HJ. 2000: Life in the Southern Ocean. *Polar Wildlife*. 200–215

21: Marchant H et al. 2000: The distribution and abundance of viruses in the Southern Ocean during spring. *Antarctic Science*. 12(4), 414–417

22: Marchant HJ, Davidson MA & Wright SW. 2000: Antarctic marine micro-organisms and climate change: Impacts and feedbacks. In press

23: Hessler RR & Kaharl VA. 1995: The deep-sea hydrothermal vent community: An overview. In *Seafloor Hydrothermal Systems. Geophysical Monograph 91*, 72–84. Am. Geophys. Union

24: Van Dover CL. 1999: Photosynthesis at deep-sea vents. *Science Frontiers* 124, Jul.-Aug. Sourcebook.

25: USGS. 2000: Exploring the deep ocean floor: Hot springs and strange creatures. *The Dynamic Earth*. USGS

26: Parr J. 2000: Deep sea aliens or long lost relatives ? *Proc. 15th Aust. Geol. Conv. Sydney*, Public Symp. 6

27: Rasmussen B. 2000: Filamentous microfossils in a 3235-million-year-old volcanogenic massive sulphide deposit. *Nature* 405, 676–679

28: Ehrlich HL. 1998: Geomicrobiology: Its significance for geology. *Earth Science Reviews* 45, 45–60

29: Doolittle WF. February 2000: Uprooting the Tree of Life. *Scientific American*. 72–77

30: Ralph W. 1984: The key to the *Leucaena* problem is in the rumen. *Rural Research* 123, 17–19

31: Elliott S & Williams KL. 1991: Modelling people using cellular slime moulds. *Aust. Nat. Hist.* 23(8), 608–616

32a: Bashan Y. 1999: Interactions of *Azospirillum* spp. in soils: a review. *Biol. Fertil. Soils* 29, 246–256, Springer-Verlag

32b: Bashan Y & Holguin G. 1997: *Azospirillum*-plant relationships: Environmental and physiological advances (1990–1996). *Can. J. Microbiol* 43, 103–121

32c: Okon Y & Labandera-Gonzalez CA. 1994: Agronomic applications of *Azospirillum*: an evaluation of 20 years worldwide field inoculation. *Soil Biol. Biochem.* 26(12), 1591–1601

33: Copley J. 2000: Ooze cruise in slime. *New Scientist*. March, 27–29

33a: Zalasiewicz J & Freedman K. 2000: The dawn of slime. *New Scientist*, March, 30–33

33b: Pain S. 2000: Slime scene. *New Scientist*. March, 34–37

33c: Crystall B. 2000: Monstrous mucus. *New Scientist*, March, 38–41

33d: Battersby S. 2000: Supergoo. *New Scientist*, March, 42–45

34: McNamara K. 1999: *Stromatolites*. WA Museum Publ.

34a: Coshell L, Roseni MR & McNamara KJ. 1998: Hydromagnesite replacement of biomineralised aragonite in a new location of Holocene stromatolites, Lake Walyungup, Western Australia. *Sedimentology* 45, 1005–1018

34b: McNamara KJ & Awramik SM. 1992: Stromatolites: A key to understanding the early evolution of life. *Sci. Progress* 76, 345–364

34c: Moore LS & Burne RV. 1994: The modern thrombolites of Lake Clifton, Western Australia. In *Phanerozoic Stromatolites II*, Crown Copyright, The Netherlands, 3–29

35: Bergeron L. 1999: ET was here: Did microbes build high-rise apartments on Mars? *New Scientist*, September, 23

36: Cohen P. 2000: The Force. *New Scientist*. February, 30–35

37: Fox D. 1999: Why we don't lay eggs. *New Scientist* 26–31

38: Proske U. 1990: The monotreme electric. *Aust. Nat. Hist.* Autumn, 23(4), 288–295

39: McGhee K. 1990: Haemoglobin in plants: When conservatism rules. *Aust. Nat. Hist.* 23(7), 554–561

40: Miller DJ & Ball EE. 2000: The coral *Acropora*: What it can contribute to our knowledge of metazoan evolution and the evolution of developmental processes. *BioEssays* 22(3), 291–296

41: Pearce F. 2000: Inventing Africa. *New Scientist* 2251, 30–33

42: MacKenzie D. 2000: Sick to death. *New Scientist* 2250, 32–35

43: Attwell B, Kriedemann P & Turnbull C. eds. 1999: *Plants in Action*. Macmillan Education Aust. Melb.

44: Burke M. 1999: Green miracle. *New Scientist* 2199, 26–30

46: Winter K & Smith JAC. 1996: An introduction to crassulacean acid metabolism. Biochemical principles and ecological diversity. In *Crassulacean Acid Metabolism*. Springer-Verlag

46a: Keeley, JE. 1996: Aquatic CAM photosynthesis. In *Crassulacean Acid Metabolism*. Springer-Verlag

47: Marchant HJ. 1994: Biological impacts of seasonal ozone depletion. *Antarctic Science*, 95–109

47a: Marchant HJ. 1993: Antarctic marine nanoplankton. *Current Topics in Bot. Res.*1, 189–201

48: Smith GD & Nga Thanh Doan. 1999: Cyanobacterial metabolites with bioactivity against photosynthesis in cyanobacteria, algae and higher plants. *J. Appld. Phycology* 11, 337–344

49: Rickards RW et al. 1999: Calothrixins A and B, novel pentacyclic metabolites from *Calothrix* cyanobacteria with potent activity against malaria parasites and human cancer cells. *Tetrahedron* 55, 13513–20

50: Smil V. 1997: *Cycles of Life: Civilisation and the Biosphere*: Scientific American Library, New York

51: Miller D & Veron C. 1990: Biochemistry of a special relationship. *New Scientist* 1719, 44–49

52: Hoffman PF & Schrag DP. 2000: Snowball Earth. *Scientific American*, January, 50–57

53: Nash JM. 1995: Life in the fast lane. *Time* Dec. 4, 70–79

54: Proceedings of Frontiers in Catchment Biogeochemistry, CSIRO, Nov. 1999. *Marine Freshwater Research* 52(1) 2001

54a: Revkin A. 2002: Earth feels heat from fires down below. *SMH* Jan. 19–20

55: Katz ME et al. 1999: The source and fate of massive carbon input during the Latest Paleocene thermal maximum. *Science* 286, 1531–1533

56: Auzende J-M et al. 2000: Thinned crust in Southwest Pacific may harbor gas hydrate. *Trans Am. Geophys. Union* 81(17), 182–185

57: Dickens GR et al. 1997: Direct measurement of *in situ* methane quantities in a large gas hydrate reservoir. *Nature* 365, Letters, 426–428

57a: Dickens GR et al. 1995: Dissociation of oceanic methane hydrate as a cause of the carbon isotope excursion at the end of the Paleocene. *Paleoceanography* 965–970

58: Dickens GR. 1999: The blast in the past. *Nature* 401, 752–755

59: Gibson JAE et al. 1990: Dimethylsulphide and the alga *Phaeocystis pouchetii* in Antarctic coastal waters. *Marine Biol.* 104, 339–346

60: Ralph W. 1986: Bacteria to boost oil recovery. *ECOS* 47, 6–8

61: Marchant HJ & Thomsen HA. 1994: Haptophytes in polar waters. In Green JC & Leadbeater BSC eds. *The Haptophyte Algae*. Clarendon Pr. 209–228

62: Rennie, J. 1992: Bug in a gilded cage: All that glitters is sometimes bacterial. *Sci. Amer.*, Sept. 82–85

63: Bischoff GCO, Coenraads RR & Lusk J. 1992: Microbial accumulation of gold: An example from Venezuela. *N. Jb. Geol. Palaont. Abh.* 185, 131–159

64: Yardley B. 1991: The successful alchemist. *New Scientist*. August. 20–24

65: Bischoff GCO & Coenraads RR. 1994: Fossil and recent traces of biodegradation on heavy minerals. *N. Jb. Geol. Palaont. Mh.* 4, 246–256

66: Behr HJ, Behr K & Watkins JJ. 2000: Cretaceous microbes: Producer of black opal at Lightning Ridge, NSW, Australia. *Geol. Soc. Aust. Abs.* 59, 28

67: Nowak R. 2000: Little gems. *New Scientist*, July, 7

68: Pearcey G. 1994: Weipa. Bauxite town in the Cape York wilderness. *Aust. Geogr.* 36, 88–109

69: McFarlane MJ. 1987: The key role of micro-organisms in the process of bauxitisation. *Modern Geology* 11, 325–344

69a: Taylor G et al. 1992: Cool climate lateritic and bauxitic weathering. *J. Geol.* 100, 669–677

70: Stott MB et al. In press: Comparative leaching of chalcopyrite by selected acidophilic bacteria and archaea.

71: Stott MB et al. 2001: Thiocyanate removal from saline CIP process water by a rotating biological contactor, with reuse of water for bioleaching. *Hydrometallurgy* 62, 93–105

72: Pearce F. 2000: Coral grief: The cause of reef health problems may be blowing in the wind. *New Scientist*, July, 22

73: Paton TR, Humphreys GS & Mitchell PB. 1995: *Soils. A New Global View*. UCL Press, London

74: Ollier CD. 1996: Chapter 1: Planet Earth. In *Companion Encyclopedia of Geography: the environment of mankind*. Routledge, London

75: Banerjie I. 1985: The wet desert. *India Today*. Oct., 150

76: Reddell P & McJannet D. 2001: Where Earth and sky meet. *CSIRO Land & Water News* 10

77: Mclaughlin MJ, Fillery IR & Till AR. 1998: Operation of phosphorus, sulphur and nitrogen cycles. *Bur. Rur.* Res. Proc. 14, 67–115

78a: Spain AV & Hutson BR. 1983: Dynamics and fauna of the litter layer. In *Soils: An Australian Viewpoint*. Division of Soils CSIRO. Melb., Chapter 38

78b: Lee KE. 1983: Soil animals and pedological processes. In *Soils: An Australian Viewpoint*. Division of Soils CSIRO. Melb., Chapter 39

78c: Greenslade PJM & P. 1983: Ecology of soil invertebrates. In *Soils: An Australian Viewpoint*. Division of Soils CSIRO. Melb., Chapter 40

78d: Foster RC. 1983: The plant root environment. In *Soils: An Australian Viewpoint*. Division of Soils CSIRO. Melb., Chapter 41

78e: Martin JK. 1983: Biology of the rhizosphere. In *Soils: An Australian Viewpoint*. Division of Soils CSIRO. Melb., Chapter 42

78f: Bowen GD. 1983: Microbial determinants of plant nutrient uptake. In *Soils: An Australian Viewpoint*. Division of Soils CSIRO. Melb., Chapter 43

78g: Date RA. 1983: Rhizobium as a factor in soil fertility. In *Soils: An Australian Viewpoint*. Division of Soils CSIRO. Melb., Chapter 44

78h: Rovira AD & Ridge EH. 1983: Soil-borne root diseases and wheat. In *Soils: An Australian Viewpoint*. Division of Soils CSIRO. Melb.. Chapter 45

78i: Warcup JH. 1983: Effects of fire and sun-baking on soil microflora and seedling growth in forest soils. In *Soils: An Australian Viewpoint*. Division of Soils CSIRO. Melb.Chapter 46

78j: Harris JR & Kimber RWL. 1983: Phytotoxins in soil. In *Soils: An Australian Viewpoint*. Division of Soils CSIRO. Melb. Chapter 47

79: Skjemstad JO & Graetz RD. 2001: The impact of burning on the nature of soil organic matter in tropical Australia. *Proc. GEOTROP 2001*. In press.

80: Margulis L, Schwartz KV & Dolan M. 1994: *The Illustrated Five*

Kingdoms. A Guide to the Diversity of Life on Earth. Harper Collins, NY

81: Pyper W & Davidson S. 2001: Bubble, Bubble. *Ecos* 106, 28–31

82: Ladd JN & Foster RC. 1996: Soil Structure and Biological Activity. In *Soil Biochemistry Vol. 9*, Stotzky G & Bollag J-M eds. Chapter 2. Marcel Dekker Inc.

83: McNamara K. 1995: *Pinnacles*. WA Museum Publ.

84: Patton PC, Pickup G & Price DM. 1993: Holocene palaeofloods of the Ross River, Central Australia. *Quat. Res.* 40, 1–11

85: Isbell RF, McDonald WS & Ashton LJ. 1997: *Concepts and Rationale of the Australian Soil Classification*. ACLEP. CSIRO Land & Water, Canberra

86: Eisenberg E. 1999: Soil City. *Whole Earth*, Spring 1999, 12–17

87: Gupta VVSR & Yeates GW. 1998: Soil microfauna as bioindicators of soil health. In *Biological Indicators of Soil Health*. CAB International

88: Roper MM. 1993: Biological diversity of micro-organisms: An Australian perspective. In *Pacific Conservation Biology*. Surrey Beatty, Sydney

89: Haynes P. 1991: Understanding soil biological changes with conservation farming. *Proc. WCFA 6th Annual Seminar, Longerenong*, 8–12

90: Pankhurst CE et al. 1996: Biodiversity of soil microbial communities in agricultural systems. *Biodiversity & Conservation* 5, 197–209

91: Brussard L et al. 1997: Biodiversity and ecosystem functioning in soil. *Ambio* 26(8), 563–570

92: Neate SM & Rovira AD. 1993: Pesticide disease interactions in conservation tillage systems. In *Pesticide Interactions in Crop Production*. CRC Press

94: Roper MM & Gupta VVSR. 1995: Management practices and soil biota. *Aust. J. Soil Res.* 33, 321–339

94a: Roper MM, Turpin JE & Thompson JP. 1994: Nitrogenase activity (C_2H_2 reduction) by free-living bacteria in soil in a long-term tillage and stubble management experiment on a vertisol. *Soil Biol. Biochem.* 26(5), 1087–1091

94b: Roper MM & Halsall DM. 1986: Use of products of straw decomposition by N_2-fixing (C_2H_2-reducing) populations of bacteria in three soils from wheat-growing areas. *Aust. J. Agric. Res.* 37, 1–9

94c: Roper MM, Marschke GW & Smith NA. 1989: Nitrogenous activity (C_2H_2 reduction) in soils following wheat straw retention: Effects of straw management. *Aust. J. Agric. Res.* 40, 241–253

94d: Roper MM. 1983: Field measurements of nitrogenase activity in soils amended with wheat straw. *Aust. J. Agric. Res.* 34, 725–739

94e: Roper MM. 1985: Straw decomposition and nitrogenase activity (C_2H_2 reduction): Effects of soil moisture and temperature. *Soil Biol. Biochem.* 17(1), 65–71

94f: Gupta VVSR. 1994: The impact of soil and crop management practices on the dynamics of soil microfauna and mesofauna. In *Soil Biota*. CSIRO

94g: Roper MM & Smith NA. 1991: Straw decomposition and nitrogenase activity (C_2H_2 reduction) by free-living microorganisms from soil: Effects of pH and clay content. *Soil Biol. Biochem.* 23(3), 275–283

94h: Roper MM & Ladha JK. 1995: Biological N_2 fixation by heterotrophic and phototrophic bacteria in association with straw. *Plant and Soil* 174, Kluwer Academ. Publ. Netherlands

94: Roper MM, Gault RR & Smith NA. 1995: Contribution to the N status of soil by free-living N_2-fixing bacteria in a lucerne stand. *Soil Biol. Biochem.* 27(4/5), 467–471

95: Rogers SL et al. 1998: Degradation of strychnine by pure bacterial cultures isolated from South Australian agricultural soils. *Soil Biol. Biochem.* 30(12), 1623–1626

96: Kookana RS, Rogers SL & Oliver DP. 1997: Sorption and desorption behaviour of strychnine rodenticide in soils. *Aust. J. Soil Res.* 35, 491–502

97: McLoughlin MJ et al. 2000: Review: A bioavailability-based rationale for controlling metal and metalloid contamination of agricultural land in Australia and New Zealand. *Aust. J. Soil Res.* 38, 1037–1086

98: Kookana RS & Rogers SL. 1995: Effects of pulp mill effluent disposal on soil. *Reviews of Environmental Contamination and Toxicology*, 142

99: Rogers SL et al. 1997: Microbial degradation of strychnine rodenticide in South Australian agricultural soils: Laboratory studies. *Soil Biol. Biochem.* 30(2), 129–132

100: Oliver DP et al. 2000: Strychnine uptake by cereals and legumes grown in glasshouse and field. *J. Environ. Quality* 29(1), 1329–1336

101: Tang W, Yang H & Ryder M. 2001: Research and application of *Trichoderma* spp. in biological control of pathogens. In *Bio-exploitation of Filamentous Fungi*. Fungal Diversity Research Series 6, 403–435

102: Ryder MH & Correll RL.1997: Assessing the potential benefits and risks of introducing natural and genetically manipulated bacteria for the control of soil-borne root diseases. In *Biological Control: Benefits and Risks*. Camb. Univ. Pr.

103: Ryder MH & Rovira AD. 1993: Biological control of Take-all of glasshouse-grown wheat using strains of *Pseudomonas corrugate* isolated from wheat field soil. *Soil Biol. Biochem.* 25, 311–320

103a: Ross IL et al. 2000:Genetic diversity and biological control activity of novel species of closely related pseudomonads isolated from wheat field in South Australia. *Appl. & Experim. Microbiol.* Apr. 1609–1616

104: Stephens PH et al. 1994: Ability of the lumbricid earthworms *Aporrectodea rosea* and *Aporrectodea trapezoides* to reduce the severity of Take-all under greenhouse or field conditions. *Soil Biol. Biochem.* 26(10), 1291–7

105: Roget DK & Rovira AD. 1991: The relationship between incidence of infection by the take-all fungus (*Gaeumannomyces graminis* var. *tritici*), rainfall and yield of wheat in South Australia. *Aust. J. Exp. Ag.* 31, 509–513

106: Roget H & DK. 2000: Prediction of Take-all disease risk in field soils using a rapid and quantitative DNA soil assay. *Plant & Soil* 227, 87–98

107: Doube BM et al. 1994: Interactions between earthworms, beneficial soil microorganisms and root pathogens. *Applied Soil Ecology* 1, 3–10

108: Ryder MH et al. 1999: Use of strains of *Bacillus* isolated in China to suppress Take-all and *Rhizoctonia* root rot, and promote seedling growth of glasshouse-grown wheat in Australian soils. *Soil Biol. Biochem.* 31, 19–29

109: Harvey PR et al. 2000: Genetic variation among populations of *Pythium irregulare* in southern Australia. *Plant Pathology* 49, 619–627

110: MacNish GC & Neate SM. 1996: *Rhizoctonia* bare patch of cereals. *Plant Diseases*, Sept., 965–971

111: Neate SM. 1991: Narrow sowing points to improve soil properties, reduce erosion and reduce root diseases. *Proc. WCFA 6th Annual Seminar, Longerenong*, 21–22

112: Roget DK, Neate SM & Rovira AD. 1996: Effect of sowing point design and tillage practice on the incidence of *Rhizoctonia* root rot, Take-all and cereal cyst nematode in wheat and barley. *Aust. J. Exper. Ag.* 36, 683–693

113: Crook IC. 1991: The influence of root diseases on cereal yield and farm performance. Proc. *WCFA 6th Annual Seminar, Longerenong*, 23–25

114: Roget DK, Venn NR & Rovira AD. 1987: Reduction of *Rhizoctonia* root rot of direct-drilled wheat by short-term fallow. *Aust. J. Exp. Ag.* 27, 425–437

115: Ross IL et al. 2000: Genetic diversity and biological control activity in novel species of closely related pseudomonads isolated from wheat field soils in South Australia. *Appld. & Environmental Microbiol.* Apr., 1609–1616

116: Coventry RJ & Williams J. 1984: Quantative relationships between morphology and current soil hydrology in some alfisols in semiarid tropical Australia. *Geoderma* 33, 191–218

116a: Gillard P, Williams J & Monypenny R. 1989: Clearing trees from Australia's semi-arid tropics: Production, economic and long-term hydrological responses. *Ag. Sci.* 2, 34–39

116b: Bui EN, Smettem KRJ, Moran CJ & Williams J. 1996: Use of soil survey information to assess regional salinisation risk using GIS. *J. Environmental Quality* 25(3), 433–439

116c: Williams J, Bui EN, Gardner EA, Littleboy M & Probert ME. 1997: Tree clearing and dryland salinity hazard in the Upper Burdekin catchment of north Queensland. *Aust. J. Soil Res.* 35, 785–801

116d: Coventry RJ & Williams J. 1990: Soil resources for sustainable agriculture: A Queensland perspective. In *Sustainability in Queensland Agriculture: The Role of Landcare in Securing the Future. Proc. Land Care Symp.* Brisbane Oct. 1989. AIAS Occ. Publ. 44, Melbourne, 5–18

117: Williams J & Coventry RJ. 1979: The contrasting soil hydrology of the red and yellow earths in a landscape of low relief. In *The Hydrology of Areas of Low Precipitation. Proc. Int. Symp. Canberra*, 1979. IAHS-AISH Publ. 128, 385–395

117a: Coventry RJ, Moss AJ & Verster E. 1988: Thin surface soil layers attributable to rain-flow transportation on low-angle slopes: An example from semi-arid tropical Queensland. *Earth Surface Processes and Landforms* 13, 421–430

117b: Coventry RJ. 1978: Late Cainozoic geology, soils and landscape evolution of the Torrens Creek area, North Queensland. *J. Geol. Soc. Aust.* 25(7), 415–427

117c: Coventry RJ. 1982: The distribution of red, yellow and grey earths in the Torrens Creek area, central North Queensland. *Aust. J. Soil Res.* 20, 1–14

117d: Williams J & Coventry RJ. 1980: The potential for groundwater recharge through red, yellow and grey earth profiles in central North Queensland. *Proc. AWRC Conf. Townsville.*

117e: Coventry RJ. 1979: The age of a red earth profile in central North Queensland. *Aust. J. Soil Res.* 17, 505–510

118: Coventry RJ, Taylor RM & Fitzpatrick RW. 1983: Pedological significance of the gravels in some red and grey earths of central North Queensland. *Aust. J. Soil Res.* 21, 219–240

119: Coventry RJ. 1985: Pedology of the red, yellow and grey earths of Northeastern Australia. *New Zealand Soil News* 33(3), 91–99

119a: Bonell M, Coventry RJ & Holt JA. 1986: Erosion of termite mounds under natural rainfall in semiarid tropical northeastern Australia. *Catena* 13, 11–28

120: Williams MAJ. 1978: Termites, soils and landscape equilibrium in the Northern Territory of Australia. In *Landform Evolution in Australasia*. Davies & Williams eds. ANU Press, 128–141

121: Watson JP. 1961: Some observations on soil horizons and insect activity in granite soils. *Proc. 1st Fed. Sci. Cong. 1960, Salisbury*, 1–6

122: Rozefelds AC & De Baar M. 1991: Silicified Kalotermitidae (Isoptera) frass in conifer wood from a mid-Tertiary rainforest in central Queensland, Australia. *Lethaia* 24, 439–442

123: Williams MAJ. 1975: Late Pleistocene tropical aridity syndromes in both hemispheres? *Nature* 253, 617–618

123a: Wyrwoll KH & Milton D. 1976: Widespread late Quaternary aridity in Western Australia. *Nature* 264, 429–430

124: Foster RC, Rovira AD & Cock TW. 1983: *Ultrastructure of the Root-Soil Interface.* Am. Phytopath. Soc.

125: Tisdall J. 1991: Effect of roots and fungal hyphae on soils. *Proc. WCFA 6th Annual Seminar, Longerenong*, 13–16

126: Greenslade P. 1991: Soil microarthropods: What they are and what they do. *Proc. WCFA 6th Annual Seminar, Longerenong*, 17–19

127: Spain A. 1999: Tropical Australian soil fauna: Termites, ants and earthworms, three groups of major pedological importance. *Trans. Acad. Sci. Symposium 'Fixing the Foundations'*. Adelaide Nov. 11–12

127a: Eldridge DJ. 1993: Cryptogam cover and soil surface condition: effects on hydrology on a semi-arid woodland soil. *Arid Soil Res. & Rehab.* 7, 203–217

128: Darwin C. 1881: *The Formation of Vegetable Mould Through the Action of Worms.* John Murray, London

129: Buckerfield JC et al. 1997: Earthworms as indicators of sustainable production in dryland cropping in southern Australia. *Soil Biol. Biochem.* 29(3/4), 547–554

130: Humphreys GS. 1981: The role of ant mounding and earthworm casting near Sydney NSW. *Search* 12, 129–131

131a/z: In *Soil Biota.* 1994: Pankhurst CE et al editors. *Proc. Symp. Soil Biota in Sustainable Farming Systems.* Adelaide. CSIRO Publ.

132: Doube BM et al. 1995: Earthworms: A down-under delivery service for biocontrol agents of root disease. *Acta Zool. Fennica* 196, 219–223

133: Coventry RJ, Holt JA & Sinclair DF. 1988: Nutrient cycling by mound-building termites in low-fertility soils of semi-arid tropical Australia. *Aust. J. Soil Res.* 26, 375–390

134: Holt JA, Coventry RJ & Sinclair DF. 1980: Some aspects of the biology and pedological significance of mound-building termites in a red and yellow earth landscape near Charters Towers, North Queensland. *Aust. J. Soil Res.* 18, 97–109

135: Eldridge DJ. 1994: Nests of ants and termites influence infiltration in a semi-arid woodland. *Pedobiologia* 38, 48–492

136: Braithwaite R. 1990: More than a home for white ants. *Aust. Nat. Hist.* 23(4), 306–313

137: Holt JA & Coventry RJ. 1990: Nutrient cycling in Australian savannas. *J. Biogeog.* 17, 427–432

137a: Cook GD. 1994: The fate of nutrients during fire in a tropical savanna. *Aust. J. Ecol.* 19, 359–365

138: Watson JAL, Lendon C & Low BS. 1973: Termites in mulga lands. *Trop. Grasslands* 7(1), 121–126

139: Foti F. 1990: Aboriginal drugstores. *Aust. Nat. Hist.* 25(4), 309

140: Eldridge DJ & Picard J. 1994: Effects of ants on sandy soils in semi-arid Eastern Australia: II. Relocation of nest entrances and consequences for bioturbation. *Aust. J. Soil Res.* 32, 323–333

141: Morton SR. 1982: Granivory in the Australian arid zone: diversity of harvester ants and structure of their communities. In *Evolution of the Flora and Fauna of Arid Australia.* Barker & Greenslade eds. Peacock Publ.

142: Eldridge DJ. 1993: Effect of ants on sandy soils in semi-arid Eastern Australia: Local distribution of nest entrances and their effect on the infiltration of water. *Aust. J. Soil Res.* 31, 509–518

144: Andersen AN. 1993: Ants as indicators of restoration success at a uranium mine in tropical Australia. *Restoration Ecology*, Sept., 154–167

145: Andersen AN. 1994: Ants as indicators of restoration success following mining in northern Australia. *Proc. Aus. IMM Conf.* Darwin, 465–467

146: Andersen AN. 1990: The use of ant communities to evaluate change in Australian terrestrial ecosystems: A review and a recipe. *Proc. Ecol. Soc. Aust.* 16, 347–357

147: Feehan J. 1994: A hard-earned pat for a true digger. *Aust. Geogr.* 36, 82–87

148: Waterhouse DF. 1998: The biological control of dung. *Sci. Amer.* 101–108

149: Monteith G. 1999: Wingless dungbeetle. *Aust. Nature* 26(7), 22–23

150: Woinarski JCZ, Milne DJ & Wanganeen G. 2001: Declines in mammal populations in relatively intact landscapes of the monsoonal tropics of the Northern Territory, Australia. In press

150a: Royal Commission into the Condition of the Crown Tenants. 1901: *Proceedings of the Legislative Assembly* 1901, Vol. 4, Government Printer, Sydney

151: Woinarski JCZ et al. 2000: Report to Kakadu Research Advisory Committee. In press

151a: Woinarski JCZ, Milne DJ & Wanganeen G. 2000: Declines in mammal populations in relatively intact landscapes of the monsoonal tropics of the Northern Territory, Australia. In press

152: Short J & Smith A. 1994: Mammal decline and recovery in Australia. *J. Mammalogy* 75(2), 288–297

153: Short J et al. 1997: Control of feral cats for nature conservation. II. Population reduction by poisoning. *CSIRO Wildlife Res.* 24, 703–714

154: Krefft G. 1866: On the verterbrated animals of the lower Murray and Darling, their habits, economy, and geographical distribution. *Trans. Phil. Soc. NSW* 1, 1–38

155: Short J & Calaby H. 2001: The status of Australian mammals in 1922: Collections and field notes of museum collector Charles Hoy. *Aust. Zoologist* 31(4) 533–560

156: Reardon M. 1999: Quoll on the run. *Aust. Geogr.* 54, 88–105

157: Friend T. 1997: Numbats on a junk food diet. *Aust. Nature* 25(9), 40–49

158: Reardon M. 2000: The bilby: Fighting for survival under the desert moon. *Aust. Geogr.* 57, 64–79

159: Van Dyck S. 1997: Queensland pebble mound mice. *Aust. Nature* Spring, 40–47

160: Short J. 1998: The extinction of rat-kangaroos (Marsupialia: Potoroidae) in New South Wales, Australia. *Biological Conservation* 86, 365–377

161: Noble JC. 1994: Bettong (*Bettongia* spp.) biocontrol of shrubs in semi-arid mulga (*Acacia aneura*) rangelands: An hypothesis. In *Ecological Research and Management in the Mulgalands*, Ed. Page & Bentel, U. Qld Gatton College, 139–145

162: Noble JC. 1993: Relict surface-soil features in semi-arid mulga (*Acacia aneura*) woodlands. *Rangel. J.* 15(1), 48–70

163: Noble JC. 1996: Mesomarsupial ecology in Australian rangelands: Burrows, bettongs (*Bettongia* spp.) and biocontrol of shrubs. *Proc. 5th IRC*, Salt Lake City, 1995

164: Short J & Turner B. 2000: Reintroduction of the burrowing bettong *Bettongia lesueur* (Marsupialia: Potoroidae) to mainland Australia. *Biological Conservation* 96, 185–196

165: Richards J. 1998: Return of the natives. *Aust. Geogr.* 50, 82–97

165a: Richards J & Short J. 1997: Western barred bandicoot. *Nature Aust.* Autumn. 19–20

165b: Richards J , Copley P & Morris K. 2001: The wopilkara's return. *Nature Aust.* 52–61

166: Strahan R. ed. 1995: *The Mammals of Australia.* Reed Books, Sydney

167: McCreery S. 1999: Fenced in and free. *Aust. Geogr.* 53, 31

168: Martin G. 2001: Small Australian ground foraging mammals: Rebuilders of soil health. In press

169: Martin G. 2001: The role of small native mammals in soil building and water balance. *Stipa Native Grasses Newsletter* 16, Autumn

170: Martin G. 2000: The effect of small Australian mammals on community dynamics. Unpubl.

171: Eldridge DJ & Greene RSB. 1994: Assessment of sediment yield by splash erosion on a semi-arid soil with varying cryptogam cover. *J. Arid Envir.* 26, 221–232

171a: Eldridge DJ. 1993: Cryptogams, vascular plants, and soil hydrological relations: Some preliminary results from the semi-arid woodlands of eastern Australia. *Great Basin Naturalist* 53(1), 48–58

172: Eldridge DJ & Greene RSB. 1994: Microbiotic soil crusts: a review of their roles in soil and ecological processes in the rangelands of Australia. *Aust. J. Soil Res.* 32, 389–415

172c: Eldridge DJ & Bradstock RA. 1994: The effect of time since fire on the cover and composition of cryptogamic soil crusts on a eucalypt shrubland soil. *Cunninghamia* 3(3), 521–527

172d: Rogers RW. 1972: Soil surface lichens in arid and sub-arid SE Australia. III. The relationship between distribution and environment. *Aust. J. Bot.* 20, 301–316

172e: Rogers RW. 1972: Soil surface lichens in arid and subarid SE Australia. II. Phytosociology and geographic zonation. *Aust. J. Bot.* 20, 215–227

172f: Rogers RW. 1971: Distribution of the lichen *Chondropsis semiviridis* in relation to its heat and drought resistance. *New Phytol.* 70, 1069–1077

172g: Greene RSB, Chartres CJ & Hodgkinson KC. 1990: The effects of fire on the soil in degraded semi-arid woodland. I. Cryptogam cover and physical and micromorphological properties. *Aust. J. Soil. Res.* 28, 755–77

172h: Noble JC et al. 1996: Fire studies in mallee (*Eucalyptus* spp.) communities of western NSW: Spatial and temporal fluxes in soil chemistry and soil biology following prescribed fire. In *Pacific Conservation Biology*. Surrey Beatty, Sydney

174: Eldridge D. 1997: Caretakers of the desert. *Aust. Nature.* Spring, 48–55

175: Bowman DMJS. 2000: *Australian Rainforests: Islands of Green in a Land of Fire.* Cambr. Univ. Press.

176: Horton D. 2000: *The Pure State of Nature.* Allen & Unwin, Sydney

177: Webb LJ & Tracey JG. 1981: The rainforests of Northern Australia. In Groves RH ed. *Australian Vegetation* Camb. Univ. Pr.

178: Nix H. 1994: The Brigalow. In Dovers S. ed. *Australian Environmental History: Essays and Cases.* OUP, Chapter 10

179: Gasteen WJ. 1987: The Brigalow lands of eastern Australia: Agricultural impact and land-use potential versus biological representation and stability. In *The Rainforest Legacy*. Aust. Herit. Comm. Publ. 7, Chapter 9

179a: Gasteen WJ. Care of our soil resources; and Some notes on artesian bores and bore drains. Unpublished

179b: Gasteen J. 2001: Some observations on soft-wood scrubs. Pers. comm.

179c: Pegler L & Bentley D. 2000. Assessing natural resource condition associated with the Great Artesian Basin. *Proc. Nat. Res. Condition & Trend Inf. Forum, Qld.* MDB

180: Flannery T. 1994: *The Future Eaters.* Reed Books, Sydney

181: Braithwaite RW & Muller WJ. 1997: Rainfall, groundwater and refuges: Predicting extinctions of Australian tropical mammal species. *Aust. J. Ecol.* 22, 57–67

182: Andersen AN & Lonsdale WM. 1990: Herbivory by insects in Australian tropical savannas: A review. *J. Biogeog.* 17, 433–444

183: Andersen AN. 1991: Responses of ground-foraging ant communities to three experimental fire regimes in a savanna forest of tropical Australia. *Biotropica* 23(4b), 575–585

184: Beard JS, Chapman AR & Gioia P. 2000: Species richness and endemism in the Western Australian flora. *J. Biogeogr.* 27, 1257–1268

185: Lamont B. 1982: Mechanisms for enhancing nutrient uptake in plants, with particular reference to Mediterranean South Africa and Western Australia. *Bot. Rev.* July-Sept. 597–689

186: Lamont B. 1983: Root hair dimensions and surface volume/weight ratios of roots with the aid of scanning electron microscopy. *Plant & Soil* 74, 149–152

187: Lamont BB, Brown G & Mitchell DT. 1984: Structure, environmental effects on their formation, and function of proteoid roots. In *Leucadendron laureolum* (Proteaceae). *New Phytol.* 97, 381–390

188: Flannery T. 1992: The diversity enigma. *Aust. Nat. Hist.* 24(1), 24–33

189: Davidson NJ, True KC & Pate JS. 1989: Water relations of the parasite: Host relationship between the mistletoe *Amyema linophyllum* (Frenzl) Tieghem and *Casuarina obesa* Miq. *Oecologia* 80, 321–330

190: Lamont BB. 2002: Green cotyledons of two *Hakea* species control seedling mass and morphology by supplying mineral nutrients rather than organic compounds. *New Phytol.* 153(1). In press

191: Kuo J, Hocking PJ & Pate JS. 1982: Nutrient reserves in seeds of selected Proteaceous species from south-western Australia. *Aust. J. Bot.* 30, 231–249

192: Dixon KW, Kuo J & Pate JS. 1983: Storage reserves of the seed-like aestivating organs of geophytes inhabiting granite outcrops in south-western Australia. *Aust. J. Bot.* 31, 85–103

193: Pate JS. 1989: Australian micro stilt plants. *Tree* 4(2), 45–49

194: Boulton AJ, Humphreys WF & Eberhard SM. 2001: Imperilled subsurface waters in Australia: Biodiversity-threatening processes and conservation. *Aquat. Ecosyst. Health & Management*

195: Franzmann PD et al. The role of microbial populations in the containment of aromatic hystocarbons in the subsurface. In press

195a: Franzmann PD et al. 1999: Microbial mineralisation of benzene and characterisation of biomass in soil above hydrocarbon-contaminated groundwater. *FEMS Microbio. Ecol.* 30, 67–76

196: Franzmann PD et al. 1998: Mineralisation of low concentrations of organic compounds and microbial biomass in surface and vadose soils from the Swan Coastal Plain, WA. *Aust. J. Soil Res.* 36, 921–939

197: Franzmann PD et al. 2000: Bioaugmentation of atrazine and fenamiphos impacted groundwater: Laboratory evaluation. *Bioremediation J.* 4(3), 237–248

198: Boulton AJ. 2001: 'Twixt two worlds: Taxonomic and functional biodiversity at the surface water/groundwater interface. *Rec. WA Mus. Suppl.* 64, 1–13

199: Humphreys WF. 1999: Relict stygofaunas living in sea salt, karst and calcrete habitats in arid northwestern Australia contain many ancient lineages. In Ponder W & Lunney D eds. *The Other 99%: The Conservation and Biodiversity of Invertebrates. Trans. Roy. Zoo. Soc. NSW* 219–227

200: Humphreys WF. 2000: Relict faunas and their derivation. In *Subterranean Ecosystems.* Elsevier, Chapter 22

200a: Wilson RS & Humphreys WF. 2001: *Prionspio thalanji* sp. nov. (Polychaeta Spionidae) from an anchialine cave, Cape Range, northwest Western Australia. *Rec. West. Aust. Mus.* Suppl. 64, in press

201: Humphreys WF. 2000: The hypogean fauna of the Cape Range Peninsula and Barrow Island, northwestern Australia. In *Subterranean Ecosystems.* Elsevier, Chapter 30

202: Humphreys WF & Adams M. 2001: Allozyme variation in the troglobitic millipede *Stygiochiropus communis* (Diplopoda: Paradoxosomaridae) from the arid tropical Cape Range, northwestern Australia: Population structure and implications for the management of the region. *Rec. WA Mus. Suppl.* 64, 15–36

203: Humphreys WF. 2001: *Milyeringa veritas* (Eleotridae), a remarkably versatile cave fish from the arid tropics of northwestern Australia. *Env. Biol. of Fishes* 62, 297–313

204: Humphreys WF. 1999: Physico-chemical profile and energy fixation in Bundera Sinkhole, an anchialine remiped habitat in north-western Australia. *J. Roy. Soc. WA* 82, 89–98

205: Humphreys WF. 2000: Bundera Sinkhole, a fragile fauna site, unique in the southern hemisphere. WA Museum newsletter, June, 16–17

206: Humphreys WF. 2000: Blind water beetles in the desert. WA Mus. newsletter

207: Boulton A. pers. comm. Letter 13 July 2001

208: Boulton AJ. 2000: River ecosystem health down under: Assessing ecological condition in riverine groundwater zones in Australia. *Ecosystem Health* 6(2), 108–118

209: Boulton AJ & Quinn JM. 2000: A simple and versatile technique for assessing cellulose decomposition potential in floodplain and riverine sediments. *Arch. Hydrobiol.* 150(1), 133–151

210: Boulton AJ et al. 1997: Land-use effects on the hyporheic ecology of five small streams near Hamilton, New Zealand. *NZ J. Marine & Freshwater Res.* 31, 609–622

211: Boulton AJ et al. 1998: The functional significance of the hyporheic zone in streams and rivers. *Annu. Res. Ecol. Syst.* 29, 59–81

212: Boulton AJ. 2000: The functional role of the hyporheos. *Verh. Internat. Verein. Limnol.* 27, 51–63

213: Boulton A. 1999: The role of subsurface filters in gravel-bed river rehabilitation strategies. *Proc. 2nd Aust. Stream Management Conf.* Adelaide SA, 81–86

214: van Oppen MJH et al. 2001: Patterns of coral-dinoflagellate associations in Acropora: Significance of local availability and physiology of *Symbiodinium* strains and host-symbiont selectivity. *J. Roy. Soc.*

215: Baillie BK et al. 2000: Conspecificity and Indo-Pacific distribution of *Symbiodinium* genotypes (Dinophyceae) from giant clams. *J. Phycol.* 36, 1153–1161

215a: Belda-Baillie CA et al. 1999: Evidence for changing symbiotic algae in juvenile tridacnids. *J. Exper. Marine Biol. & Ecol.* 241, 207–221

215a: Ghiold J. 1990: White death: The fate of deserted coral. *New Scientist* 1719, 46

216: Woodford J. 2001: Embarrassing: How we're killing our own reef. *SMH*, Feb. 17

217: Franzmann PD. 1996: Examination of Antarctic prokaryotic diversity through molecular comparisons. *Biodiversity & Conserv.* 5, 1295–1305

218: Marchant HJ, Watanabe K & Kawachi M. 1996: Marine snow in Antarctic coastal waters. *Proc. NIPR Symp. Polar Biol. Tokyo*, 75–83

219: Marchant HJ & Nash GV. 1986: Electron microscopy of gut contents and faeces of *Euphausia superba* Dana. *Mem. Nat. Inst. Polar Res.* Special Issue 40, 167–177

220: Whitman WB, Coleman DC & Wiebe WJ. 1998: Prokaryotes: The unseen majority. *Proc. Natl. Acad. Sci. USA* 95, 6578–6583

221: McCully M. 1995: How do real roots work? *Plant Physiol.* 109, 1–6

222: McCully ME. 1999: Roots in soil: Unearthing the complexities of roots and their rhizospheres. *Annu. Rev. Plant Physiol. Plant. Mol. Biol.* 50, 695–718

223: Nisbet EG & Sleep NR. 2001: The habitat and nature of early life. *Nature* 409, 1083–1091

224: Rothschild LJ & Mancinelli RL. 2001: Life in extreme environments. *Nature* 409, 1092–1101

225: McCully ME. 1995: Water efflux from the surface of field-grown grass roots: Observations by cryo-scanning microscopy. *Physiol. Plantarum* 95, 217–224

226: Harrison PL & Eward S. 2001: Elevated levels of nitrogen and phosphorus reduce fertilisation success of gametes from scleractinian reef corals. *Marine Biol.* 139, 1057–1068

227: Reichelt-Brushett AJ & Harrison PL. 1999: The effect of copper, zinc and cadmium on fertilisation success of gametes from scleractinian reef corals. *Marine Pollution Bull.* 38(3), 182–187

228: Reichelt-Brushett AJ & Harrison PL. 2000: The effect of copper on the settlement success of larvae from the scleractinian coral *Acropora tenuis*. *Marine Pollution Bull.* 41(7-12), 385–391

229: Ward S & Harrison PL. 2000: Changes in gametogenesis and fecundity of acroporid corals that were exposed to elevated nitrogen and phosphorus during the ENCORE experiment. *J. Exper. Marine Biol. & Ecol.* 246, 179–221

230: Ward S & Harrison PL. 1997: The effects of elevated nutrient levels on settlement of coral larvae during the ENCORE experiment, Great Barrier Reef, Australia. *Proc. 8th Int. Coral Reef Sym.* 1, 891–896

231: Koop K et al. 2001: ENCORE: The effect of nutrient enrichment on coral reefs: Synthesis of results and conclusions. *Marine Pollution Bull.* 42(2), 91–120

232: Harrison PL & Wallace CC. 1990: Reproduction, Dispersal and recruitment of scleractinian corals. In *Coral Reefs*, Z. Dubinsky ed. Elsevier, Chapter 7

233: *Sydney Morning Herald*, Weekend Edition May 4–5, 2002: Newsday Press Association: Smallest living organism is a very big discovery

234: Lawrence JF. 1999: The Australian Ommatidae (Coleoptera: Archostemata): New species, larva and discussion of relationships. *Invertebrate Taxonomy* 13, 369–390

235: Osborne RAL & Cooper IB. 2001: Sulfide-bearing palaeokarst deposits at Lune River Quarry, Ida Bay, Tasmania. *Aust. J. Earth Sci.* 48, 409–416

236: Bonham KJ, Mesibov R & Bashford R. 2002: Diversity and abundance of some ground-dwelling invertebrates in plantation vs. native forests in Tasmania, Australia. *Forest Ecol. & Management* 158, 237–247

237: Mesibov, Bob. 1999: Litter: The good sort. *Forest Practices News* 2(1)

Index

(Figures in italics indicate pictures, maps or diagrams)

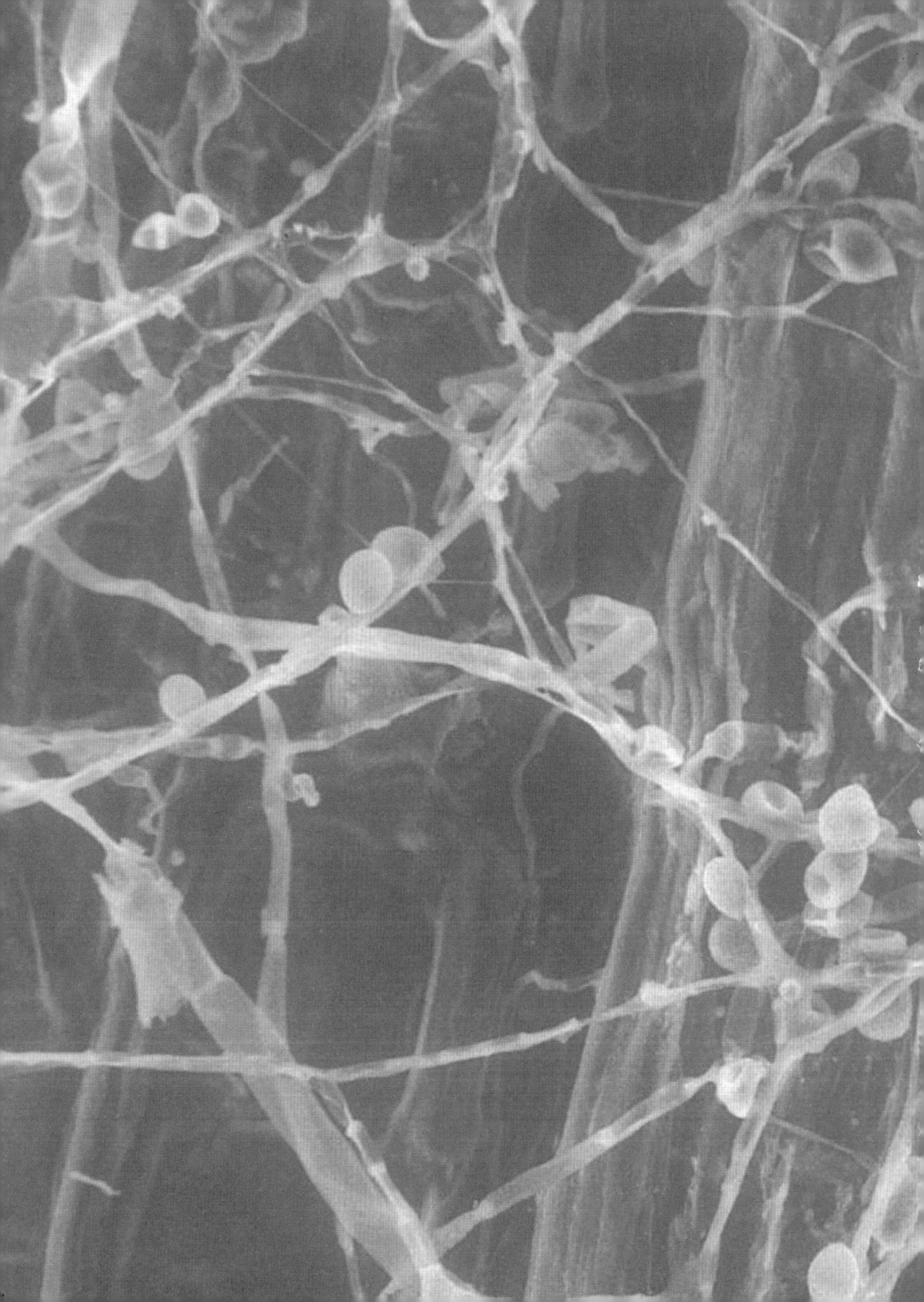